The Philosophy of Mathematical Practice

The Philosophy of Mathematical Practice

Paolo Mancosu

OXFORD
UNIVERSITY PRESS

Great Clarendon Street, Oxford OX2 6DP

Oxford University Press is a department of the University of Oxford.
It furthers the University's objective of excellence in research, scholarship,
and education by publishing worldwide in

Oxford New York

Auckland Cape Town Dar es Salaam Hong Kong Karachi
Kuala Lumpur Madrid Melbourne Mexico City Nairobi
New Delhi Shanghai Taipei Toronto

With offices in

Argentina Austria Brazil Chile Czech Republic France Greece
Guatemala Hungary Italy Japan Poland Portugal Singapore
South Korea Switzerland Thailand Turkey Ukraine Vietnam

Oxford is a registered trademark of Oxford University Press
in the UK and in certain other countries

Published in the United States
by Oxford University Press Inc., New York

© the several contributors 2008

The moral rights of the authors have been asserted
Database right Oxford University Press (maker)

First published 2008

All rights reserved. No part of this publication may be reproduced,
stored in a retrieval system, or transmitted, in any form or by any means,
without the prior permission in writing of Oxford University Press,
or as expressly permitted by law, or under terms agreed with the appropriate
reprographics rights organization. Enquiries concerning reproduction
outside the scope of the above should be sent to the Rights Department,
Oxford University Press, at the address above

You must not circulate this book in any other binding or cover
and you must impose the same condition on any acquirer

British Library Cataloguing in Publication Data

Data available

Library of Congress Cataloging in Publication Data

Data available

Typeset by Laserwords Private Limited, Chennai, India
Printed in Great Britain
on acid-free paper by
CPI Antony Rowe, Chippenham, Wiltshire

ISBN 978–0–19–929645–3

10 9 8 7 6 5 4 3 2 1

Preface

When in the spring of 2005 I started planning the present book, I had two aims in mind. First, I wanted to unify the efforts of many philosophers who were making contributions to a philosophy of mathematics informed by a desire to account for many central aspects of mathematical practice that, by and large, had been ignored by previous philosophers and logicians. Second, I wished to produce a book that would be useful to a large segment of the community, from interested undergraduates to specialists. I like to think that both goals have been met.

Concerning the first aim, I consider the book to provide a representative sample of the best work that is being produced in this area. The eight topics selected for inclusion encompass much of contemporary philosophical reflection on key aspects of mathematical practice. An overview of the topics is given in my introduction to the volume.

The second goal dictated the organization of the book and my general introduction to it. Each topic is discussed in an introductory chapter and in a research article in the very same area. The rationale for this division is that I conceive the book both as a pedagogical tool and as a first-rate research contribution. Thus, the aim of the introductory chapters is to provide a general and accessible overview of an area of research. I hope that, in addition to the experts, these will be useful to undergraduates as well as to non-specialists. The research papers obviously have the aim of pushing the field forwards.

As for the introduction to the book, my aim was to provide the context out of which, and sometimes against which, most of the contributions to the volume have originated. Once again, the idea was to give a fair account of the landscape that could be useful also, but not only, to the non-initiated. Each author has been in charge of writing both the introduction and the research paper to the area that was commissioned from him. The only exception is the subject area of 'purity of methods' where two specialists on the topic teamed up, Mic Detlefsen and Michael Hallett. In addition, Johannes Hafner has been brought in as co-author of the research paper on explanation jointly written with me.

I would like to thank all the contributors for their splendid work. Not only did they believe in the project from the very start and accept enthusiastically my invitation to participate in it, but they also performed double duties (introduction and research paper). That a project of this size could be brought to completion within two years from its inception is a testimony to their energy, enthusiasm, and commitment.

I am also very grateful to Peter Momtchiloff, editor at Oxford University Press, for having believed in the project from the very beginning, for having encouraged me to submit it to OUP, and for having followed its progress all along.

The production of the manuscript was the work of Fabrizio Cariani, a graduate student in the Group in Logic and the Methodology of Science at U.C. Berkeley. With great patience and expertise he turned a set of separate essays (some with lots of diagrams) in different formats into a beautiful and uniform LaTex document. I thank him for his invaluable help. His work was supported by a Faculty Research Grant at U.C. Berkeley.

Other individual acknowledgements will be given after the individual contributions. But I would like to take advantage of my position as editor of the volume to thank my wife, Elena Russo, for her loving patience and support throughout the project.

Berkeley, 17 May 2007

Contents

Biographies	viii
Introduction	1
1. Visualizing in Mathematics	22
2. Cognition of Structure	43
3. Diagram-Based Geometric Practice	65
4. The Euclidean Diagram (1995)	80
5. Mathematical Explanation: Why it Matters	134
6. Beyond Unification	151
7. Purity as an Ideal of Proof	179
8. Reflections on the Purity of Method in Hilbert's *Grundlagen der Geometrie*	198
9. Mathematical Concepts and Definitions	256
10. Mathematical Concepts: Fruitfulness and Naturalness	276
11. Computers in Mathematical Inquiry	302
12. Understanding Proofs	317
13. What Structuralism Achieves	354
14. 'There is No Ontology Here': Visual and Structural Geometry in Arithmetic	370
15. The Boundary Between Mathematics and Physics	407
16. Mathematics and Physics: Strategies of Assimilation	417
Index of Names	441

Biographies

Jeremy Avigad is an Associate Professor of Philosophy at Carnegie Mellon University. He received a B.A. in Mathematics from Harvard in 1989, and a Ph.D. in Mathematics from the University of California, Berkeley in 1995. His research interests include mathematical logic, proof theory, automated reasoning, formal verification, and the history and philosophy of mathematics. He is particularly interested in using syntactic methods, in the tradition of the Hilbert school, towards obtaining a better understanding of mathematical proof.

Michael Detlefsen is Professor of Philosophy at the University of Notre Dame and long time editor of the *Notre Dame Journal of Formal Logic*. His scholarly work includes a number of projects concerning (i) Gödel's incompleteness theorems (and related theorems) and their philosophical implications, (ii) Hilbert's ideas in the foundations of mathematics, (iii) Brouwer's intuitionism, (iv) Poincaré's conception of proof, and (v) the history and philosophy of formalist thinking from the 17th century to the present. Recently, he has been thinking about the classical distinction between problems and theorems and the role played by algebra in shaping the modern conception of problem-solving. Throughout his work, he has sought to illuminate meaningful historical points of connection between philosophy and mathematics. His current projects include a book on formalist ideas in the history of algebra, another on constructivism, and a third (with Tim McCarthy) on Gödel's theorems.

Marcus Giaquinto studied philosophy for a B.A. at University College London (UCL), then mathematical logic for an M.Sc. taught by John Bell, Christopher Ferneau, Wilfrid Hodges, and Moshé Machover. Further study of logic at Oxford under the supervision of Dana Scott was followed by turning to philosophy of mathematics for a Ph.D. supervised by Daniel Isaacson. Giaquinto is a Professor in UCL's Philosophy Department and an associate member of UCL's Institute of Cognitive Neuroscience. He has written two books, *The Search for Certainty: a Philosophical Account of Foundations of Mathematics* (OUP 2002), and *Visual Thinking in Mathematics: an Epistemological Study* (OUP 2007). The research for this work and for Giaquinto's articles in this volume was funded by a British Academy two-year readership.

Johannes Hafner is Assistant Professor of Philosophy at North Carolina State University and formerly lecturer at the University of Vienna. After receiving his Magister degree in philosophy he pursued graduate studies in philosophy and logic at CUNY and at U.C. Berkeley (Ph.D. in Logic, 2005). He has a strong interest in the history of logic and has in particular worked on the emergence of model-theoretic methods within the Hilbert school and on Bolzano's concept of indirect proof. Other research interests include ontological issues within the philosophy of mathematics in particular Putnam's argument for antirealism in set theory and the concept of mathematical explanation.

Michael Hallett is Associate Professor of Philosophy at McGill University in Montreal. His main interests are in the philosophy and history of mathematics. He is the author of *Cantorian Set Theory and Limitation of Size* (Oxford, Clarendon Press, 1984), a study of Cantor's development of set theory and of the subsequent axiomatization. Much more of his recent work has centred on Hilbert's treatment of the foundations of mathematics and what distinguishes it from other major foundational figures, e.g. Frege and Gödel. He is a General Editor (along with William Ewald, Ulrich Majer, and Wilfried Sieg) of a six-volume series (to be published by Springer) containing many important and hitherto unpublished lecture notes of Hilbert on the foundations of mathematics and physics. *Volume 1: David Hilbert's Lectures on the Foundations of Geometry, 1891–1902*, co-edited by Hallett and Ulrich Majer, appeared in 2004. *Volume 3: David Hilbert's Lectures on Arithmetic and Logic, 1917–1933*, edited by William Ewald and Wilfried Sieg, will appear in 2008.

Colin McLarty is the Truman P. Handy Associate Professor of Philosophy, and of Mathematics, at Case Western Reserve University. He is the author of *Elementary Categories, Elementary Toposes* (OUP 1996) and works on category theory especially in logic and the foundations of mathematics. His current project is a philosophical history of current methods in number theory and algebraic geometry, which largely stem from Grothendieck. The history includes Poincaré's topology, Noether's abstract algebra and her influence on topology, and Eilenberg and Mac Lane's category theory. He has published articles on these and on Plato's philosophy of mathematics.

Paolo Mancosu is Professor of Philosophy at U.C. Berkeley. His main interests are in logic, history and philosophy of mathematics, and history and philosophy of logic. He is the author of *Philosophy of Mathematics and Mathematical Practice in the Seventeenth Century* (OUP 1996) and editor of *From Brouwer to Hilbert. The debate on the foundations of mathematics in the 1920s*

(OUP 1988). He has recently co-edited the volume *Visualization, Explanation and Reasoning Styles in Mathematics* (Springer 2005). He is currently working on mathematical explanation and on Tarskian themes (truth, logical consequence, logical constants) in philosophy of logic.

Kenneth Manders (Ph.D., U.C. Berkeley, 1978) is Associate Professor of philosophy at the University of Pittsburgh, with a secondary appointment in History and Philosophy of Science, and fellow of the Center for Philosophy of Science. He was a fellow of the Institute for Advanced Study in the Behavioral Sciences and NEH fellow, and has held a NATO postdoctoral fellowship in science (at Utrecht), an NSF mathematical sciences postdoctoral fellowship (at Yale), and a Howard Foundation Fellowship. His research interests lie in the philosophy, history, and foundations of mathematics; and in general questions on relations between intelligibility, content, and representational or conceptual casting. He is currently working on a book on geometrical representation, centering on Descartes. He has published a number of articles on philosophy of mathematics, history of mathematics, model theory, philosophy of science, measurement theory, and the theory of computational complexity.

Jamie Tappenden is an Associate Professor in the Philosophy Department of the University of Michigan. He completed a B.A. (Mathematics and Philosophy) at the University of Toronto and a Ph.D. (Philosophy) at Princeton. His current research interests include the history of 17th century mathematics, especially geometry and complex analysis, both as subjects in their own right and as illustrations of themes in the philosophy of mathematical practice. As both an organizing focus for this research and as a topic of independent interest, Tappenden has charted Gottlob Frege's background and training as a mathematician and spelled out the implications of this context for our interpretation of Frege's philosophical projects. Representative publications are: 'Extending knowledge and "fruitful concepts": Fregean themes in the philosophy of mathematics' (*Noûs* 1995) and 'Proof Style and Understanding in Mathematics I: Visualization, Unification and Axiom Choice', in P. Mancosu *et al.* (eds.), *Visualization, Explanation and Reasoning Styles in Mathematics* (Springer 2005). His book on 19th century mathematics with special emphasis on Frege is to be published by Oxford University Press.

Alasdair Urquhart is a Professor in the Departments of Philosophy and Computer Science at the University of Toronto. He studied philosophy as an undergraduate at the University of Edinburgh, and obtained his doctorate at the University of Pittsburgh under the supervision of Nuel D. Belnap. He has published articles and books in the areas of non-classical logics, algebraic logic, lattice theory, universal algebra, complexity of proofs, complexity of algorithms, philosophy of logic, and history of logic. He is the co-author (with Nicholas Rescher) of *Temporal Logic* and the editor of Volume 4 of the *Collected Papers of Bertrand Russell*. He is currently the managing editor of the reviews section of the *Bulletin of Symbolic Logic*.

Introduction

The essays contained in this volume have the ambitious aim of bringing some fresh air to the philosophy of mathematics. Contemporary philosophy of mathematics offers us an embarrassment of riches. Anyone even partially familiar with it is certainly aware of the recent work on neo-logicism, nominalism, indispensability arguments, structuralism, and so on. Much of this work can be seen as an attempt to address a set of epistemological and ontological problems that were raised with great lucidity in two classic articles by Paul Benacerraf. Benacerraf's articles have been rightly quite influential, but their influence has also had the unwelcome consequence of crowding other important topics off the table. In particular, the agenda set by Benacerraf's writings for philosophy of mathematics was that of explaining how, if there are abstract objects, we could have access to them. And this, by and large, has been the problem that philosophers of mathematics have been pursuing for the last fifty years. Another consequence of the way in which the discussion has been framed is that no particular attention to mathematical practice seemed to be required to be an epistemologist of mathematics. After all, the issue of abstract objects confronts us already at the most elementary levels of arithmetic, geometry, and set theory. It would seem that paying attention to other branches of mathematics is irrelevant for solving the key problems of the discipline. This engendered an extremely narrow view of mathematical epistemology within mainstream philosophy of mathematics, due partly to the over-emphasis on ontological questions.

The authors in this collection believe that the single-minded focus on the problem of 'access' has reduced the epistemology of mathematics to a torso. They believe that the epistemology of mathematics needs to be extended well beyond its present confines to address epistemological issues having to do with fruitfulness, evidence, visualization, diagrammatic reasoning, understanding, explanation, and other aspects of mathematical epistemology which are orthogonal to the problem of access to 'abstract objects'. Conversely, the ontology of mathematics could also benefit from a closer look at how interesting ontological issues emerge both in connection to some of the epistemological problems mentioned above (for instance, issues concerning the existence of 'natural kinds' in mathematics) and from mathematical practice itself (issues of individuation of objects and structuralism in category theory).

The contributions presented in this book are thus joined by the shared belief that attention to mathematical practice is a necessary condition for a renewal of the philosophy of mathematics. We are not simply proposing new topics for investigation but are also making the claim that these topics cannot be effectively addressed without extending the range of mathematical practice one needs to look at when engaged in this kind of philosophical work. Certain philosophical problems become salient only when the appropriate area of mathematics is taken into consideration. For instance, geometry, knot theory, and algebraic topology are bound to awaken interest in (and philosophical puzzlement about) the issue of diagrammatic reasoning and visualization, whereas other areas of mathematics, say elementary number theory, might have much less to offer in this direction. In addition, for theorizing about structures in philosophy of mathematics it seems wise to go beyond elementary algebra and take a good look at what is happening in advanced areas, such as cohomology, where 'structural' reasoning is pervasive. Finally, certain areas of mathematics can actually provide the philosophy of mathematics with useful tools for addressing important philosophical problems.

There is an interesting analogy to be drawn here with the philosophy of the natural sciences, which has flourished under the combined influence of both general methodology and classical metaphysical questions (realism vs. antirealism, space, time, causation, etc.) interacting with detailed case studies in the special sciences (physics, biology, chemistry, etc.). Revealing case studies have been both historical (studies of Einstein's relativity, Maxwell's electromagnetic theory, statistical mechanics, etc.) and contemporary (examinations of the frontiers of quantum field theory, etc.). By contrast, with few exceptions, philosophy of mathematics has developed without the corresponding detailed case studies.

In calling for renewed attention to mathematical practice, we are the inheritors of several traditions of work in philosophy of mathematics. In the rest of this introduction, I will describe those traditions and the extent to which we differ from them.

1 Two traditions

Many of the philosophical directions of work mentioned at the outset (neo-logicism, nominalism, structuralism, and so on) were elaborated in close connection to the classical foundational programs in mathematics, in particular logicism, Hilbert's program, and intuitionism. It would not be possible to make sense of the neo-logicism of Hale and Wright unless seen as a proposal for overcoming the impasse into which the original Fregean logicist program fell as a consequence of Russell's discovery of the paradoxes. It would be even harder to understand Dummett's anti-realism without appropriate knowledge of intuitionism as a foundational position. Obviously, it would take more space than I have to trace here the sort of genealogy I have in mind; but in a way it would also be useless. For it cannot be disputed that already in the 1960s, first with Lakatos and later through a group of 'maverick' philosophers of mathematics (Kitcher, Tymoczko, and others),[1] a strong reaction set in against philosophy of mathematics conceived as foundation of mathematics. In addition to Lakatos' work, the philosophical opposition took shape in three books: Kitcher's *The Nature of Mathematical Knowledge* (1984), Aspray and Kitcher's *History and Philosophy of Modern Mathematics* (1988) and Tymoczko's *New Directions in the Philosophy of Mathematics* (Tymoczko, 1985) (but see also Davis and Hersh (1980) and Kline (1980) for similar perspectives coming from mathematicians and historians). What these philosophers called for was an analysis of mathematics that was more faithful to its historical development. The questions that interested them were, among others: How does mathematics grow? How are informal arguments related to formal arguments? How does the heuristics of mathematics work and is there a sharp boundary between method of discovery and method of justification? Evaluating the analytic philosophy of mathematics that had emerged from the foundational programs, Aspray and Kitcher (1988) put it this way:

> Philosophy of mathematics appears to become a microcosm for the most general and central issues in philosophy—issues in epistemology, metaphysics, and philosophy of language—and the study of those parts of mathematics to which philosophers of mathematics most often attend (logic, set theory, arithmetic) seems designed to test the merits of large philosophical views about the existence of abstract entities or the tenability of a certain picture of human knowledge. There is surely nothing wrong with the pursuit of such investigations, irrelevant though they may be to the concerns of mathematicians and historians of mathematics.

[1] I borrow the term 'maverick' from the 'opinionated introduction' by Aspray and Kitcher (1988).

Yet it is pertinent to ask whether there are not also other tasks for the philosophy of mathematics, tasks that arise either from the current practice of mathematics or from the history of the subject. A small number of philosophers (including one of us) believe that the answer is yes. Despite large disagreements among the members of this group, proponents of the minority tradition share the view that philosophy of mathematics ought to concern itself with the kinds of issues that occupy those who study the other branches of human knowledge (most obviously the natural sciences). Philosophers should pose such questions as: How does mathematical knowledge grow? What is mathematical progress? What makes some mathematical ideas (or theories) better than others? What is mathematical explanation? (p. 17)

They concluded the introduction by claiming that the current state of the philosophy of mathematics reveals two general programs, one centered on the foundations of mathematics and the other centered on articulating the methodology of mathematics.

Kitcher (1984) had already put forward an account of the growth of mathematical knowledge that is one of the earliest, and still one of the most impressive, studies in the methodology of mathematics in the analytic literature. Starting from the notion of a mathematical practice,[2] Kitcher's aim was to account for the rationality of the growth of mathematics in terms of transitions between mathematical practices. Among the patterns of mathematical change, Kitcher discussed generalization, rigorization, and systematization.

One of the features of the 'maverick' tradition was the polemic against the ambitions of mathematical logic as a canon for philosophy of mathematics. Mathematical logic, which had been essential in the development of the foundationalist programs, was seen as ineffective in dealing with the questions concerning the dynamics of mathematical discovery and the historical development of mathematics itself. Of course, this did not mean that philosophy of mathematics in this new approach was reduced to the pure description of mathematical theories and their growth. It is enough to think that Lakatos' *Proofs and Refutations* rests on the interplay between the 'rational reconstruction' given in the main text and the 'historical development' provided in the notes. The relation between these two aspects is very problematic and remains one of the central issues for Lakatos scholars and for the formulation of a dialectical philosophy of mathematics (see Larvor (1998)). Moreover, in addition to providing an empiricist philosophy of mathematics, Kitcher proposed a theory of mathematical change that was based on a rather idealized model (see Kitcher 1984, Chapters 7–10).

---

[2] A quintuple consisting of five components: 'a language, a set of accepted statements, a set of accepted reasonings, a set of questions selected as important, and a set of metamathematical views' (Kitcher, 1984).

A characterization in broad strokes of the main features of the 'maverick' tradition could be given as follows:

a. anti-foundationalism, i.e. there is no certain foundation for mathematics; mathematics is a fallible activity;
b. anti-logicism, i.e. mathematical logic cannot provide the tools for an adequate analysis of mathematics and its development;
c. attention to mathematical practice: only detailed analysis and reconstruction of large and significant parts of mathematical practice can provide a philosophy of mathematics worth its name.

Quine's dissolution of the boundary between analytic and synthetic also helped in this direction, for setting mathematics and natural science on a par led first to the possibility of a theoretical analysis of mathematics in line with natural science and this, in turn, led philosophers to apply tools of analysis to mathematics which had meanwhile become quite fashionable in the history and philosophy of the natural sciences (through Kuhn, for instance). This prompted questions by analogy with the natural sciences: Is mathematics revisable? What is the nature of mathematical growth? Is there progress in mathematics? Are there revolutions in mathematics?

There is no question that the 'mavericks' have managed to extend the boundaries of philosophy of mathematics. In addition to the works already mentioned I should refer the reader to Gillies (1992), Grosholz and Breger (2000), van Kerkhove and van Bengedem (2002, 2007), Cellucci (2002), Krieger (2003), Corfield (2003), Cellucci and Gillies (2005), and Ferreiros and Gray (2006) as contributions in this direction, without of course implying that the contributors to these books and collections are in total agreement with either Lakatos or Kitcher. One should moreover add the several monographs published on Lakatos' philosophy of mathematics, which are often sympathetic to his aims and push them further even when they criticize Lakatos on minor or major points (Larvor (1998), Koetsier (1991); see also Bressoud (1999)).

However, the 'maverick tradition' has not managed to substantially redirect the course of philosophy of mathematics. If anything, the predominance of traditional ontological and epistemological approaches to the philosophy of mathematics in the last twenty years proves that the maverick camp did not manage to bring about a major reorientation of the field. This is not *per se* a criticism. Bringing to light important new problems is a worthy contribution in itself. However, the iconoclastic attitude of the 'mavericks' *vis-à-vis* what had been done in foundations of mathematics had as a consequence a reduction of their sphere of influence. Logically trained philosophers of mathematics and traditional epistemologists and ontologists of mathematics felt that the 'mavericks' were throwing away the baby with the bathwater.

Within the traditional background of analytic philosophy of mathematics, and abstracting from Kitcher's case, the most important direction in connection to mathematical practice is that represented by Maddy's naturalism. Roughly, one could see in Quine's critique of the analytic/synthetic distinction a decisive step for considering mathematics methodologically on a par with natural science. This is especially clear in a letter to Woodger, written in 1942, where Quine comments on the consequences brought about by his (and Tarski's) refusal to accept the Carnapian distinction between the analytic and the synthetic. Quine wrote:

> Last year logic throve. Carnap, Tarski and I had many vigorous sessions together, joined also, in the first semester, by Russell. Mostly it was a matter of Tarski and me against Carnap, to this effect. (a) C[arnap]'s professedly fundamental cleavage between the analytic and the synthetic is an empty phrase (cf. my "Truth by convention"), and (b) *consequently* the concepts of logic and mathematics are as deserving of an empiricist or positivistic critique as are those of physics. (quoted in Mancosu (2005); my emphasis)

The spin Quine gave to the empiricist critique of logic and mathematics in the early 1940s was that of probing how far one could push a nominalistic conception of mathematics. But Quine was also conscious of the limits of nominalism and was led, reluctantly, to accept a form of Platonism based on the indispensability, in the natural sciences, of quantifying over some of the abstract entities of mathematics (see Mancosu (Forthcoming) for an account of Quine's nominalistic engagement).

However, Quine's attention to mathematics was always directed at its logical structure and he showed no particular interest in other aspects of mathematical practice. Still, there were other ways to pursue the possibilities that Quine's teachings had opened. In Section 3 of this introduction I will discuss the consequences Maddy has drawn from the Quinean position. Let me mention as an aside that the analogy between mathematics and physics was also something that emerged from thinkers who were completely opposed to logical empiricism or Quinean empiricism, most notably Gödel. We will see how Maddy combines both the influence of Quine and Gödel. Her case is of interest, for her work (unlike that of the 'mavericks') originates from an active engagement with the foundationalist tradition in set theory.

The general spirit of the tradition originating from Lakatos as well as Maddy's naturalism requires extensive attention to mathematical practice. This is not to say that classical foundational programs were removed from such concerns. On the contrary, nothing is further from the truth. Developing a formal language, such as Frege did, which aimed at capturing formally all valid forms of reasoning occurring in mathematics, required a keen understanding of the reasoning patterns to be found in mathematical practice.[3] Central to Hilbert's program was, among other things, the distinction between real and ideal elements that also originates in mathematical practice. Delicate attention to certain aspects of mathematical practice informs contemporary proof theory and, in particular, programs such as reverse mathematics. Finally, Brouwer's intuitionism takes its origin from the distinction between constructive vs. non-constructive procedures, once again a prominent distinction in, just to name one area, the debates in algebraic number theory in the late 19th century (Kronecker vs. Dedekind). Moreover, the analytical developments in philosophy of mathematics are also, to various extents, concerned with certain aspects of mathematical practice. For instance, nominalistic programs force those engaged in reconstructing parts of mathematics and natural science to pay special attention to those branches of mathematics in order to understand whether a nominalistic reconstruction can be obtained.

This will not be challenged by those working in the Lakatos tradition or by Maddy or by the authors in this collection. But in each case the appeal to mathematical practice is different from that made by the foundationalist tradition as well as by most traditional analytic philosophers of mathematics in that the latter were limited to a central, but ultimately narrow, aspect of the variety of activities in which mathematicians engage. This will be addressed in the following sections.

My strategy for the rest of the introduction will be to discuss in broad outline the contributions of Corfield and Maddy, taken as representative philosophers of mathematics deeply engaged with mathematical practice, yet who come from different sides of the foundational/maverick divide. I will begin with Corfield, who follows in the Lakatos lineage, and then move to Maddy, taken as an exemplar of certain developments in analytic philosophy. It is within this background, and by contrast with it, that I will present, in Section 4, the contributions contained in this volume and articulate, in Section 5, how they differ from, and relate to, the traditions being currently described. Regretfully, I will have to refrain from treating many other contributions that would deserve extensive discussion, most notably Kitcher (1984), but completeness is not what I am aiming at here.

---

[3] For a reading of Frege which stresses the connection to mathematical practice, see Tappenden (2008).

2 Corfield's *Towards a Philosophy of Real Mathematics* (2003)

A good starting point is Corfield's recent book *Towards a Philosophy of Real Mathematics* (2003). Corfield's work fits perfectly within the frame of the debate between foundationalists and 'maverick' philosophers of mathematics I described at the outset. Corfield attributes his desire to move into philosophy of mathematics to the discovery of Lakatos' *Proofs and Refutations* (1976) and he takes as the motto for his introduction Lakatos' famous paraphrasing of Kant:

> The history of mathematics, lacking the guidance of philosophy, has become *blind*, while the philosophy of mathematics, turning its back on the most intriguing phenomena in the history of mathematics, has become *empty*. (Lakatos, 1976, p. 2)

Corfield's proposal for moving out of the impasse is to follow in Lakatos' footsteps, and he proposes a philosophy of 'real' mathematics. A succinct description of what this is supposed to encompass is given in the introduction:

> What then is a *philosophy* of real mathematics? The intention of this term is to draw a line between work informed by the concerns of mathematicians past and present and that done on the basis of at best token contact with its history or practice. (Corfield, 2003, p. 3)

Thus, according to Corfield, neo-logicism is not a philosophy of real mathematics, as its practitioners ignore most of 'real' 20th century mathematics and most historical developments in mathematics with the exception of the foundational debates. In addition, the issues raised by such philosophers are not of concern to mathematicians. For Corfield, contemporary philosophy of mathematics is guilty of not availing itself of the rich trove of the history of the subject, simply dismissed as 'history' (you have to say that with the right disdainful tone!) in the analytic literature, not to mention a first-hand knowledge of its actual practice. Moreover,

> By far the larger part of activity in what goes by the name *philosophy of mathematics* is dead to what mathematicians think and have thought, aside from an unbalanced interest in the 'foundational' ideas of the 1880–1930 period, yielding too often a distorted picture of that time. (Corfield, 2003, p. 5)

It is this 'foundationalist filter', as Corfield calls it, which he claims is responsible for the poverty of contemporary philosophy of mathematics. There are two major parts to Corfield's enterprise. The first, the *pars destruens*, consists in trying to dismantle the foundationalist filter. The second, the *pars construens*, provides philosophical analyses of a few case studies from mainstream mathematics of the last seventy years. His major case studies come from the interplay between mathematics and computer science and from n-dimensional algebras and algebraic topology.

The *pars destruens* shares with Lakatos and some of his followers a strong anti-logical and anti-foundational polemic. This has unfortunately damaged the reception of Corfield's book and has drawn attention away from the good things contained in it. It is not my intention here to address the significance of what Corfield calls the 'foundationalist filter' or to rebut the arguments given by Corfield to dismantle it (on this see Bays (2004) and Paseau (2005)). Let me just mention that a very heated debate on this topic took place in October 2003 on the FOM (Foundations of Mathematics) email list. The *pars destruens* in Corfield's book is limited to some arguments in the introduction. Most of the book is devoted to showing by example, as it were, what a philosophy of mathematics could do and how it could expand the range of topics to be investigated. This new philosophy of mathematics, a philosophy of 'real mathematics' aims at the following goals:

> Continuing Lakatos' approach, researchers here believe that a philosophy of mathematics should concern itself with what leading mathematicians of their day have achieved, how their styles of reasoning evolve, how they justify the course along which they steer their programmes, what constitute obstacles to their programmes, how they come to view a domain as worthy of study and how their ideas shape and are shaped by the concerns of physicists and other scientists. (p. 10)

This opens up a large program which pursues, among other things, the dialectical nature of mathematical developments, the logic of discovery in mathematics, the applicability of mathematics to the natural sciences, the nature of mathematical modeling, and what accounts for the fruitfulness of certain concepts in mathematics.

More precisely, here is a list of topics that motivate large chunks of Corfield's book:

1) Why are some mathematical entities important, natural, and fruitful while others are not?
2) What accounts for the connectivity of mathematics? How is it that concepts developed in one part of mathematics suddenly turn out to be connected to apparently unrelated concepts in other areas?
3) Why are computer proofs unable to provide the sort of understanding at which mathematicians aim?

4) What is the role of analogy and other types of inductive reasoning in mathematics? Can Bayesianism be applied to mathematics?
5) What is the relationship between diagrammatic thinking and formal reasoning? How to account for the fruitfulness of diagrammatic reasoning in algebraic topology?

Of course, several of these issues had already been discussed in the literature before Corfield, but his book was the first to bring them together. Thus, Corfield's proposed philosophy of mathematics displays the three features of the mavericks' approach mentioned at the outset. In comparison with previous contributions in that tradition, he expands the set of topics that can be fruitfully investigated and seems to be less concerned than Lakatos and Kitcher with providing a grand theory of mathematical change. His emphasis is on more localized case studies. The foundationalist and the analytic tradition in philosophy of mathematics are dismissed as irrelevant in addressing the most pressing problems for a 'real' philosophy of mathematics. In Section 5, I will comment on how Corfield's program relates to the contributions in this volume.

3 Maddy on mathematical practice

Faithfulness to mathematical practice is for Maddy a criterion of adequacy for a satisfactory philosophy of mathematics (Maddy, 1990, p. 23 and p. 28). In her 1990 book, *Realism in Mathematics*, she took her start from Quine's naturalized epistemology (there is no first philosophy, natural science is the court of arbitration even for its own methodology) and forms of the indispensability argument. Her realism originated from a combination of Quine's Platonism with that of Gödel. But Maddy is also critical of certain aspects of Quine's and Gödel's Platonisms, for she claims that both fail to capture certain aspects of the mathematical experience. In particular, she finds objectionable that unapplied mathematics is not granted right of citizenship in Quine's account (see Quine, 1984, p. 788) and, *contra* Quine, she emphasizes the autonomy of mathematics from physics. By contrast, the Gödelian brand of Platonism respects the autonomy of mathematics but its weakness consists in the postulation of a faculty of intuition in analogy with perception in the natural sciences. Gödel appealed to such a faculty of intuition to account for those parts of mathematics which can be given an 'intrinsic' justification. However, there are parts of mathematics for which such 'intrinsic', intuitive, justifications cannot be given and for those one appeals to 'extrinsic' justifications; that is, a justification in terms of their consequences. *Realism in Mathematics* aims at providing both a naturalistic epistemology that replaces Gödel's intuition as well as a detailed study of the practice of extrinsic justification. It is this latter aspect of the project that leads Maddy, in Chapter 4, to quite interesting methodological studies which involve, among other things, the study of the following notions and aspects of mathematical methodology: verifiable consequence; powerful new methods for solving pre-existing problems; simplifying and systematizing theories; implying previous conjectures; implying 'natural' results; strong intertheoretic connections; and providing new insights into old theorem (see Maddy, 1990, pp. 145–6). These are all aspects of great importance for a philosophy of mathematics that wants to account for mathematical practice. Maddy's study in Chapter 4 focuses on justifying new axioms for set theory ($V = L$ or SC [there exists a supercompact cardinal]). In the end, her analysis of the contemporary situation leads to a request for a more profound analysis of 'theory formation and confirmation':

> What's needed is not just a description of non-demonstrative arguments, but an account of why and when they are reliable, an account that should help set theorists make a rational choice between competing axiom candidates. (Maddy, 1990, p. 148)

And this is described as an open problem not just for the 'compromise platonist' but for a wide spectrum of positions. Indeed, on p. 180, she recommends engagement with such problems of rationality 'even to those philosophers blissfully uninvolved in the debate over Platonism' (p. 180).

In *Naturalism in Mathematics* (1997), the realism defended in *Realism in Mathematics* is abandoned. But certain features of how mathematical practice should be accounted for are retained. Indeed, what seemed a self-standing methodological problem in the first book becomes for Maddy the key problem of the new book and a problem that leads to the abandonment of realism in favor of naturalism. This takes place in two stages. First, she criticizes the cogency of indispensability arguments. Second, she positively addresses the kinds of considerations that set-theorists bring to bear when considering new axioms, the status of statements independent of ZFC, or when debating new methods, and tries to abstract from them more general methodological maxims.

Her stand on the relation between philosophy and mathematics is clear and it constitutes the heart of her naturalism:

> If our philosophical account of mathematics comes into conflict with successful mathematical practice, it is the philosophy that must give. This is not, in itself, a philosophy of mathematics; rather, it is a position on the proper relations between the philosophy of mathematics and the practice of mathematics. Similar

sentiments appear in the writings of many philosophers of mathematics who hold that the goal of philosophy of mathematics is to account for mathematics as it is practiced, not to recommend reform. (Maddy, 1997, p.161)

Naturalism, in the Maddian sense, recognizes the autonomy of mathematics from natural science. Maddy applies her naturalism to a methodological study of the considerations leading the mathematical community to the acceptance or rejection of various (set-theoretical) axioms. She envisages the formulation of 'a naturalized model of practice' (p. 193) that will provide 'an accurate picture of the actual justificatory practice of contemporary set theory and that this justificatory structure is fully rational' (pp. 193–4). The method will proceed by identifying the goals of a certain practice and by evaluating the methodology employed in that branch of mathematics (set theory, in Maddy's case) in relation to those goals (p. 194). The naturalized model of practice is both purified and amplified. It is purified in that it eliminates seemingly irrelevant (i.e. philosophical) considerations in the dynamics of justification; and it is amplified in that the relevant factors are subjected to more precise analysis than what is given in the practice itself and they are also applied to further situations:

> Our naturalist then claims that this model accurately reflects the underlying justificatory structure of the practice, that is, that the material excised is truly irrelevant, that the goals identified are among the actual goals of the practice (and that the various goals interact as portrayed), and that the means-ends reasoning employed is sound. If these claims are true, then the practice, in so far as it approximates the naturalist's model, is rational. (Maddy, 1997, p. 197)

Thus, using the example of the continuum hypothesis and other independent questions in descriptive set theory, she goes on to explain how the goal of providing 'a complete theory of sets of real numbers' gives rational support to the investigation of CH (and other questions in descriptive set theory). The tools for such investigations will be mathematical and not philosophical. While a rational case for or against CH cannot be built out of the methodology that Maddy distils from the practice, she provides a case against $V = L$ (an axiom that Quine supported).

We need not delve into the details of Maddy's analysis of her case studies and the identification of several methodological principles, such as *maximize* and *unify*, that in her final analysis direct the practice of set theorists and constitute the core of her case against $V = L$. Rather, let us take stock.

Comparing Maddy's approach to that of the 'maverick' tradition, we can remark that just as in the 'maverick' tradition, there is a shift in what problems Maddy sets out to investigate. While not denying that ontological and epistemological problems are worthy of investigation she has decided to focus on an aspect of the methodology of mathematics completely ignored in previous analytic philosophy of mathematics. This is a large subject area concerning the sort of arguments that are brought to bear in the decision in favor or against certain new axioms in set theory. Previous analytic philosophy of mathematics would have relegated this to the 'context of discovery' and as such not worthy or suitable for rigorous investigation. Maddy counters that these decisions are rational and can be accounted for by a naturalistic model that spells out the principles and maxims directing the practice. Maddy's project can be seen as a contribution to the general problem of how evidence and justification functions in mathematics. This can be seen as related to a study of 'heuristics', although this has to be taken in the appropriate sense as her case studies cannot be confused with, or reduced to, traditional studies on 'problem solving'. Another feature of Maddy's work that ties her approach to that of the mavericks is the appeal to the history of logic and mathematics as a central component in her naturalized account. This is not surprising: mathematical practice is embodied in the concrete work of mathematicians and that work has taken place in history. Although Maddy, unlike Kitcher 1984, is not proposing an encompassing account of the rationality in the changes in mathematical practice, or a theory of mathematical growth, the case studies she investigated have led her to consider portions of the history of analysis and of set theory. The history of set theory (up to its present state) is the 'laboratory' for the distillation of the naturalistic model of the practice. Finally, a major difference in attitude between Maddy and the 'mavericks' is the lack on Maddy's part of any polemic against logic and foundations. Rather, her ambition is one of making sense of the inner rationality of foundational work in set theory.

4 This collection

The rather stark contrast used to present different directions of philosophical work on mathematical practice in Sections 2 and 3 would not be appropriate to characterize some of the most recent contributions in this area, in which a variety of approaches often coexist together. This is especially true of the volumes 'Perspectives on Mathematical Practices' (van Kerkhove and van Bengedem 2002 and 2007) which contain a variety of contributions, some of which find their inspiration in the maverick tradition and others in Maddy's work, while others yet point the way to independent developments. Similar considerations apply to Mancosu *et al.* (2005), although in contrast to the two former collections, this book does not trace its inspiration back to Lakatos. It contains a wide range of contributions on visualization, explanation and reasoning styles in mathematics carried out both by philosophers and historians of mathematics. The above-mentioned volumes contain contributions that overlap in topic and/or inspiration with those of the present collection. However, this collection is more systematic and more focused in its aims.

The eight topics studied here are:

1) Visualization
2) Diagrammatic reasoning
3) Explanation
4) Purity of methods
5) Concepts and definitions
6) Philosophical aspects of uses of computer science in mathematics
7) Category theory
8) Mathematical physics

Taken all together, they represent a broad spectrum of contemporary philosophical reflection on different aspects of mathematical practice. Each author (with one exception to be mentioned below) has written a general introduction to the subject area and a research paper in that area. I will not here summarize the single contributions but rather point out why each subject area is topical.

The first section is on **Visualization**. Processes of visualization (e.g. by means of mental imagery) are central to our mathematical activity and recently this has become once again a central topic of concern due to the influence of computer imagery in differential geometry and chaos theory and to the call for visual approaches to geometry, topology, and complex analysis. But in what sense can mental imagery provide us with mathematical knowledge? Shouldn't visualization be relegated to heuristics? Marcus Giaquinto (University College London) argues in his introduction that mathematical visualization can play an epistemic role. Then, in his research paper, he proceeds to examine the role of visual resources in cognitive grasp of structures.

The second section is entitled **Diagrammatic reasoning**. In the last twenty years there has been an explosion of interest in this topic due also to the importance of such diagrammatic systems for artificial intelligence and their extended use in certain branches of contemporary mathematics (knot theory, algebraic topology, etc.). Kenneth Manders (University of Pittsburgh) focuses in his introduction on some central philosophical issues emerging from diagrammatic reasoning in geometry and in his research paper—an underground classic that finally sees publication—he addresses the problem of the stability of diagrammatic reasoning in Euclidean geometry.

If mathematicians cared only about the truth of certain results, it would be hard to understand why after discovering a certain mathematical truth they often go ahead to prove the result in several different ways. This happens because different proofs or different presentations of entire mathematical areas (complex analysis etc.) have different epistemic virtues. **Explanation** is among the most important virtues that mathematicians seek. Very often the proof of a mathematical result convinces us *that* the result is true but does not tell us *why* it is true. Alternative proofs, or alternative formulations of entire theories, are often given with this explanatory aim in mind. In the introduction, Paolo Mancosu (U.C. Berkeley) shows that the topic of mathematical explanation has far-reaching philosophical implications and then he proceeds, in the joint paper with Johannes Hafner (North Carolina State), to test Kitcher's model of mathematical explanation in terms of unification by means of a case study from real algebraic geometry.

Related to the topic of epistemic virtues of different mathematical proofs is the ideal of **Purity of methods** in mathematics. The notion of purity has played an important role in the history of mathematics—consider, for instance, the elimination of geometrical intuition from the development of analysis in the 19th century—and in a way it underlies all the investigations concerning issues of conservativity in contemporary proof theory. That purity is often cherished in mathematical practice is made obvious by the fact that Erdös and Selberg were awarded the Fields Medal for the elementary proof of the prime number theorem (already demonstrated with analytical tools in the late 19th century. But why do mathematicians cherish purity? What is epistemologically to be gained by proofs that exclude appeal to 'ideal' elements? Proof theory has given us a rich analysis of when ideal elements can be eliminated in principle (conservativity results), but what proof theory leaves open is the philosophical question of why and whether we should seek either the use or the elimination of such ideal elements. Michael Detlefsen (University of Notre Dame) provides a general historical and conceptual introduction to the topic. This is followed by a study of purity in Hilbert's work on the foundations of geometry written by Michael Hallett (McGill University). In addition to emphasizing the epistemic role of purity, he also shows that in mathematical practice the dialectic between purity and impurity is often very subtle indeed.

Mathematicians seem to have a very good sense of when a particular mathematical concept or theory is fruitful or 'natural'. A certain concept might provide the 'natural' setting for an entire development and reveal this by its fruitfulness in unifying a wide group of results or by opening unexpected new vistas. But when mathematicians appeal to such virtues of concepts (fruitfulness, naturalness, etc.) are they simply displaying subjective tastes or are there objective features, which can be subjected to philosophical analysis, which can account for the rationality and objectivity of such judgments? This is the topic of the introductory chapter on **Concepts and Definitions** written by Jamie Tappenden (University of Michigan) who has already published on these topics in relation to geometry and complex analysis in the 19th century. His research paper ties the topic to discussions on naturalness in contemporary metaphysics and then discusses Riemann's conceptual approach to analysis as an example from mathematical practice in which the natural definitions give rise to fruitful results.

The influence of **Computer Science** on contemporary mathematics has already been mentioned in connection with visualization. But uses of the computer are now pervasive in contemporary mathematics, and some of the aspects of this influence, such as the computer proof of the four-color theorem, have been sensationally popularized. Computers provide an aid to mathematical discovery; they provide experimental, inductive confirmation of mathematical hypotheses; they carry out calculations that are needed in proofs; and they enable one to obtain formal verification of proofs. In the introduction to this section, Jeremy Avigad (Carnegie Mellon University) addresses the challenges that philosophy will have to meet when addressing these new developments. In particular, he calls for an extension of ordinary epistemology of mathematics that will address issues that the use of computers in mathematics make urgent, such as the problem of characterizing mathematical evidence and mathematical understanding. In his research paper he then goes on to focus on mathematical understanding and here, in an interesting reversal of perspective, he shows how formal verification can assist us in developing an account of mathematical understanding, thereby showing that the epistemology of mathematics can inform and be informed by research in computer science.

Some of the most spectacular conceptual achievements in 20th century mathematics are related to the developments of **Category Theory** and its role in areas such as algebraic geometry, algebraic topology, and homological algebra. Category theory has interest for the philosopher of mathematics both on account of the claim made on its behalf as an alternative foundational framework for all of mathematics (alternative to Zermelo–Fraenkel set theory) as well as for its power of unification and its fruitfulness revealed in the above-mentioned areas. Colin McLarty (Case Western Reserve University) devotes his introduction to spelling out how the structuralism involved in much contemporary mathematical practice threatens certain reductionist projects influenced by set-theoretic foundations and then proceeds to argue that only detailed attention to the structuralism embodied in the practice (unlike other philosophical structuralisms) can account for certain aspects of contemporary mathematics, such as the 'unifying spirit' that pervades it. In his research paper he looks at *schemes* as a tool for pursuing Weil's conjectures in number theory as a case study for seeing how 'structuralism' works in practice. In the process he draws an impressive fresco of how structuralist and categorial ideas developed from Noether through Eilenberg and Mac Lane to Grothendieck.

Finally, the last chapter is on the philosophical problems posed by some recent developments in **Mathematical Physics** and how they impact pure mathematics. In the third quarter of the 20th century what seemed like an inevitable divorce between physics and pure mathematics turned into an exciting renewal of vows. Developments in pure mathematics turned out to be incredibly fruitful in mathematical physics and, vice versa, highly speculative developments in mathematical physics turned out to bear extremely fruitful results in mathematics (for instance in low-dimensional topology). However, the standards of acceptability between the two disciplines are very different. Alasdair Urquhart (University of Toronto) describes in his introduction some of the main features of this renewed interaction and the philosophical problems posed by a variety of physical arguments which, despite their fruitfulness, turn out to be less than rigorous. This is pursued in his research paper where several examples of 'non-rigorous' proofs in mathematics and physics are discussed with the suggestion that logicians and mathematicians should not dismiss these developments but rather try to make sense of these unruly parts of the mathematical universe and to bring the physicists' insights into the realm of rigorous argument.

5 A comparison with previous developments

The time has come to articulate how this collection differs from previous traditions of work in philosophy of mathematical practice. Let us begin with the Lakatos tradition.

There are certainly a remarkable number of differences. First of all, Lakatos and many of the Lakatosians (for instance, Lakatos (1976), Kitcher (1984)) were quite concerned with metaphilosophical issues such as: How do history and philosophy of mathematics fit together? How does mathematics grow? Is the process of growth rational? The aim of the authors in the collection is much more restricted. While not dismissing these questions, we think a good amount of humility is needed to avoid the risk of theorizing without keeping our feet on the ground. It is interesting to note that also a recent volume written by historians and philosophers of mathematics (Ferreiros and Gray 2006) displays the same modesty with respect to these metaphilosophical issues. At the same time, the number of topics we touch upon is immeasurably vaster than the ones addressed by the Lakatos tradition. Visualization, diagrammatic reasoning, purity of methods, category theory, mathematical physics, and many other topics we investigate here are remarkably absent from a tradition which has made attention to mathematical practice its call to arms. One exception here is Corfield (2003), which does indeed touch upon many of the topics we study. However, and this is another important point, we differ from Corfield in two essential points. First of all, the authors of this collection do not engage in polemic with the foundationalist tradition and, as a matter of fact, many of them work, or have worked, also as mathematical logicians (of course, there are differences of attitude, vis-à-vis foundations, among the contributors). We are, by and large, calling for an extension to a philosophy of mathematics that will be able to address topics that the foundationalist tradition has ignored. But that does not mean that we think that the achievements of this tradition should be discarded or ignored as being irrelevant to philosophy of mathematics. Second, unlike Corfield, we do not dismiss the analytic tradition in philosophy of mathematics but rather seek to extend its tools to a variety of areas that have been, by and large, ignored. For instance, to give one example among many, the chapter on explanation shows how the topic of mathematical explanation is connected to two major areas of analytic philosophy: indispensability arguments and models of scientific explanation. But this conciliatory note should not hide the force of our message: we think that the aspects of mathematical practice we investigate are absolutely vital to an understanding of mathematics and that having ignored them has drastically impoverished analytic philosophy of mathematics.

In the Lakatos tradition, it was Kitcher in particular who attempted to build a bridge with analytic philosophy. For instance, when engaged in his work on explanation in philosophy of science he also made sure that mathematical explanation was also taken into account. We are less ambitious than Kitcher, in that we do not propose a unified epistemology and ontology of mathematics and a theory of how mathematical knowledge grows rationally. But we are much more ambitious in another respect, in that we cover a broad spectrum of case studies arising from mathematical practice which we subject to analytic investigation. Thus, in addition to the case of explanation already mentioned, Giaquinto investigates whether synthetic *a priori* knowledge can be obtained by appealing to experiences of visualization; Tappenden engages recent work in metaphyics when discussing fruitfulness and naturalness of concepts; and MacLarty shows how structuralism in mathematical practice can help us evaluate structuralist philosophies of mathematics. Indeed, the whole book is an attempt to expand the boundaries of epistemology of mathematics well beyond the problem of how we can have access to abstract entities.

But coming now to the analytic developments related to Maddy's work, I should point out that whereas Maddy has limited her investigations to set theory we take a much wider perspective on mathematical practice, drawing our case studies from geometry, complex analysis, real algebraic geometry, category theory, computer science, and mathematical physics. Once again, we believe that while set theory is a very important subject of methodological investigation, there are central phenomena that will be missed unless we cast our net more broadly and extend our investigations to other areas of mathematics. Moreover, while Maddy's study of set-theoretic methodology has some points of contact with our investigations (evidence, fruitfulness, theory-choice) we look at a much broader set of issues that never come up for discussion in her work (visualization, purity of methods, explanation, rigor in mathematical physics). The closest point of contact between her investigations and this book is probably the discussion of evidence in Avigad's introduction. There is also no explicit commitment in our contributions to the form of mathematical naturalism advocated by Maddy; actually, the spirit of many of our contributions seems to go against the grain of her philosophical position.

Let me conclude by coming back to the comparison with the situation in philosophy of science. I mentioned at the beginning that recent philosophy of science has thrived under the interaction of traditional problems (realism vs. instrumentalism, causality, etc.) with more localized studies in the philosophies of the special sciences. In general, philosophers of science are happy with claiming that both areas are vital for the discipline. Corfield takes as the model for his approach to the philosophy of mathematics the localized studies in the philosophy of physics, but decrees that classical philosophy of mathematics is a useless pursuit (see Pincock (2005)). As for Maddy, she gets away from traditional ontological and epistemological issues (realism, nominalism, etc.) by means of her naturalism. What is distinctive in this volume is that we integrate local studies with general philosophy of mathematics, *contra* Corfield, and we also keep traditional ontological and epistemological topics in play, *contra* Maddy.

Hopefully, the reader will realize that my aim has not been to make any invidious comparisons but only to provide a fair account of what the previous traditions have achieved and why we think we have achieved something worth proposing to the reader. We are only too aware that we are making the first steps in a very difficult area and we hope that our efforts might stimulate others to do better.

Acknowledgements. I would like to thank Jeremy Avigad, Paddy Blanchette, Marcus Giaquinto, Chris Pincock, Thomas Ryckman, José Sagüillo, and Jamie Tappenden for useful comments on a previous draft.

Bibliography

ASPRAY, William and KITCHER, Philip (1988), *History and Philosophy of Modern Mathematics* (Minneapolis: University of Minnesota Press).

BAYS, Timothy (2004), 'Review of Corfield's *Towards a Philosophy of Real Mathematics*', <http://ndpr.nd.edu/review.cfm?id=1042>.

BRESSOUD, David (1999), *Proofs and Confirmations: The Story of the Alternating Sign Matrix Conjecture* (Cambridge: Cambridge University Press).

CELLUCCI, C. and GILLIES, D. (eds.) (2005), *Mathematical Reasoning and Heuristics* (London: King's College Publications).

CELLUCCI, Carlo (2002), *Filosofia e Matematica* (Bari: Laterza).

CORFIELD, David (2003), *Towards a Philosophy of Real Mathematics* (Cambridge: Cambridge University Press).

DAVIS, Philip and HERSH, Reuben (1980), *The Mathematical Experience* (Basel: Birkhäuser).

FERREIROS, J. and GRAY, J. (eds.) (2006), *The Architecture of Modern Mathematics* (Oxford: Oxford University Press).

GILLIES, D. (ed.) (1992), *Revolutions in Mathematics* (Oxford: Oxford University Press).

GROSHOLZ, E. and BREGER, H. (eds.) (2000), *The Growth of Mathematical Knowledge* (Dordrecht: Kluwer).

HERSH, R. (ed.) (2006), *18 Unconventional Essays on the Nature of Mathematics* (New York: Springer).

KITCHER, Philip (1984), *The Nature of Mathematical Knowledge* (Oxford: Oxford University Press).

KLINE, Morris (1980), *Mathematics. The Loss of Certainty* (Oxford: Oxford University Press).

KOETSIER, Teun (1991), *Lakatos' Philosophy of Mathematics: A Historical Approach* (Amsterdam: North Holland).

KRIEGER, Martin (2003), *Doing Mathematics: Convention, Subject, Calculation, Analogy* (Singapore: World Scientific Publishing).

LAKATOS, Imre (1976), *Proofs and Refutations* (Cambridge: Cambridge University Press).

LARVOR, Brendan (1998), *Lakatos: An Introduction* (London: Routledge).

MADDY, Penelope (1990), *Realism in Mathematics* (Oxford: Oxford University Press).

MADDY, Penelope (1997), *Naturalism in Mathematics* (Oxford: Oxford University Press).

MANCOSU, Paolo (2005), 'Harvard 1940–41: Tarski, Carnap and Quine on a Finitistic Language of Mathematics for Science', *History and Philosophy of Logic*, 26, 327–357.

—— (Forthcoming), 'Quine and Tarski on Nominalism', *Oxford Studies in Metaphysics*.

MANCOSU, P., JØRGENSEN, K., and PEDERSEN, S.(eds.) (2005), *Visualization, Explanation and Reasoning Styles in Mathematics* (Dordrecht: Springer).

PASEAU, Alex (2005), 'What the Foundationalist Filter Kept Out', *Studies in History and Philosophy of Science*, 36, 191–201.

PINCOCK, Chris (2005), 'Review of Corfield's *Towards a Philosophy of Real Mathematics*', *Philosophy of Science*, 72, 632–634.

QUINE, Willard Van Orman (1984), 'Review of Parsons' Mathematics in Philosophy', *Journal of Philosophy*, 81, 783–794.

TAPPENDEN, Jamie (2008), *Philosophy and the Origins of Contemporary Mathematics: Frege in his Mathematical Context* (Oxford: Oxford University Press).

TYMOCZKO, Thomas (1985), *New Directions in the Philosophy of Mathematics* (Basel: Birkhäuser).

VAN KERKHOVE, B. and VAN BENGEDEM, J. P. (eds.) (2002), *Perspectives on Mathematical Practices*, Special Issue of Logique and Analyse, 179–180 [published in 2004].

—— (2007), *Perspectives on Mathematical Practices* (Dordrecht: Springer-Verlag).

1

Visualizing in Mathematics

MARCUS GIAQUINTO

Visual thinking in mathematics is widespread; it also has diverse kinds and uses. Which of these uses is legitimate? What epistemic roles, if any, can visualization play in mathematics? These are the central philosophical questions in this area. In this introduction I aim to show that visual thinking does have epistemically significant uses. The discussion focuses mainly on visual thinking in proof and discovery and touches lightly on its role in understanding.

1.1 The context

'Mathematics can achieve nothing by concepts alone but hastens at once to intuition' wrote Kant (1781/9, A715/B743), before describing the geometrical construction in Euclid's proof of the angle sum theorem (Euclid, Book 1, proposition 32). The Kantian view that visuo-spatial thinking is essential to mathematical knowledge and yet consistent with its *a priori* status probably appealed to mathematicians of the late 18th century. By the late 19th century a different view had emerged: Dedekind, for example, wrote of an overpowering feeling of dissatisfaction with appeal to geometric intuitions in basic infinitesimal analysis (Dedekind, 1872, Introduction). The grounds were felt to be uncertain, the concepts employed vague and unclear. When such concepts were replaced by precisely defined alternatives that did not rely on our sense of space, time, and motion, our intuitive expectations turned out to be unreliable: an often cited example is the belief that a continuous function on an interval of real numbers is everywhere differentiable except at isolated points. Even in geometry the use of figures came to be regarded as unreliable: 'the theorem is only truly demonstrated if the proof is completely independent of the figure'

Pasch said, and he was echoed by Hilbert and Russell (Pasch, 1882; Hilbert, 1894; Russell, 1901).

In some quarters this turn led to a general disdain for visual thinking in mathematics: 'In the best books' Russell pronounced 'there are no figures at all.' (Russell, 1901). Although this attitude was opposed by some prominent mathematicians, others took it to heart. Landau, for example, wrote a calculus textbook without a single diagram (Landau, 1934). But the predominant view was not so extreme: thinking in terms of figures was valued as a means of facilitating grasp of formulae and linguistic text, but only reasoning expressed by means of formulae and text could bear any epistemological weight.

By the late 20th century the mood had swung back in favour of visualization: Mancosu (2005) provides an excellent survey. We find books that advertise their defiance of anti-visual puritanism in their titles, such as *Visual Geometry and Topology* (Fomenko, 1994) and *Visual Complex Analysis* (Needham, 1997); mathematics educators turn their attention to pedagogical uses of visualization (Zimmermann and Cunningham, 1991); the use of computer-generated imagery begins to bear fruit at research level (Hoffman, 1987; Palais, 1999); and diagrams find their way into research papers in abstract fields: see for example the papers on higher dimensional category theory by Joyal *et al.* (1996), Leinster (2004), and Lauda (2005). But attitudes to the epistemology of visual thinking remain mixed. The discussion is almost entirely confined to the role of diagrams in proofs. In some cases, it is claimed, a picture alone is a proof (Brown, 1999, Ch. 3). But that view is rare. Even the editor of *Proofs without Words: Exercises in Visual Thinking*, writes 'Of course, "proofs without words" are not really proofs' (Nelsen, 1993, p. vi). At the other extreme is the stolid attitude of Tennant (1986, p. 304):

> [the diagram] has no proper place in the proof as such. For the proof is a syntactic object consisting only of sentences arranged in a finite and inspectable array.

Between the extremes, others hold that, even if no picture alone is a proof, visual images can have a non-superfluous role in reasoning that constitutes a proof (Barwise and Etchemendy, 1996a; Norman, 2006). Visual representations, such as geometric diagrams, graphs, and maps, all carry information. Taking valid deductive reasoning to be the reliable extraction of information from information already obtained, Barwise and Etchemendy (1996a) pose the following question: Why cannot the representations composing a proof be visual as well as linguistic? The sole reason for denying this role to visual representations is the thought that, with the possible exception of very restricted cases, visual thinking is unreliable, hence cannot contribute to proof. In the next section I probe this matter by considering visualization in proving, where that excludes what is involved in the constructive phase, such as getting the main ideas for a proof, but covers thinking through the steps in a proof, either for the first time or following a given proof, in such a way that the soundness of the argument is apparent to the thinker.

1.2 Proving

We should distinguish between a proof and a presentation of a proof. A proof can be presented in different ways. How different can distinct presentations be and yet be presentations of the same proof? There is no context-invariant answer to this, and even within a context there may be some indeterminacy. Usually mathematicians are happy to regard two presentations as presenting the same proof if the central idea is the same in both cases. But if one's main concern is with what is involved in thinking through a proof, its central idea is not enough to individuate it: the overall structure, the sequence of steps and perhaps other factors affecting the cognitive processes involved will be relevant. Even so, not every cognitive difference in the processes of following a proof will entail distinctness of proofs: in some cases, presumably, the same bits of information in the same order can be given in ink and in Braille.

Once individuation of proofs has been settled, we can distinguish between replaceable thinking and superfluous thinking. In the process of thinking through a proof, a given part of the thinking is replaceable if thinking of some other kind could stand in place of the given part in a process that would count as thinking through the same proof. A given part of the thinking is superfluous if its excision without replacement would be a process of thinking through the same proof. Let us agree that there can be superfluous diagrammatic thinking in thinking through a proof, thinking which serves merely to facilitate or reinforce understanding of the text. This leaves several possibilities.

(a) All thinking that involves a diagram in thinking through a proof is superfluous.
(b) Not all thinking that involves a diagram in thinking through a proof is superfluous; but if not superfluous it will be replaceable by non-diagrammatic thinking.
(c) Some thinking that involves a diagram in thinking through a proof is neither superfluous nor replaceable by non-diagrammatic thinking.

The negative view stated earlier that diagrams can have no role in proof entails claim (a). The idea behind (a) is that, because visual reasoning is unreliable, if a process of thinking through an argument contains some non-superfluous visual thinking, that process lacks the epistemic security to be a case of thinking through a proof. This view, claim (a) in particular, is threatened by cases in which the reliability of visual thinking is demonstrated non-visually. The clearest kind of example would be provided by a formal system which has diagrams in place of formulas among its syntactic objects, and types of inter-diagram transition for inference rules. Suppose you take in such a formal system and an interpretation of it, and then think through a proof of the system's soundness with respect to that interpretation; suppose you then inspect a sequence of diagrams, checking along the way that it constitutes a derivation in the system; suppose finally that you recover the interpretation to reach a conclusion. That entire process would constitute thinking through a proof of the conclusion; and the visual thinking involved would not be superfluous. Such a case has in fact been realized. A formal diagrammatic system of Euclidean geometry called 'FG' has been set out and shown to be sound by Nathaniel Miller (2001). Figure 1.1 presents Miller's derivation in FG of Euclid's first theorem that on any given finite line segment an equilateral triangle can be constructed.

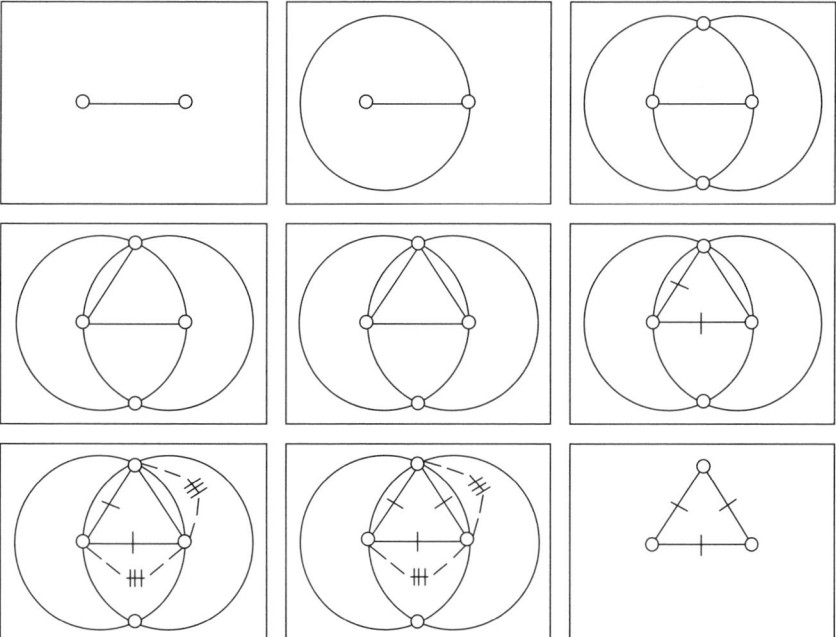

Fig. 1.1.

Miller himself has surely gone through exactly the kind of process described above. Of course the actual event would have been split up in time, and Miller would have already known the conclusion to be true; still, the whole thing would have been a case of thinking through a proof, a highly untypical case of course, in which visual thinking occurred in a non-superfluous way.

This is enough to refute claim (a), the claim that all diagrammatic thinking in thinking through a proof is superfluous. What about Tennant's claim that a proof is 'a syntactic object consisting only of sentences' as opposed to diagrams? A proof is *never* a syntactic object. A formal derivation on its own is a syntactic object but not a proof. Without an interpretation of the language of the formal system the end-formula of the derivation says nothing; and so nothing is proved. Without a demonstration of the system's soundness with respect to the interpretation, one lacks reason to believe that derived conclusions are true. A formal derivation *plus* an interpretation and soundness proof can be a proof of the derived conclusion. But one and the same soundness proof can be given in syntactically different ways, so the whole proof, i.e. derivation + interpretation + soundness proof, is not a syntactic object. Moreover, the part of the proof which really is a syntactic object, the formal derivation, need not consist solely of sentences; it can consist of diagrams, as Miller's example shows.

The visual thinking in this example consists in going through a sequence of diagrams and at each step seeing that the next diagram results from a permitted alteration of the previous diagram. It is a non-superfluous part of the process of thinking through a proof that on any straight line segment an equilateral triangle is constructible. It is clear too that in a process that counts as thinking through *this* proof, the visual thinking is not replaceable by non-diagrammatic thinking. That knocks out (b), leaving only (c): some thinking that involves a diagram in thinking through a proof is neither superfluous nor replaceable by non-diagrammatic thinking.

This is not an isolated example. In the 1990s Barwise led a programme aimed at the development of formal systems of reasoning using diagrams and establishing their soundness. There was renewed interest in Peirce's graphical systems for propositional and quantifier logic, and systems employing Euler diagrams and Venn diagrams were developed and investigated, culminating in the work of Sun-Joo Shin (1994). Barwise was interested in systems which better model how we reason than these, and to this end he turned his attention to heterogeneous systems, systems deploying both formulas and diagrams: he and Etchemendy developed such a system for teaching logic, *Hyperproof*, and began to investigate its metalogical properties (Barwise and Etchemendy, 1996b). This was part of a surge of research interest in the use of diagrams, encompassing computer science, artificial intelligence, formal logic, and philosophy. For a representative sample see the papers in Blackwell (2001).[1]

All that is for the record. Mathematical practice almost never proceeds by way of formal systems. For most purposes there is no need to master a formal system and work through a derivation in the system. In fact there is reason to avoid going formal: in a formalized version of a proof, the original intuitive line of thought is liable to be obscured by a multitude of minute steps. While formal systems may eventually prove useful for modelling actual reasoning with diagrams in mathematics, much more investigation of actual practice is needed before we can develop formal systems that come close to real mathematical reasoning. This prior investigation has two branches: a close look at practices in the history of mathematics, such as Manders's work on the use of diagrams in Euclid's *Elements* (see his contributions in this volume), and cognitive study of individual thinking using diagrams in mathematics. Cognitive scientists have not yet paid much attention to this; but there is a large literature now on visual perception and visual imagery that epistemologists of mathematics can draw on. A very useful short discussion of both the Barwise programme and the relevance of cognitive sudies (plus bibliography) can be found in (Mancosu, 2005).[2] So let us set aside proofs by means of formal systems and restrict attention to normal cases.

Outside the context of formal diagrammatic systems, the use of diagrams is widely felt to be unreliable. There are two major sorts of error:

1. relevant mismatch between diagrams and captions;
2. unwarranted generalization from diagrams.

In errors of sort (1), a diagram is unfaithful to the described construction: it represents something with a property that is ruled out by the description, or without a property that is entailed by the description. This is exemplified by diagrams in the famous argument for the proposition that all triangles are isosceles: the meeting point of an angle bisector and the perpendicular bisector of the opposite side is represented as falling inside the triangle, when it has to be outside—see Rouse Ball (1939). Errors of sort (1) are comparatively rare, usually avoidable with a modicum of care, and not inherent in the nature of diagrams; so they do not warrant a general charge of unreliability.

---

[1] The most philosophically interesting questions concern the essential differences between sentential and diagrammatic representation and the properties of diagrams that explain their special utility and pitfalls. Especially illuminating in this regard is the work of Atsushi Shimojima: see the slides for his conference presentation, of which Shimojima (2004) is the abstract, on his web page.

[2] Two other pertinent references are Pylyshyn (2003) and Grialou *et al.* (2005).

Typically diagrams (and other non-verbal visual representations) do not represent their objects as having a property that is actually ruled out by the intention or specification of the object to be represented. But diagrams very frequently do represent their objects as having properties that, though not ruled out by the specification, are not demanded by it. In fact this is often unavoidable. Verbal descriptions can be discrete, in that they supply no more information than is needed.[3] But visual representations are typically indiscrete, because for many properties or kinds F, a visual representation cannot represent something as being F without representing it as being F *in a particular way*. Any diagram of a triangle, for instance, must represent it as having three acute angles or as having just two acute angles, even if neither property is required by the specification, as would be the case if the specification were 'Let ABC be a triangle'. As a result there is a danger that in using a diagram (or other visual representation) to reason about an arbitrary instance of class K, we will unwittingly rely on a feature represented in the diagram that is not common to all instances of the class K. Thus the risk of errors of sort (2), unwarranted generalization, is a danger inherent in the use of diagrams.

The indiscretion of diagrams is not confined to geometrical figures. The dot diagrams of ancient mathematics used to convince one of elementary truths of number theory necessarily display particular numbers of dots, though the truths are general. Figure 1.2 is an example, used to justify the formula for the nth triangular number, i.e. the sum of the first n positive integers.

Some people hold that the figure alone constitutes a proof (Brown, 1999); others would say that some accompanying text is required,to indicate how the image is to be interpreted and used. If there is no text, some background of conventions of interpretation must be assumed. But often more than one set of conventions is available. In fact Grosholz (2005) shows how the multiple interpretability of diagrams has sometimes been put to good use in mathematical

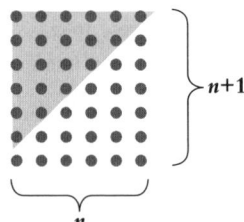

Fig. 1.2.

---

[3] This happy use of 'discrete' and its cognates is due to Norman, 2006.

thinking, a topic I will have to leave aside. In the following discussion I will assume that appropriate text is present.

The conclusion drawn is that the sum of integers from 1 to n is $[n \times (n+1)]/2$ for any positive integer n, but the diagram presents the case only for $n = 6$. We can perhaps avoid representing a particular number of dots when we merely imagine a display of the relevant kind; or if a particular number is represented, our experience may not make us aware of the number—just as, when one imagines the sky on a starry night, for no particular number k are we aware that exactly k stars are represented. Even so, there is likely to be some extra specificity. For example, in imagining an array of dots of the form just illustrated, one is unlikely to imagine just two columns of three dots, the rectangular array for $n = 2$. Typically the subject will be aware of imagining an array with more than two columns. This entails that an image is likely to have unintended exclusions. In this case it would exclude the three-by-two array. An image of a triangle representing all angles as acute would exclude triangles with an obtuse angle or a right angle. The danger is that the visual reasoning will not be valid for the cases that are unintentionally excluded by the visual representation, with the result that the step to the conclusion is an unwarranted generalization.

What should we make of this? First, let us note that in a few cases the image or diagram will not be over-specific. When in geometry all instances of the relevant class are similar to one another, for instance all circles or all squares, the image or diagram will not be over-specific for a generalization about that class; so there will be no unintended exclusions and no danger of unwarranted generalization. Here then are possibilities for non-superfluous visual thinking in proving.

To get clear about the other cases, where there is a danger of overgeneralization, it helps to look at generalization in ordinary non-visual reasoning. Schematically put, in reasoning about things of kind K, once we have shown that from certain premises it follows that such-and-such a condition is true of arbitrary instance c, we can validly infer from those same premises that that condition is true of all Ks, with the proviso that neither the condition nor any premiss mentions c. (If a premiss or the condition does mention c, the reasoning may depend on a property of c that is not shared by all other Ks, and so the generalization would be unsafe.) A question we face is whether in following an argument involving generalization on an arbitrary instance,[4] the thinking must include a conscious, explicit check that the proviso[5] is met. It is

---

[4] Other terms for this are 'universal generalization' and 'universal quantifier introduction'.
[5] The proviso is that neither the condition nor any premiss mentions c.

clearly not enough that the proviso is in fact met. For in that case it might just be the thinker's good luck that the proviso is met; hence the thinker would not know that the generalization is valid and so would not have genuinely thought through the proof at that step. This leaves two options. The strict option is that without a conscious, explicit check one has not really thought through the proof. The relaxed option is that one *can* properly think through the proof without checking that the proviso is met, but only if one is sensitive to the potential error and would detect it in otherwise similar arguments. For then one is not just lucky that the proviso is met. Being sensitive in this context consists in being alert to dependence on features of the arbitrary instance not shared by all members of the class of generalization, a state produced by a combination of past experience and current vigilance. Without a compelling reason to prefer one of these options, decisions on what is to count as proving must be conditional.

How does all this apply to generalizing from visual thinking about an arbitrary instance? Take the example of the visual route to the formula for triangular numbers just indicated. The image used reveals that the formula holds for the 6th triangular number. The generalization to all triangular numbers is justified only if the visuo-spatial method used is applicable to the nth triangular number for all positive integers n, that is, provided that the method used does not depend on a property not shared by all positive integers. A conscious, explicit check that this proviso is met requires making explicit the method exemplified for 6 and proving that the method is applicable for all positive integers in place of 6. For a similar idea in the context of automating visual arguments, see Jamnik (2001). This is not done in practice when thinking visually, so if we accept the strict option for thinking through a proof involving generalization, we would have to accept that the visual route to the formula for triangular numbers does not amount to thinking through a proof of it; and the same would apply to the familiar visual routes to other general positive integer formulas, such as that $n^2 = $ the sum of the first n odd numbers.

But what if the strict option for proving by generalization on an arbitrary instance is too strict, and the relaxed option is right? When arriving at the formula in the visual way indicated, one does not pay attention to the fact that the visual display represents the situation for the 6th triangular number; it is as if the mind had somehow extracted a general schema of visual reasoning from exposure to the particular case, and had then proceeded to reason schematically, converting a schematic result into a universal proposition. What is required, on the relaxed option, is sensitivity to the possibility that the schema is not applicable to all positive integers; one must be so alert to ways a schema of the given kind can fall short of universal applicability that if one had been presented with a schema that did fall short, one would have detected the failure.

In the example at hand, the schema of visual reasoning involves at the start taking a number k to be represented by a column of k dots, thence taking the triangular array of n columns to represent the sum of the first n positive integers, thence taking that array combined with an inverted copy to make a rectangular array of n columns of $n + 1$ dots. For a schema starting this way to be universally applicable, it must be possible, given any positive integer n, for the sum of the first n positive integers to be represented in the form of a triangular array, so that combined with an inverted copy one gets a rectangular array. This actually fails at the extreme case: $n = 1$. The formula $[n \times (n + 1)]/2$ holds for this case; but that is something we know by substituting '1' for the variable in the formula, not by the visual method indicated. That method cannot be applied to $n = 1$, because a single dot does not form a triangular array, and combined with a copy it does not form a rectangular array. The fact that this is frequently missed by commentators suggests that the required sensitivity is often absent. This and similar 'dot' arguments are discussed in more detail in Giaquinto (1993b).

Missing an untypical case is a common hazard in attempts at visual proving. A well-known example is the proof of Euler's formula $V - E + F = 2$ for polyhedra by 'removing triangles' of a triangulated planar projection of a polyhedron. One is easily convinced by the thinking, but only because the polyhedra we normally think of are convex, while the exceptions are not convex. But it is also easy to miss a case which is not untypical or extreme when thinking visually. An example is Cauchy's attempted proof (Cauchy, 1813) of the claim that if a convex polygon is transformed into another polygon keeping all but one of the sides constant, then if some or all of the internal angles at the vertices increase, the remaining side increases, while if some or all of the internal angles at the vertices decrease, the remaining side decreases. The argument proceeds by considering what happens when one transforms a polygon by increasing (or decreasing) angles, angle by angle. But in a trapezoid, changing a single angle can turn a convex polygon into a concave polygon, and this invalidates the argument (Lyusternik, 1963).

The frequency of such mistakes indicates that visual arguments often lack the transparency required for proof; even when a visual argument is in fact sound, its soundness may not be clear, in which case the argument is better thought of as a way of discovering rather than proving the truth of the conclusion. But this is consistent with the claim that visual thinking can be and often is a way of proving something. That is, visual thinking can form a non-superfluous part of a process of thinking through a proof that is not replaceable by some non-visual thinking (without changing the proof.) Euclid's proof of the proposition that the internal angles of a triangle sum to two right angles provides an example. Norman (2006) makes a strong case that following this proof requires visual thinking, and that the visual thinking is not replaceable by non-visual thinking. Returning to our visual route to the formula for triangular numbers, we can check that the method works for all positive integers after the first, using visual reasoning to assure ourselves that it works for 2 and that if the method works for k it works for $k+1$. Together with this reflective thinking, the visual thinking sketched earlier constitutes following a proof of the formula for the nth triangular number for all integers $n > 1$, at least if the relaxed view of thinking through a proof is correct. In some cases, then, the danger of unwarranted generalization in visual reasoning can be overcome.

1.3 Discovering

Though philosophical discussion of visual thinking in mathematics has concentrated on its role in proof, visual thinking more often shows its worth in discovery than in proof. By 'discovering' a truth I will mean coming to believe it by one's own lights (as opposed to reading it or being told) in way that is reliable and involves no violation of epistemic rationality (given one's epistemic state). Priority is not the point: discovering something, in this sense, entails seeing it for oneself rather than seeing it before anyone else. The difference between merely discovering a truth and proving it is a matter of transparency: for proving or following a proof the subject must be aware of both the way in which the conclusion is reached and the soundness of that way; this is not required for discovery.

The oldest and best known discussion of visual discovery is to be found in Plato's *Meno* (82b–86b). Using the diagram of Fig. 1.3, it appears quite easy to

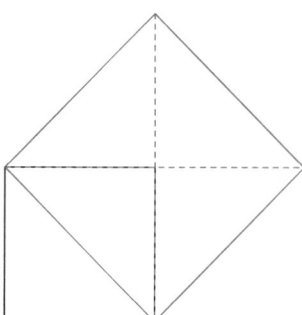

Fig. 1.3.

discover the general truth of Euclidean geometry that a square on a diagonal of a given square has twice the area of the given square. The many visual ways of reaching Pythagoras's Theorem provide similar examples.

While it is easy to reach the theorem Plato presents by means of the visual image, it is very difficult to show that the mode of belief-acquisition is reliable, because much of the process is fast, unconscious, and sub-personal. While the way of reaching the belief that there are twice as many small triangles in the square on the diagonal as in the square with the horizontal base is clear and open, the way of reaching the belief that the small triangles are congruent is hidden. For an initial speculative account of the process, see Giaquinto (2005). (The relevant passage in the *Meno* is examined in Giaquinto (1993a) and a similar example is discussed at length in Giaquinto (1992).)

My hypothesis is that the hidden process involves the activation of dispositions that come with possession of certain geometrical concepts (e.g. for square, diagonal, congruent). What triggers the activation of these dispositions is conscious, indeed attentive, visual experience; but the presence and operation of these dispositions is hidden from the subject. Because the process is fast and hidden, the resulting belief seems to the subject immediate and obvious. The feeling of obviousness occurs even in some more complicated cases. As we occasionally turn out to be fooled by these feelings, and because (at least part of) the process of belief-acquisition in visual discovery is hidden, we need in addition some transparent means of reaching the discovered proposition as a check: this is proof.

But in other cases there is nothing hidden. One makes a discovery by means of explicit visual thinking using background knowledge. Figure 1.4 illustrates an open way of discovering that the geometric mean of two numbers is less than or equal to their arithmetic mean (Eddy, 1985).

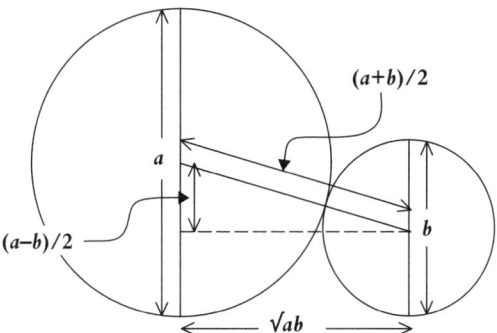

Fig. 1.4.

Pythagoras's theorem is used to infer that the base of the right-angled triangle has length $\sqrt{ab}$, where a and b are the diameters of the larger and smaller of the osculating circles respectively; then visualizing what happens to the triangle when the diameter of the smaller circle varies between 0 and the diameter of the larger circle, one infers that $\sqrt{ab} \leq (a+b)/2$ and that $\sqrt{ab} = (a+b)/2$ just when $a = b$. Reflecting on this way of reaching the inequality one can retrieve the premisses and the steps, and by expressing them in sentences construct a sound non-visual argument. This is quite different from the visual genesis of our belief that the triangles of a square either side of a diagonal are congruent. In that case reflection on our thinking does not provide us with any argument, and we are reduced to saying 'It's obvious!'

In some cases visual thinking inclines one to believe something but only on the basis of assumptions suggested by the visual representation that remain to be justified given the subject's current knowledge. In such cases there is always the danger that the subject takes the visual representation to show the correctness of the assumptions and ends up with an unwarranted belief. In such a case, even if the belief is true, the subject has not discovered the truth, as the means of belief-acquisition is not reliable and demands of rationality have been transgressed. This can be illustrated by means of an example taken from Nelsen's *Proof Without Words* (Montuchi and Page, 1988) presented in Fig. 1.5.

Reflection on this can incline one to think that when the positive real number values of x and y are constrained to satisfy the equation $x.y = k$ (where k is a constant), the positive real number values of x and y for which $x + y$ is minimal are $x = \sqrt{k} = y$. (Let '#' denote this claim.) Suppose that a person knows the conventions for representing functions by graphs in a Cartesian coordinate system, and knows also that the diagonal represents the function $y = x$, and that a line segment with gradient -1 from $(0, b)$ to $(b, 0)$ represents

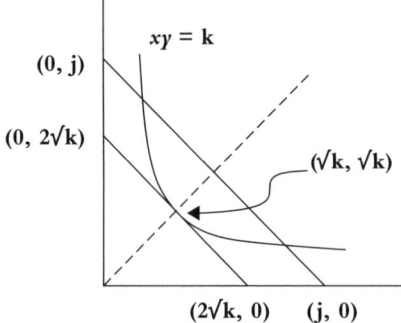

Fig. 1.5.

the function $x + y = b$. Then looking at the diagram may incline the subject to think that for no positive value of x does the value of y in the function $x.y = k$ fall below the value of y in $x + y = 2\sqrt{k}$, and that these functions coincide just at the diagonal. From these beliefs the subject may (correctly) infer the conclusion #. But mere attention to the diagram (or a visual image of it) cannot warrant believing that the y-value of $x.y = k$ never falls below that of $x + y = 2\sqrt{k}$ and that the functions coincide just at the diagonal; for the conventions of representation do not rule out that the curve of $x.y = k$ meets the curve of $x + y = 2\sqrt{k}$ at two points extremely close to the diagonal, and that the former curve falls below the latter in between those two points. So the visual thinking is not in this case a means of discovering proposition #. But it is useful because it leads the subject to propositions which, once she has established their truth,[6] would provide her with what is needed to transform an inclination to believe into knowledge.

We can say about the case just discussed that visual thinking is a heuristic aid rather than a means of discovery. When it comes to analysis mathematicians of the 19th and 20th centuries were right to hold that visual thinking rarely delivers knowledge. Visualizing cannot reveal what happens in the tail of an infinite process. So visual thinking is unreliable for situations in which limits are involved, and so it is not a means of discovery in those situations, let alone a means of proof. A more optimistic view is presented in Brown (1999), where it is proposed that reflection on a single diagram (Fig. 1.6) suffices to prove the Intermediate Zero Theorem.

The idea is that reflection on the diagram warrants the conviction that a perceptually continuous graphical line with endpoints above and below the horizontal axis must cross the axis, and that in turn justifies the theorem: if a

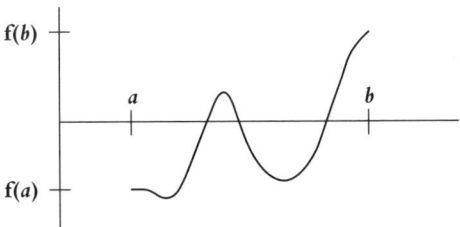

Fig. 1.6.

---

[6] This is easy enough. (i) For each equation it is trivial that if $x = y$, their common value is $\sqrt{k}$. So the functions expressed by those equations meet at the diagonal. (ii) To show that the y-values of $x.y = k$ never fall below the y-values of $x + y = 2\sqrt{k}$, we need only show that for positive x, $2\sqrt{k} - x \leq k/x$. As a geometric mean is less than or equal to the corresponding arithmetic mean, $\sqrt{[x.(k/x)]} \leq [x + (k/x)]/2$. So $2\sqrt{k} \leq x + (k/x)$. So $2\sqrt{k} - x \leq k/x$.

function f (in the real numbers) is $\epsilon - \delta$ continuous on an interval $[a, b]$ and $f(a) < 0 < f(b)$, f must take the value 0 at some point between a and b. A central point of contention is whether the perceptual concept of graphical line continuity is so related to the analytical ($\epsilon - \delta$) concept of function continuity that the perceptual conviction justifies the analytical conclusion. One reason for doubt is that not every $\epsilon - \delta$ continuous function has a visualizable curve, e.g. Weierstrass's continuous but nowhere differentiable function,[7] or f defined on [0, 1] thus:

$$f(x) = x . \sin(1/x) \text{ if } 0 < x \leqslant 1; f(0) = 0$$

So the visual thinking does not provide us with a form of reasoning that is applicable to all $\epsilon - \delta$ continuous functions. The matter is discussed further in Giaquinto (1994).

1.4 Other uses of visual thinking

Even if visual thinking in analysis cannot be a means of proof or discovery (in the sense in which I am using that term), it is extremely useful in other ways. One way in which it is useful is in augmenting understanding. This has two dimensions. First, there is understanding in the sense of one's grasp of a definition, a formula, an algorithm, a valid inference-type, and so on. Then there is the understanding involved in not only grasping the correctness of a claim, method or proof, but also appreciating why it is correct. Understanding of this kind falls within the ambit of Mancosu's research programme on mathematical explanation (Mancosu, 2000, 2001, and this volume).[8] The following discussion is confined to understanding in the first sense.

Students of calculus and analysis will be aware that visual representations can help one grasp analytically defined concepts; of course the danger of hasty generalization lurks, but one can also have visual representations of

---

[7] This claim is contentious. Mancosu (2005) discusses the question in historical context. My view in brief is this. We do have a visual way of reducing the puzzling character of a continuous nowhere differentiable function: we visualize first a curve with sharp peaks and sharp valleys; then we imagine zooming in on an apparently smooth part between a peak and a valley, only to find that this part itself contains sharp peaks and valleys; and we suppose that this is repeatable without end. But at every stage the curve visualized appears to have smooth parts and so is the image of a function that is differentiable over small intervals. The curve of a continuous function which is non-differentiable at a point makes a sharp turn at that point, and a curve consisting of sharp turns at every point, without any smooth segments between sharp points, is unvisualizable.

[8] See also Tappenden (2005) for a nuanced, informative and illuminating discussion of understanding and visualization in mathematics.

exceptional cases. When a function $f(x)$ does have a visualizable curve, visualization can help us grasp symbolically given operations on the function, e.g. $f(x) + k, f(x + k), |f(x)|, f(|x|), f^{-1}(x)$ (Eisenberg, 1991), and operations on functions that cannot be grasped pointwise, such as:

$$I(x) = \int_0^x f(t)d(t)$$

Visual representations can also help explain the rationale of a definition. For an example consider Euler's formula:

$$e^{i\theta} = \cos\theta + i\sin\theta$$

This is often introduced as a definition for extending the exponential function to complex numbers.[9] We can help to make sense of this definition visually, in terms of its geometric significance. Consider the point on the unit circle at angle θ anticlockwise from the unit vector on the x-axis (Fig. 1.7 left). That point has coordinates $\langle\cos\theta, \sin\theta\rangle$. So it represents the complex number $\cos\theta + i\sin\theta$. Thinking of this as the vector from the origin to the point $\langle\cos\theta, \sin\theta\rangle$, Euler's formula tells us that $e^{i\theta}$ is that vector.

If we expand (or contract) the x and y coordinates of that vector by real magnitude r to $r\cos\theta$ and $r\sin\theta$, it is clear that the corresponding vector must also expand or contract by a factor of r. This gives an immediate geometrical significance to the following trivial consequence of Euler's formula:

$$re^{i\theta} = r\cos\theta + ri\sin\theta$$

It tells us that $re^{i\theta}$ is the vector with length r at angle θ (Fig. 1.7 right). Thus we have a notation for vectors that makes explicit its determining geometric properties, its length r and its angle θ, properties hidden by the pairs-of-reals notation. This is what explains the aptness of Euler's formula. For confirmation,

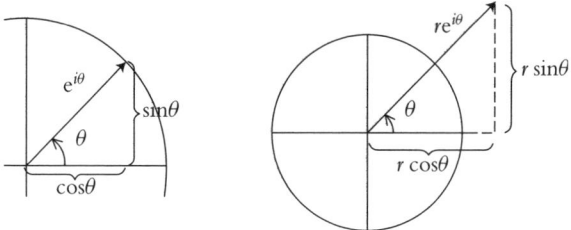

Fig. 1.7.

---

[9] Given the law for multiplying by adding exponents, Euler's formula is trivially equivalent to the equation $e^{x+i\theta} = e^x(\cos\theta + i\sin\theta)$. Sometimes this equation is given as the definition.

recall the puzzlement one feels when first introduced to vector multiplication in terms of pairs-of-reals. Given that $i^2 = -1$, it is clear that

$$(x + iy)(u + iv) = (xu - yv) + i(xv + yu)$$

But why does the term on the right denote the vector whose length is the *product* of the lengths of the multiplied vectors and whose angle is the *sum* of the angles of the multiplied vectors? Given the law for multiplication by adding exponents, the answer is immediate using the Euler notation for vectors:

$$re^{i\theta} se^{i\eta} = rse^{i(\theta+\eta)}$$

For further light on Euler's formula I recommend Needham (1997), 10–14. Another way in which visualization can be useful is as an aid to discovery or proof. We have already seen an example in the use of the diagram in Fig. 1.5 in getting to the theorem that, if $x.y = k$, the values of x and y for which $x + y$ is smallest are $x = \sqrt{k} = y$. The visual thinking does not get us all the way to the theorem, but it reduced the problem to simple algebraic manipulations. Often a visual representation will seem to show properties that might be useful in solving a problem, a maximum here or a symmetry there; thus the investigators are supplied with plausible hypotheses which they can then attempt to prove. An excellent example is the successful attempt to show that the Costa minimal surface (Fig. 1.8) is embeddable[10] by David Hoffman, William Meeks and James Hoffman.

Fig. 1.8.

[10] A surface is embeddable if and only if it can be placed in R^3 without self-intersections. More generally, a surface is *n*-embeddable if and only if it can be placed in R^n without self-intersecting, but for all $m < n$ cannot be placed in R^m without self-intersecting.

David Hoffman's own report (Hoffman, 1987) is worth quoting:

> ...Jim Hoffman and I could see after one long night of staring at orthogonal projections of the surface from a variety of viewpoints that it was free of self-intersections. Also, it was highly symmetric. This turned out to be the key to getting a proof of embeddedness. Within a week, the way to prove embeddedness was worked out. During that time we used computer graphics as a guide to "verify" certain conjectures about the geometry of the surface. We were able to go back and forth between the equations and the images. The pictures were extremely useful as a guide to the analysis.

In discussing such uses of visual representations it is sometimes said that they are 'merely heuristic'. This use of 'merely' is a symptom of our preoccupation with proof and our tendency to undervalue heuristic devices of all sorts. Heuristic devices vary in value, and they can be used in good or bad ways. The evaluation of heuristic aspects of inquiry is a subject of epistemic importance that we epistemologists have so far failed to investigate. The heuristic use of visual representations is clearly significant, indeed vital, in mathematics, and deserves greater attention in future studies of mathematical thinking.

Finally, visual thinking is useful in calculation. What I have in mind is the rearrangement of symbol arrays in imagination or on paper. This is a topic I do not have space to go into here. But its importance should be clear: even the algorithms we are taught at junior school for basic multidigit calculation are visuo-spatial in nature.[11] So it is not that surprising to find that people with certain visuo-spatial deficits (Williams syndrome and Turner syndrome) or agraphia (Gerstmann syndrome) also have calculation difficulties, though the nature of the connection between the deficits, if they are connected, is unclear. Symbol manipulation also plays a significant role in thinking in algebra, though contrary to popular belief there is more to algebra than that. Matrix algebra provides a significant example of visual operations on symbol displays, and the importance of matrices in applications again attests to the utility of symbol manipulation in calculation.

1.5 Conclusion

Visual thinking can occur as a non-superfluous part of thinking through a proof and it can at the same time be irreplaceable, in the sense that one could not think through the same proof by a process of thought in which the visual

---

[11] See Sawyer (1964) for the potential of visual means at elementary level.

thinking is replaced by some thinking of a different kind. Often, however, when visual thinking is a non-superfluous and irreplaceable part of thinking through an argument, the soundness of the process and sometimes even the process itself are not apparent to the thinker; in that case the thinking does not constitute proving. But it may be sound and the thinker may believe the conclusion without any violation of epistemic rationality. That is discovery ('seeing for oneself' rather than 'finding first'), as I have been using that term. So visual thinking can have an epistemically significant role in proving and, as it often does in practice, in discovering. Visual thinking more frequently plays other roles in practice. It may lead one to plausible hypotheses which, if proved, lead to the solution of a problem under investigation. This is its role as a heuristic aid, a role no less important than its role in proving and discovering. Visual thinking also augments understanding and enables calculation.

That is a summary of what I hope to have made plausible in this introduction. What is unavoidably missing is a sense of the richness of visual thinking in mathematics, the diversity of types of visual representation and visual transformations of them, and the ways they are used. This is extensive and still largely unexplored terrain, potentially fruitful for the cognitive science and epistemology of mathematics.

Bibliography

BARWISE, Jon and ETCHEMENDY, John (1996a), 'Heterogeneous logic', in G. Allwein and J. Barwise (eds.), *Logical Reasoning with Diagrams* (Oxford: OUP).

—— (1996b), 'Visual information and valid reasoning', in G. Allwein and J. Barwise (eds.), *Logical Reasoning with Diagrams* (Oxford: OUP).

BLACKWELL, Alan (2001), *Thinking with Diagrams* (Dordrecht: Kluwer Academic Publishers).

BROWN, James R. (1999), *Philosophy of Mathematics: an Introduction to the World of Proofs and Pictures* (London: Routledge).

CAUCHY, Augustin (1813), *Sur Les Polygones et Les Polyèdres* (Paris: Gauthier-Villars (1882–1938)).

DEDEKIND, Richard (1872), *Stetigkeit und irrationale Zahlen* (Braunschweig: Vieweg). Translated in *Essays on the Theory of Numbers*, Beman W. (trans.).

EDDY, Roland (1985), 'The arithmetic mean–geometric mean inequality III', *College Mathematics Journal*, 16(208), (reprinted in Nelsen (1993)).

EISENBERG, Theodore (1991), 'Functions and associated learning difficulties', in D. Tall (ed.), *Advanced Mathematical Thinking* (Dordrecht: Kluwer Academic Publishers).

EUCLID, *Elements*, published as *Euclid's Elements: all Thirteen Books Complete in One Volume* Heath T. (trans.), D. Densmore (ed.) (Santa Fe: Green Lion Press).

FOMENKO, Anatolij (1994), *Visual Geometry and Topology*, M. Tsaplina (trans.) (New York: Springer-Verlag).

GIAQUINTO, Marcus (1992), 'Visualizing as a means of geometrical discovery', *Mind & Language*, 7, 382–401.

—— (1993a), 'Diagrams: Socrates and Meno's slave', *International Journal of Philosophical Studies*, 1, 81–97.

—— (1993b), 'Visualizing in arithmetic', *Philosophy and Phenomenological Research*, 53, 385–396.

—— (1994), 'Epistemology of visual thinking in elementary real analysis', *British Journal for the Philosophy of Science*, 45, 789–813.

—— (2005), 'From symmetry perception to basic geometry', in P. Mancosu, K. Jørgensen, and S. Pedersen (eds.), *Visualization, Explanation and Reasoning Styles in Mathematics* (New York: Springer-Verlag).

GRIALOU, Pierre, LONGO, Giuseppe, and OKADA, Mitsuhiro (2005), *Images and Reasoning* (Tokyo: Keio University Press).

GROSHOLZ, Emily (2005), 'Constructive ambiguity in mathematical reasoning', in D. Gillies and D. Cellucci (eds.), *Mathematical Reasoning and Heuristics* (London: King's College Publications).

HILBERT, David (1894), 'Die Grundlagen der Geometrie', in M. Hallett and U. Majer (eds.), *David Hilbert's Lectures on the Foundations of Geometry* (Berlin: Springer-Verlag).

HOFFMAN, David (1987), 'The computer-aided discovery of new embedded minimal surfaces', *Mathematical Intelligencer*, 9, 8–21.

JAMNIK, Mateja (2001), *Mathematical Reasoning with Diagrams: From Intuition to Automation* (Stanford: CSLI Publications).

JOYAL, André, STREET, Ross, and VERITY, Dominic (1996), 'Traced monoidal categories', *Mathematical Proceedings of the Cambridge Philosophical Society*, 119(3), 447–468.

KANT, Immanuel (1781/9), *Kritik der reinen Vernunft*, P. Guyer and A. Wood (trans. & eds.), (Cambridge: Cambridge University Press (1998)).

LANDAU, Edmund (1934), *Differential and Integral Calculus*, M. Hausner and M. Davis (trans.) (New York: Chelsea (1950)).

LAUDA, Aaron (2005), 'Frobenius algebras and planar open string topological field theories', <http://uk.arxiv.org/PS_cache/math/pdf/0508/0508349.pdf>.

LEINSTER, Tom (2004), 'Operads in higher-dimensional category theory', *Theory and Applications of Categories*, 12(3), 73–194.

LYUSTERNIK, Lazar (1963), *Convex Figures and Polyhedra*, T. Smith (trans.) (New York: Dover).

MANCOSU, Paolo (2000), 'On mathematical explanation', in E. Grosholz and H. Breger (eds.), *Growth of Mathematical Knowledge* (Dordrecht: Kluwer Academic Publishers).

—— (2001), 'Mathematical explanation: problems and prospects', *Topoi*, 20, 97–117.

MANCOSU, Paolo (2005), 'Visualization in logic and mathematics', in P. Mancosu, K. Jørgensen, and S. Pedersen (eds.), *Visualization, Explanation and Reasoning Styles in Mathematics* (Dordrecht: Springer-Verlag).

MILLER, Nathaniel (2001), *A Diagrammatic Formal System for Euclidean Geometry*, Ph.D. thesis, Cornell University.

MONTUCHI, Paolo and PAGE, Warren (1988), 'Two extremum problems', *College Mathematics Journal* (reprinted 1993), 19, 347.

NEEDHAM, Tristan (1997), *Visual Complex Analysis* (Oxford: Clarendon).

NELSEN, Roger (1993), *Proofs Without Words: Exercises in Visual Thinking* (Washington DC: The Mathematical Association of America).

NORMAN, Jesse (2006), *After Euclid* (Stanford: CSLI Publications).

PALAIS, Richard (1999), 'The visualization of mathematics: towards a mathematical exploratorium', *Notices of the American Mathematical Society*, 46, 647–658.

PASCH, Moritz (1882), *Vorlesungen über neuere Geometrie* (Springer (1926)).

PLATO, *Meno*, Sharples R. (ed. & trans.) (Warminster: Aris and Phillips (1985)).

PYLYSHYN, Zenon (2003), *Seeing and Visualizing* (Cambridge: Cambridge University Press).

ROUSE BALL, W. W. (1939), *Mathematical Recreations and Essays* (New York: MacMillan).

RUSSELL, Bertrand (1901), 'Recent work on the principles of mathematics', *International Monthly*, 4 (reprinted as 'Mathematics and the Metaphysicians' in Russell (1918)), 83–101.

SAWYER, Warwick (1964), *Vision in Elementary Mathematics* (New York: Dover (2003)).

SHIMOJIMA, Atsushi (2004), 'Inferential and expressive capacities of graphical representations: survey and some generalizations', in *Diagrammatic Representation and Inference: Third International Conference, Diagrams 2004, Proceedings* (Berlin: Springer-Verlag).

SHIN, Sun-Joo (1994), *The Logical Status of Diagrams* (Cambridge: Cambridge University Press).

TAPPENDEN, Jamie (2005), 'Proof-style and understanding in mathematics I: Visualization, unification and axiom choice', in P. Mancosu, K. Jørgensen, and S. Pedersen (eds.), *Visualization, Explanation and Reasoning Styles in Mathematics* (Dordrecht: Springer-Verlag).

TENNANT, Neil (1986), 'The withering away of formal semantics?', *Mind & Language*, 1(4), 302–318.

ZIMMERMANN, Walter and CUNNINGHAM, Steve (1991), *Visualization in Teaching and Learning Mathematics* (Washington DC: Mathematical Association of America).

2
Cognition of Structure

MARCUS GIAQUINTO

2.1 Introduction

What is the nature of our cognitive grasp of mathematical structures? A *structured set* is a set considered under one or more relations, functions, distinguished elements (constants) or some combination of these. Examples are the ring of integers, the group of isometries of the Euclidean plane, the lattice of subsets of natural numbers, and the Banach space of continuous functions on the closed unit interval I with norm $\|f\| = sup\{f(x) : x \in I\}$. The *structure* of a structured set, as I will be using the word, is that property it shares with all and only those structured sets isomorphic to it, in other words its isomorphism type.[1]

By 'cognitive grasp' I mean the kind of knowledge that has a non-propositional complement, such as one's knowledge of the taste of cinnamon, the Cyrillic alphabet or Fauré's requiem.[2] With this kind of knowledge in mind, Russell (1912) drew attention to a distinction between knowledge by description and knowledge by acquaintance: you may know Rome by acquaintance but Carthage only by description. As structures are highly abstract, it seems clear that we cannot know them by acquaintance. If we know them at all, we must know them by description; and in many cases this is just how it seems to be. If we know that all the models of a mathematical theory are isomorphic, we can know a structure as *the structure common to all models of the theory*, e.g. the structure of all models of the second-order axioms for Dedekind–Peano arithmetic. This is knowledge by description, and it seems to be the only kind of knowledge of structure available to us. Nonetheless, I will try to show that, on the contrary, this is not the only way

---

[1] Mathematicians sometimes use the word 'structure' this way; but they sometimes use it to refer to what I am calling a structured set, and sometimes to a common structural kind, as in 'group structure'.
[2] There is no English word for just this kind of knowing. I sometimes use the frankly metaphorical 'grasp', but also sometimes 'discern' or even 'cognize', when the object is a structure.

of knowing structures; in some cases we can know structures in more intimate ways, by means of our visual capacities.

We can say, broadly, that our ability to discern the structure of simple structured sets is a power of abstraction; but we do not at present have an adequate account of the operative faculties or processes. That we have an ability to discern structure is made evident by our manifest ability to spot a structural analogy. A striking case is provided by the history of biology. The pattern of *gene distribution* in dihybrid reproduction postulated by Mendelian theory can be found in the *behaviour of chromosomes* in meiosis (division of cells into four daughter cells, each with half the number of chromosomes of the parent) and subsequent fusion of pairs of daughter cells from different parents. The hypothesis that chromosomes are agents of inherited characteristics issued from this observation, and was later confirmed.[3]

Somehow biologists spotted the common pattern. But how? For an answer one might look to the large cognitive science literature on analogy. This subject is still at an early stage, with lots of speculation and many theories. Among the most promising are those based on the structure-mapping idea of Dedre Gentner.[4] This, however, is no good for present purposes, because it takes cognition of structure as given, and cognition of structure is precisely what we want to understand.

I will proceed by suggesting some possibilities consistent with what I know of current cognitive science, in particular the science of vision and visual imagination. I will restrict consideration to just a few structures, mostly very simple finite structures. But I will also say something about our grasp of infinite structures.

2.2 Visual grasp of simple finite structures

Here is a simpler biological example of a structured set. Consider the set consisting of a certain cell and two generations of cells formed from that initial cell by mitosis, division of cells into two daughter cells, under the relation 'x is a parent of y'. In this structured set there is a unique initial cell (initial, in that no cell in the set is a parent of it), a first generation of two daughters and a second generation of four grand-daughters, which are terminal (as they are not parents of any cell in the set). How can we have knowledge of the structure of this structured set?

---

[3] Priority is attributed to W. Sutton (1902). Others were hard on his heels.   [4] Gentner (1983).

2.2.1 Visual templates

One possibility, highlighted by Michael Resnik (1997, Ch. 11 §2), is that we use a perceptible template. This might be a particular configuration of marks on a page, an instance of a diagram, which itself can be construed as a structured set, one that has the very same structure as the cell set. Let Fig. 2.1 be our diagram.

The structured set here is the set of square nodes of Fig. 2.1 under the relation 'there is an arrow from x to y'. (One has to be careful here. We get a different structured set if the chosen relation is 'there is an arrow or *chain of arrows* from x to y'.) Now we let the top square node represent the initial cell, the square nodes at the tips of arrows from the top represent the daughter cells, and the four terminal nodes represent the grand-daughter cells, with no two nodes representing the same cell. We can tell that these conventions provide an isomorphism: cell c is a parent of cell d if and only if there is an arrow from the node representing c to the node representing d. In this case we are using a *particular visible structured set* to represent another structured set. In the same way we can use the visible structured set to represent all structured sets that share its structure. This representative can thus serve as a visual template for the structure. We can see the visual template, and we can see it as a structured set. Seeing it as a structured set requires a non-visual factor, to determine what is the relevant set (e.g. square nodes or arrows) and what is the relevant relation (e.g. 'there is an arrow from x to y' or 'there is chain of one or more arrows from x to y'). But the result is still a kind of visual cognition.

Suppose now we are interested in some other structured set (like the cells under the 'parent-of' relation). By giving names to its elements and labelling the nodes of the visual template with these names, we can establish that the given structured set is isomorphic to the visual template. The labelled configuration is something we see, and checking that the labelling gives a one–one

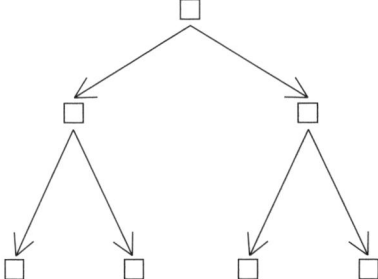

Fig. 2.1.

order-preserving correlation (an isomorphism) is a mechanical-perceptual task. In this way we have epistemic access to a structure.

Here I have been supposing that a visual template is one particular configuration of actual physical marks, e.g. those in Fig. 2.1 in the copy of the book you are now reading! Just as you can recognize different physical inscriptions as instances of an upper-case letter A, so you can recognize different configurations of physical marks as instances of the same *type*. You would have no difficulty in recognizing, as an instance of that type, the instance of Fig. 2.1 in another copy of the book. Moreover, you could recognize as instances of that type other configurations that are geometrically similar but differing in size and orientation.

To account for this recognitional capacity it is thought that the visual system stores representations of *types* of visible configurations. A representation of a visual type is not itself a visual image; but

(i) a visual stimulus can activate a representation of a type. This is required for recognition as opposed to mere perception.
(ii) Also, we can activate a type representation to produce an image of an instance of the type by voluntary visual imagining.

In this situation it seems reasonable to take the configuration *type* to have the structure-fixing role that we had been assigning to a particular physical instance of the type. This is still within the ambit of the template idea. In place of a single physical template, however, we allow as templates any configuration of marks that we can recognize as instances of the type.

2.2.2 Beyond visual templates

We can have an awareness of structure that is more direct than the visual template mode, at least in such very simple cases as this, two generations of binary splitting. The awareness I have in mind is not tied to a particular configuration type. Although a particular configuration of marks, viewed as a set of elements visibly related in a particular way, may serve a person as an initial instance of a structure, one may later think of the structure without thinking of it as the structure of configurations of just that visual type. Once one has perceived a configuration of marks as a structured set, one can acquire the ability to perceive configurations of *other visual types* as structured in the same way. For example, one has no difficulty in seeing the configurations of Fig. 2.2 as structured in the same way, even though they appear geometrically quite dissimilar.

It is not clear how we do this. In these cases there is a visualizable spatio-temporal transformation of one configuration into another that preserves number of members and relevant relations between them. (Relevant relations

COGNITION OF STRUCTURE 47

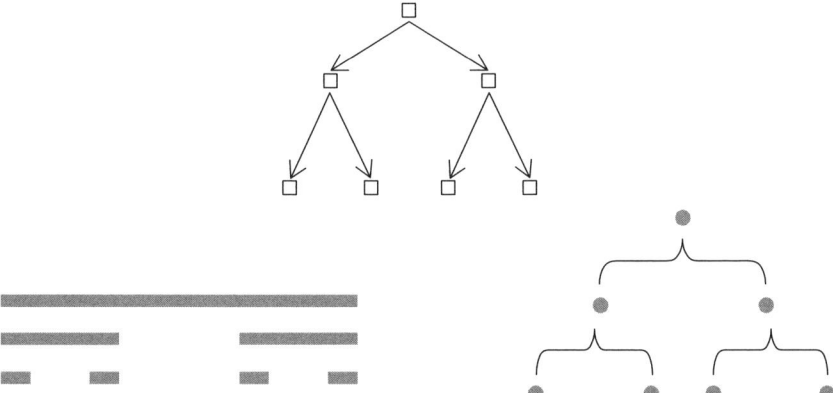

Fig. 2.2.

are those picked out in conceiving the configurations as structured sets.) So, seeing two dissimilar configurations, conceived as structured sets, as instances of the same structure *may* involve visualizing an appropriate spatial transformation.

But I doubt that it is necessary to visualize a spatial transformation. We may instead be able to see, quite directly, diverse configurations as instances of the same structure. How is this possible? We can directly recognize objects of different shapes, sizes, colours, surface textures, etc. as members of a single class, such as the category of *hand*. We sometimes think of all hands, or at least all left hands, as having the same shape. But the actual shape of a hand varies with degree of openness and positions of the fingers, as well as their relative thickness. What we actually mean by 'the shape' of a hand is something more abstract, a spatial property that is preserved not only under shape-preserving transformations, but also under transformations involved in normal changes of palm shape and finger positions. Yet we can visually recognize something as a hand directly, regardless of palm shape and finger positions. We certainly do not *deduce* that something we see is a hand from other perceptible properties; rather, the visual system has acquired a category representation for *hand*; and when visual inputs from seeing a hand strongly activate this category representation, one visually recognizes what one is seeing as a hand. In the same way, the visual system can acquire a representation for a category of visual configurations of marks that provide instances of a common structure. Let us call these representations *category specifications*.[5] My suggestion is that a visual category specification gives us a visual means of grasping structure that is more flexible than a visual template.

[5] This is what is called a 'category pattern' in Kosslyn (1994).

A category specification is a nexus of related feature specifications that cannot be 'read off' the subjectively accessible features of the visual experience of an instance of the category. Just as with face recognition, the subject may have no way of knowing precisely which congregations of features and relations leads to recognition of the category when presented with an instance. A visual category specification is a kind of representation in the visual system, but it is very unlike a visual image or percept. An image or percept is a transient item of experience of specific phenomenological type, whereas a category specification is a relatively enduring representation that is not an item of experience, though its activation affects experience.

We can recognize a perceived configuration of marks as an instance of a certain structure, by activation of an appropriate visual category specification. Thus, I suggest, we can have a kind of visual grasp of structure that does not depend on the particular configuration we first used as a template for the structure. We may well have forgotten that configuration altogether. Once we have stored a visual category specification for a structure, we have no need to remember any particular configuration as a means of fixing the structure in mind. We can know it without thinking of it as 'the structure of this or that configuration'. There is no need to make an association. So this is more direct than grasp of structure by a visual template.

What about identifying the structure of a non-visual structured set given by verbal description? Recall the method mentioned earlier using a visual template: (1) we first name the members of the set, (2) then label elements of the template with those names, and (3) then check that the labelling provides an isomorphism. Though this can happen, we often do not need any naming and labelling. Consider the following structured set: {Mozart, his parents, his grandparents} under the relation 'x is a child of y'. Surely one can tell without naming and labelling that it is isomorphic to the structured sets given earlier. We know this for any set consisting of a person, her parents and grandparents under the 'child-of' relation (assuming no incest). It is as if our grasp of these sets as structured sets already involves activation of a visual category specification for two generations of binary splitting. Exactly the same applies to the set obtained from the first two stages in the construction of the Cantor set by excluding open middle thirds, starting from the closed unit interval, under set inclusion. Naming and labelling are not needed to recognize this as a case of two generations of binary splitting. I offer as a tentative hypothesis that we can recognize the structure in this case as a result of activation of a visual category specification for two generations of binary splitting.

2.3 Extending the approach

2.3.1 More complicated structures

The structured sets considered so far have all been very small finite sets under a single binary relation. You may reasonably harbour the suspicion that sets structured by a plurality of relations or operations lie beyond any visual means of cognizing structure. I will now try to allay that suspicion. Figure 2.3 is a labelled visual template for the structure of the power set of a three-membered set $\{a, b, c\}$ under inclusion.

As before, this is a very small set structured by one binary relation. The configuration of Fig. 2.3 provides a template, namely, the set H of nodes under the relation '$n = m$, or, there is an upward path from m to n', which we can shorten to '$n \leqslant m$'. This structured set is easily seen to be isomorphic to the power set of S under inclusion by means of the labelling in the figure. In symbols,

$$\langle H; \leqslant \rangle \cong \langle \mathcal{P}(S); \subseteq \rangle$$

However, there is another way in which the set of nodes of Fig. 2.3 is structured. In place of the binary relation $\leqslant$ there are operations and constants defined as follows:

(∧) $x \wedge y =$ the highest node n such that $n \leqslant x$ and $n \leqslant y$.
(∨) $x \vee y =$ the lowest node n such that $x \leqslant n$ and $y \leqslant n$.

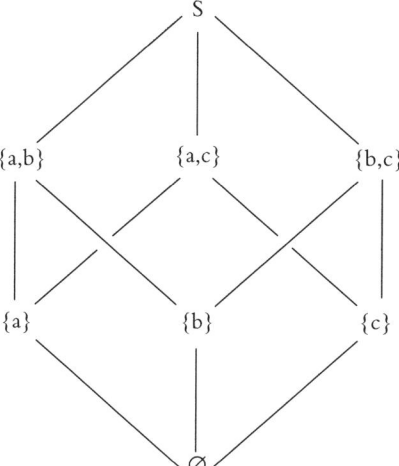

Fig. 2.3.

We name the bottom and top nodes, our constants, Min and Max respectively.

(−) $-x$ = the node n such that $x \wedge n$ is Min and $x \vee n$ is Max.

The set of nodes structured by $\wedge$, $\vee$, $-$, Max and Min is isomorphic to the power set algebra of S, which is the power set of S under the operations of intersection, union, and relative complement, and the constants S and the empty set. In symbols,

$$\langle H; \wedge, \vee, -, \text{Max}, \text{Min} \rangle \cong \langle \mathcal{P}(S); \cap, \cup, \sim, S, \emptyset \rangle$$

The isomorphism can be checked visually. A structured set of this kind is known as a *three-atom Boolean algebra*. Thus we have a visual template for the structure of a three-atom Boolean algebra, and this is a structure involving three operations, known as *meet* ($\wedge$), *join* ($\vee$), and *complement* ($-$), and two constants.

With a small amount of practice it is easy to acquire the visual ability to see the meet, join, and complement of nodes in Fig. 2.3 right away. The least straightforward is complement, but here is how it is done. Viewing the configuration as a cube, we see that every node is at one end of a unique diagonal of the cube; the complement of a node is the node at the other end the diagonal. So, although it is an exaggeration to say that we can simply see the configuration of Fig. 2.3 as the structured set $\langle H; \wedge, \vee, -, \text{Max}, \text{Min} \rangle$, it is strictly correct to say that we can have a perceptual grasp of that structured set. But how do we grasp its structure, the structure of a three-atom Boolean algebra? First, given an instance of this structure we can map it isomorphically onto a configuration like the one shown here, construed as a structured set in the way described. This is the visual template method. But practice may take us beyond it. We may eventually acquire an ability to tell, given an instance, that it can be mapped isomorphically onto the configuration without actually having carried out the mapping, even in thought. Provided that this ability is discriminating, in that it would not also lead us to think, of a non-instance, that it too can be mapped isomorphically onto the configuration, our grasp of the structure may consist in our having this ability, or, if that is too behaviourist, in our having the cognitive basis of the ability.

2.3.2 Structural kinds

The cognitive abilities involved in discerning the structures of the last example can also be used to discern *kinds* of structure. Representing a structure by a spatial pattern can be useful when the structure is very small or very simple, but moderate size and complexity usually nullifies any gain. Try drawing a

Fig. 2.4.

diagram of the power set algebra of a four-element set: it can be done, but the result will not be easy to take in and memorize. Moreover, there seems to be no iterative procedure for obtaining a diagram or spatial model for the power set algebra of an $n+1$-element set from a diagram for the power set algebra of an n-element set. If that is right the structures of the power set algebras of finite sets do not constitute a *kind* that we can grasp by means of visual configurations. Some kinds of structure, however, are sufficiently simple that visual representations together with rules for extending them can provide awareness of the structural kind. For discrete linear orderings with endpoints the representations might be horizontal configurations of stars and lines, giving a structured set of stars under the relation 'x is left of y'. The first few of these we can see or visualize, as illustrated in Fig. 2.4.

We have a uniform rule for obtaining, from any given diagram, the diagram representing the next discrete linear ordering with endpoints: add to the star at the right end a horizontal line segment with a star at its right end. This is a visualizable operation that together with the diagrams of the initial cases provides a grasp of a species of visual templates for finite discrete linear orderings with endpoints. Through awareness of this kind of template we have knowledge of the kind of structure which they are templates for. Binary trees, and more generally n-ary trees for finite n, comprise a more interesting kind of structure. We have already seen configurations that can be used to represent a binary tree in earlier figures. Taking elements to be indicated by points at which a line splits into two and by terminal points, the three binary trees after the first (which is just a 'v') can be visualized as in Fig. 2.5.

As mentioned before, a configuration represents a structured set only given some rules governing the representation of relations, functions, and constants. Binary trees are sets ordered by a single transitive relation R with one

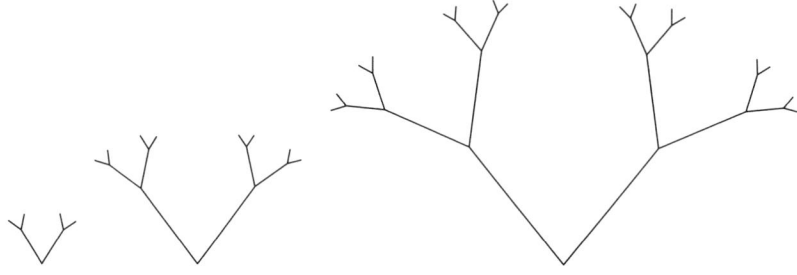

Fig. 2.5.

initial element and for every non-terminal element exactly two immediate *R*-successors.[6] I am restricting attention to binary trees such that, for some positive finite number *k*, all and only *k*th generation elements are terminal.[7] The rule is this:

> *x* bears relation *R* to *y* if and only if there is an upward path from the node (branching point) representing *x* to the node representing *y*.

This convention clearly respects the transitivity of the relation. The largest tree that we can easily visualize has very few generations. But we have a uniform rule for obtaining the next tree from any given *n*-generation tree by a visualizable operation:

> Add to each of the 2^n terminal points an appropriately sized 'v'.

This visualizable operation, together with the diagrams or images of the first few trees, provides us with a visual awareness of a kind of template for finite binary trees. But however we proportion the 'v's of each succeeding generation to their predecessors, for some *n* which is not very large the parts of a tree diagram from the *n*th generation onwards will be beyond the scope or resolution of a visual percept or image in which the initial part of the tree is clearly represented. Yet we have a visual way of thinking of a binary tree too large for its spatial representation to be completely visible at one scale. We can visualize first the result of scanning a tree diagram along any one of its branches; we can then visualize the result of zooming in on 'v's only just within visual resolution, and we can iterate this process. This constitutes a kind of visual grasp of tree diagrams beyond those we perceive entirely and at once. This visual awareness, arising from the combination of our visual experience of the first few tree diagrams and our knowledge of the visualizable operation for extending them, gives us a grasp of the structural kind comprising the finite binary trees.

The same considerations apply with regard to ternary and perhaps even quaternary trees. But trees whose nodes split into as many as seven branches defy visualization beyond the first generation, and we cannot visualize even the first generation of a tree with 29 branches from each node, except perhaps in a way in which it is indistinguishable from trees with 30-fold branching. In

---

[6] So the ordered sets represented earlier with the 'parent-of' and 'child-of' ordering relations are not binary trees as neither of these is transitive. But those same sets ordered by 'forbear-of' and 'descendant-of' respectively are binary trees.

[7] The initial element is zeroth generation, its two successors are first generation, and so on. Accordingly, a tree whose terminal elements are first generation is a 1-generation tree, and in general a tree whose terminal elements are *n*th generation is an *n*-generation tree.

these cases we operate by description and where possible by analogy with the trees we can visualize, just as we sometimes use three-dimensional objects as analogues of objects in higher dimensions. However, the sort of acquaintance we can have with the binary tree kind we can also have with many other kinds of structure, if we can give a recursive specification of the kind. What is required is just that templates of the smallest two or three structures are easy to visualize and we have a rule for obtaining a template of the 'next' structure from a template of a given structure by a visualizable process. A structural kind of this sort comprises an infinite sequence of structures that are often nested, in the sense that every structure in the sequence is a substructure of all later structures. The union of such a sequence is an infinite structure. What kind of visual grasp of infinite structures is possible, if any at all? That is our next topic.

2.4 Knowledge of infinite structures

If there can be visual cognition of any infinite structure, the simple and familiar structure of the natural numbers will be the prime example. Here I am talking about the structure of the finite cardinals under their natural 'less than' ordering. There is significant evidence that people with a standard education mentally represent positive integers as aligned in space. This is an active topic of cognitive research that goes under the title of 'mental number lines'.[8] In our culture the standard mental number line is a horizontal alignment with a left-to-right ordering. Details may vary. One possibility is a set of evenly spaced vertical marks on a horizontal line, with a single leftmost mark, continuing endlessly to the right such that every mark, however far to the right, is reachable by constant rate scanning from the leftmost mark (Fig. 2.6). One mark is taken to precede another if it is to the left of it; with respect to this ordering the leftmost mark is the only initial mark and there is no terminal mark.[9]

An obvious problem with the idea that a mental number line provides a grasp of the natural number structure is that we cannot see or visualize more than a finite part of any such line. When it comes to actual images (or percepts), something like Fig. 2.6 will be the best we can do. So the idea that we have a

Fig. 2.6.

[8] This is the subject of Chapter 6 of Giaquinto (2007).
[9] An initial element is one that has no immediate predecessor; a terminal element is one that has no immediate successor.

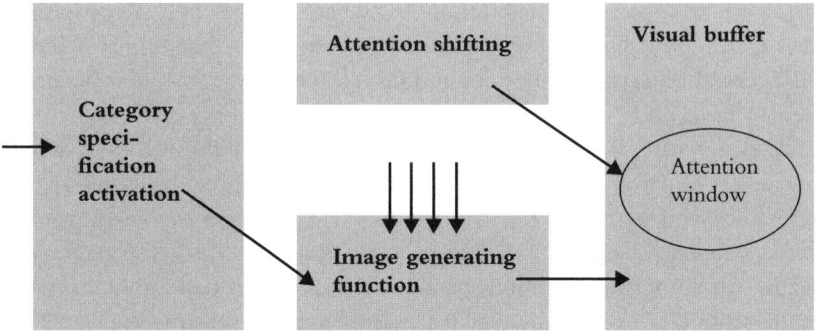

Fig. 2.7.

visual representation of the whole of such a line appears to be plain wrong. But it may not be wrong if visual representations include not only visual images but also visual category specifications. Recall the distinction: a *visual image* is a transient item of visual experience of specific phenomenological type, whereas a *visual category specification* is a stored representation, consisting of an ensemble of feature descriptions.[10] What is impossible is an infinitely extended visual image. But it is possible that a category specification specifies a line with no right end, one that continues rightward endlessly. To explain this I need to say more about the relation of category specification to image, which I will do with reference to Kosslyn's view of the functional architecture of the human visual system, a part of which is represented in Fig. 2.7. This figure is adapted from Kosslyn (1994).[11]

Category specifications are representations in the category specification activation subsystem and images are identified with patterns of activity in the visual buffer. When attention shifts to an image part, the corresponding activity (now in the attention window) is augmented. An internal 'instruction' to visualize something is input to the category specification activation system; the eventual output is a pattern of activity in the visual buffer that constitutes an image. More than one image can be generated from a given category specification; in fact, a continuously[12] changing image can be generated over an interval of time from one category specification. When the category

---

[10] Here 'image' is used broadly to include percepts. Where 'image' is used strictly for products of visual imagination, the representation results from activation of a category specification by 'top-down' processing, as opposed to 'bottom-up' processing originating with retinal stimulation.

[11] I have omitted all arrows representing bottom-up processing; only a few parts of the system are represented. See the figure on p. 383 of Kosslyn (1994). The subsystem I call 'category specification activation' Kosslyn calls 'category pattern activation'; what I call 'image generating function' he calls 'spatiotopic mapping'.

[12] Here I use 'continuously' in the ordinary non-mathematical sense.

specification is activated, the resulting image depends also on a number of parameter values corresponding to (i) viewpoint, (ii) distance, (iii) orientation, and others. These values can be continuously changed, and when that occurs the result will be continuously changing visual imagery. Imagine an ordinary cup upside down on an eye-level shelf with its handle to your right. Now imagine the cup as you take it down and bring it to a position and orientation that allows you to look into it from above. In that case, continuously changing parameters for location and orientation act on the category specification for a cup, continuously changing the output image.

Among the image-transforming operations that can result from continuous changes of parameter values are operations misleadingly called *scanning*, *zooming in*, *zooming out*, and *rotating*. Our present concern is with scanning. Imagistic scanning is not inspecting in sequence the parts of a fixed image, but continuously changing the image in a way that is subjectively like perceptual scanning. A momentary image generated by activation of that category specification will represent only a finite portion of the line; but the specification that the line has no right end ensures that rightward imagistic scanning will never produce an image of a right-ended line. In this way the category specification is a visual representation for a line that extends infinitely in one direction.

My suggestion is that our grasp of this structured set, the well-ordered set of evenly spaced marks on an endless horizontal line, issues (or *can* issue) from a stored visual category specification. But this is not achieved by 'reading off' the descriptions of the category specification, since we have no direct access to those descriptions. Rather, as a result of having the category specification, we have a number of dispositions which, taken together, give some indication of the kind of structured set it represents. These are dispositions to answer certain questions one way rather than another. For example:

> Given any two marks, must one precede the other? *Yes.*
> Do the intermark spaces vary in length? *No.*
> Is the precedence of marks transitive? *Yes.*
> Can any (non-initial) mark be reached from the initial mark by scanning to the right at a constant speed? *Yes.*

But some questions will have no answer:

> Is the intermark length more than a centimetre?

These answers tell us something about the nature of the mental number line as determined by the features specified in the category specification. The answers entail that no mark has infinitely many predecessors; as the marks form a strict linear ordering, this entails that they form a well-ordering. So we can say that the structure of the mental number line is that of a well-ordered set with a single initial element and no terminal element. I will call this structure **N**. My proposal is twofold. First, in becoming aware in the indirect way just described of the representational content of a visual category specification for the mental number line, we have a grasp of a *type* of structured set, namely, a set of number marks on a line endless to the right taken in their left-to-right order of precedence. Secondly, we can have knowledge of the structure **N** as the structure of a 'number line' of this type.

Mathematical logicians are understandably preoccupied by the fact that the set of *first-order* Dedekind–Peano axioms for number theory has non-isomorphic models. So those axioms taken together fail to determine a unique structure as the structure of the natural numbers[13] (under successor, addition, multiplication, and 0). This raises the question of how we determine that structure, how the mind succeeds in picking out models of just one isomorphism type (which we call 'the standard model' when identifying in thought isomorphic models.) One response is that our concept of the system of natural numbers is essentially second-order; if we replace the first-order induction axiom by the second-order induction axiom of Dedekind's original presentation, the result is an axiom set whose models are all isomorphic, as Dedekind showed. The worry about this response is that in order to understand the second-order induction axiom we would already need a cognitive grasp of the totality of sets of natural numbers; thus, if this response were adequate, our grasp of the set of natural numbers would depend on a prior grasp of its power set—hardly a plausible position.

I am inclined to think that there are two mutually reinforcing sources of comprehension of the natural number structure. One comes from our understanding of the natural numbers as the denotations of the number-word expressions in our natural language, and (later) as the denotations of our written numerals. We pick up algorithms for generating the number-words/numerals, and we think of a number as what such an expression stands for. The number system thus has the structure of the number-word system and the numeral system. So we can grasp the structure of the set of natural numbers under their natural 'less-than' ordering as the structure of the set of number-words (or numerals) under their order of precedence. The second comes from the visual category specification described earlier. This in turn depends on representations of space, time, and motion that cannot incorporate infinite bounded lengths,

---

[13] Models of these axioms are sets with functions for successor, plus, times and constant zero; a 'less-than' relation can be defined in terms of these. The models of the first-order axioms vary with respect to the induced 'less-than' relation: many, but not all, have infinite receding subsets.

infinite completed durations, or infinite speeds. Finiteness is in this sense 'built-in'. Under these background constraints the category specification determines that the number marks are well-ordered by their relation of precedence. This suffices to determine a unique structure. So we can grasp the structure of the natural number system as the structure of the set of number marks of the mental number line under their order of precedence.

Knowing the natural number structure in this way is much less direct than the kind of knowledge of finite structures discussed earlier. In this case we cannot experience an entire instance of the structure. So this knowledge of structure does not consist in the cognitive basis of an ability to recognize instances and to distinguish them from non-instances. We have to gather the nature of a number line from our inclinations to answer certain questions about it; although visual experience plays some role in this process, our answers are not simply reports of experience. In becoming aware in this indirect way of the content of a visual category specification for a mental number line, we acquire a grasp of a type of structured set, and we can then know the structure **N** as the structure of structured sets of this type. While this kind of knowledge is quite different from the experiential knowledge of small structures discussed earlier, it does have an experiential element that distinguishes it from knowledge by a description of the form *the structure of models of such-&-such axioms*, which of course requires knowing that the axiom set is categorical. To help appreciate how significant this difference is, it is worth examining a contrasting case.

2.4.1 An infinite structure beyond visual grasp

A contrasting case is the structure of the set of real numbers in the closed unit interval [0, 1], under the 'less than' relation. I will call this structure **U** for 'unit'. We do have visual ways of representing **U**, but I claim that they do not give us knowledge of it. I will now try to substantiate this claim.

An obvious thought is that we can think of **U** as the structure of the set of points on a straight line segment with left and right endpoints, when each point corresponds to a unique distance from one end, and the order of points is determined by the corresponding distance. We can certainly visualize a finite horizontal line segment, taking its points to be the locations of intersection of the horizontal line segment with (potential) vertical line segments; and we can visually grasp what it is for one such location to lie to the left of another. Why does not this give us a visual grasp of the structure **U**?

One reason concerns points. If the points on a line constitute a set with structure **U** they must be not merely too small to be seen by us, but absolutely invisible, having zero extension. Neither vision nor visualization gives us any acquaintance with even a single point of this kind, let alone uncountably many of them. In addition to a geometrical concept of extensionless points we have a perceptual concept of points as tiny dots. Perceptual points do have extension; some parts of a perceptual point on the line would be nearer the beginning of the line than others, so a perceptual point does not lie at exactly one distance from the beginning. For this reason no line of juxtaposed perceptual points could have the structure of [0, 1] under the 'less than' ordering. So we must make do with geometrical points. But thinking of a line as composed of geometrical points leads to numerous paradoxes. For instance, the parts of a line either side of a given point would have to be both separated (as the given point lies between them) and touching (as there is no distance between the two parts, the given point being extensionless). A related puzzle is that a line segment must have positive extension; but as all its components have zero extension, the line segment must also have zero extension. Yet another puzzle: the unit interval has symmetric left and right halves of equal length; but symmetry is contradicted by the fact that just one of the two parts has two endpoints—either the left part has a right endpoint or the right part has a left endpoint, but not both, otherwise there would be two points with none intervening. These puzzles show that the visuo-spatial idea of a continuous line cannot be coherently combined with the analytical idea of a line as a geometrical-point set of a certain kind.[14]

Setting aside these puzzles, thinking of an interval of real numbers as a set of points composing a line segment cannot reveal a crucial structural feature, Dedekind-continuity (completeness): for every set of real numbers, if it is bounded above it has a least upper bound, and if it is bounded below it has a greatest lower bound. So this visual way of thinking of an interval of real numbers is unrevealing about that feature which distinguishes it structurally from an interval of rational numbers. Hence we cannot know the structure **U** by thinking of it visually in terms of the points in a line segment.

Perhaps there are other visual ways of thinking of the closed unit interval. But none that I can think of fares any better. One possibility is the set of all branches of the infinite binary tree under the following relation of precedence:

x precedes y if and only if, when the branches x and y first diverge, x goes to the left and y goes to the right.

We can identify branches with infinite sequences of 0s and 1s. The left successor of each node is assigned 0, the right successor is assigned 1; nothing is assigned to the initial node. The branch up to a given node is identified

---

[14] This is not to deny that we can usefully flip back and forth between the two conceptions in practice.

with the sequence of 0s and 1s assigned to nodes on that branch up to and including the given node; so, for example, the branch up to the leftmost fourth generation node (after the initial node) is ⟨0, 0, 0, 0⟩. A single infinite path up the tree, a branch, represents the infinite sequence of 0s and 1s assigned to its nodes. Each infinite sequence of 0s and 1s is the binary expansion of a real number in the closed unit interval. So we can use branches to represent real numbers in the unit interval. This correlation of sequences with real numbers is not injective: some pairs of sequences represent the same real number. But we can easily rectify the matter by cutting out redundant branches.[15] Call the resulting set of branches S. Then there is a one–one correlation of S with [0, 1]: a branch is correlated with the real number of which it is the binary expansion. S ordered by '<', the relation of precedence among branches, is isomorphic to [0, 1] under 'less than'; in other words it has the structure **U**.

Why does *that* not give us a visual grasp of the structure **U**? Why is it wrong to think of S as standing to the two-generation binary tree as the infinite mental number line stands to a finite line segment marked like a ruler? And if the mental number line provides us with a grasp of **N**, why does a mental representation of S *fail* to provide a grasp of **U**? The reason, in brief, is that the two-generation binary tree, as we were conceiving of it earlier, is not a substructure of S. The elements of the two-generation binary tree, as we were conceiving of it, are its nodes; the elements of S are its infinite branches. Since we can see or visualize only a finite portion of an infinite binary tree, not even one element of S is represented in any visual image of it. We see or visualize at most finite initial segments of branches, and each finite initial segment is common to infinitely many different branches. It is true that there are some branches for which we can have a visual category specification: for instance the leftmost branch for unending zeros. But we can have only finitely many category specifications, and that leaves most branches unrepresented.[16]

One might reply that we have a visual appreciation of how the infinite binary tree continues from one generation of nodes to the next, thus giving us a grasp of the structure **U** similar to our grasp of **N**. But that is mistaken. The category specification for the infinite binary tree provides the basis for an

---

[15] Two binary expansions are correlated with the same real number if and only if they are identical up to and including their *n*th components but the remainder of one of them is 0 followed by 1 recurring while the remainder of the other is 1 followed by 0 recurring. In terms of the binary tree, two branches represent the same real number if and only if, when they first diverge, the one that goes left will go right ever after and the one that goes right will go left ever after. We take the result of cutting out that branch in every such pair that first diverges left.

[16] Logicians will add that in any language properly so called there is at most a countable infinity of category specifications (as each of these is finite), whereas there is an uncountable infinity of branches in S.

awareness of the structure of the set of nodes under the relation 'there is a path up from x to y'. That is very different from the structure of the set of *branches* in S under the relation of precedence '<' given earlier, and only the latter is the structure **U**.

Perhaps reflection on the visual representations used when we think of S ordered by '<' visually reveals that the ordering is linear and dense; but it is not clear to me how they would reveal that the ordering is Dedekind-continuous. If it cannot, that is another reason why thinking visually of an infinite binary tree does not give us the capacities that would warrant a claim to knowledge of the structure. I suspect that a stumbling block to any visually anchored grasp of **U** is continuity. If this is right, our knowledge of its structure is wholly theoretical rather than experiential. This constitutes a clear contrast between the kind of knowledge we have of the structure **U** and the kind of knowledge we have of the structure **N**.[17]

2.4.2 *Visual grasp of structures beyond* **N**

I have argued that we can have a grasp of the structure **N** that is anchored to a visuo-spatial representation, and that this gives us a kind of awareness of the nature of **N** that is unavailable for **U**. Is the kind of awareness that we have of **N** available for other infinite structures? Or is our grasp of **N** an isolated case? In a passage discussing the reach of what Hilbert called finitary mathematics, Gödel suggests by implication that there are other infinite structures knowable with the same kind of immediacy as our knowledge of **N**. Gödel begins as follows:

> Due to the lack of a precise definition of either concrete or abstract evidence there exists, today, no rigorous proof for the insufficiency (even for the consistency proof of number theory) of finitary mathematics. However, this surprising fact has been made abundantly clear through the examination of induction up to ϵ_0 used in Gentzen's consistency proof of number theory.

He continues:

> The situation may be roughly described as follows: Recursion for ϵ_0 could be proved finitarily if the consistency of number theory could. On the other hand the validity of this recursion can certainly not be made *immediately* evident, as is possible, for example in the case of ω^2. That is to say, one cannot grasp at one glance the various structural possibilities which exist for decreasing sequences, and there exists, therefore, no *immediate* concrete knowledge of the termination of every such sequence. But furthermore such *concrete* knowledge (in Hilbert's sense) cannot be realized either by a stepwise transition from smaller to larger

---

[17] S shorn of its leftmost and rightmost branches is isomorphic to the real numbers under their standard ordering. As these remarks apply to S thus shorn, they also apply to the real line.

ordinal numbers, because the concretely evident steps, such as $\alpha \to \alpha^2$, are so small that they would have to be repeated ϵ_0 times in order to reach ϵ_0.[18]

The significant implications of this passage for present concerns are that the step from ω to ω^2 is 'concretely evident'; that, as one can 'grasp at one glance' the structural possibilities for decreasing sequences in ω^2, one can have 'immediate concrete knowledge' that all such sequences terminate; hence that the validity of recursion (induction) for ω^2 can be made 'immediately evident', whereas the same is not true for ϵ_0 in place of ω^2. Is ω^2 really knowable in the implied way? Let us first step back. The ordinal ω under the membership relation has the structure $\mathbf{N}$; in fact this is normally what set theory uses to represent the set of natural numbers under '<'. So we can represent it in the same way, by a horizontal string of evenly spaced marks, with a leftmost mark, running off to the right endlessly. I will call such strings 'ω-strings'.

How do we make the step from the structure $\mathbf{N}$, exemplified by a single ω-string, to the structure of ω^2? Consider a vertical sequence of horizontal ω-strings, starting from the top, left-aligned and evenly spaced, proceeding downward endlessly, so that if, visualizing, we were to scan this sequence downward, new ω-strings would come into view, still evenly spaced, however far we continued; and if we were to scan upward from any ω-string, at constant pace from each ω-string to its predecessor, we would arrive back at the topmost ω-string. This constitutes imagining a two-dimensional array of marks that has a top edge (the first ω string) and a left edge (the column of first elements of ω strings) but is infinite rightward and downward. I will call such an array an 'ω-square', though it is not a square, as the array has no right or lower edge. The ordering is taken to be as for western script: for any two elements, if they are on the same line the leftmost element precedes, and if on different lines the element on the uppermost line precedes. Of course, we can have no image of more than a finite part of an ω-square; our sole visual representation of it is, as of an ω-string, a category specification. Thus we can have the kind of grasp of the structure of ω^2 that we have of the structure $\mathbf{N}$.

What about Gödel's further claims about knowledge of ω^2? These are that, as one can grasp at a glance the various structural possibilities for decreasing sequences, one can have immediate concrete knowledge of the termination

---

[18] Gödel (1972). The italics are Gödel's own. ω is the least infinite ordinal. ω^2 is the first ordinal after $\omega \times n$ for every finite n, and $\omega \times n$ is the ordinal consisting of a sequence of n ω-sequences. Let α be any ordinal and define $\alpha \uparrow n$ thus: $\alpha \uparrow 0$ is α; $\alpha \uparrow n+1$ is $\alpha^{(\alpha \uparrow n)}$. Then ϵ_0 is the first ordinal after every $\omega \uparrow n$ for finite n. By 'recursion for ϵ_0' Gödel meant induction up to ϵ_0: If for every $\alpha < \epsilon_0$, ϕ is true of α if ϕ is true of every $\beta < \alpha$, then ϕ is true of every $\alpha < \epsilon_0$.

of every such sequence; hence that the validity of induction up to ω^2 can be made immediately evident.

What I think Gödel had in mind here is that, using the ω-square representation, there is a visuo-spatial way of telling that every decreasing sequence of members of ω^2 terminates. From this fact the validity of induction up to ω^2 quickly follows, a fact which Gödel presumably took as background knowledge for his readers. But how can we tell that every decreasing sequence of members of ω^2 terminates? Thinking of ω^2 in terms of an ω-square as described earlier, it is obvious that for any decreasing sequence, among rows containing members of that sequence, there will be an uppermost row; and that, of the members of the sequence in that row there will be a leftmost member—call it α. Then, recalling the ordering of ω^2, it is clear that α is the least member of the sequence: so the decreasing sequence terminates. This way of acquiring the knowledge is relatively immediate, and is concrete in the sense that it has an experiential element. This may have been what Gödel had in mind, but we do not know. Either way, it does substantiate all but one of his claims. It does not support Gödel's implied claim that one can grasp at one glance all the structural possibilities for decreasing sequences in ω^2, for there are decreasing sequences with elements that occur arbitrarily far to the right in an ω-square.[19] But Gödel's other claims are untouched by this.

It is credible that we can have the same kind of grasp of the structure of ω^2 as we can have of **N**, and that this plays a role in actual mathematical thinking. Investigation is needed to determine how much further into the transfinite we may go before this kind of cognition of ordinal structure becomes impossible. At present I do not see how to extend the kind of account I have suggested for ω^2 to ω^ω. But I think that we can form a visual category specification for ω-many layers of ω-squares, an ω-cube, thus forming a representation of ω^3. There would be a top ω-square, and beneath each ω-square another one. The ordering within ω-squares is unchanged; for elements α and β in different ω-squares, α precedes β if and only if α's ω-square is above β's ω-square.

This uses our natural representation of space as extending infinitely in each of three dimensions. As our natural representation of space lacks a fourth dimension, it is clear that we cannot get to ω^4 from ω^3 in the same way as we got to ω^3 from ω^2. What we can do is to take each element of an ω-cube to represent an ω-string, an ω-square or an ω-cube, thus getting a way of thinking of ω^4, ω^5 or ω^6. But this way of using a visual category specification for ω-cubes is not a way of *extending* it to get another visual category specification. Can we *imagine* putting an ω-cube in each position in an ω-cube, as opposed

---

[19] Robert Black pointed this out in discussion.

to merely using each position in an ω-cube to *represent* an ω-cube?[20] Not if this is visuo-spatial imagination, as opposed to supposition. This is because a representation of an ω-cube is a representation of something with infinite spatial extension, while each element of an ω-cube is represented as finitely extended. If, however, 'imagine' is understood broadly to include fictional supposition, we can imagine each element of an ω-cube to be a benign black hole into which we can dive; and once in there, we find ourselves in a new infinite three-dimensional space containing another ω-cube (and of course one could iterate the story). Although this does not amount to forming a visual category specification for an ω-cube of ω-cubes, it does give us a semi-visual way of thinking of higher powers of ω. But even this semi-visual thinking runs out before ϵ_0.[21]

There may be other kinds of infinite structures, non-ordinal structures, that are knowable with the kind of visual awareness that we have for the structure of ω and ω^2 under their standard ordering. That is a matter for further research. **N**, the structure of ω, is the cognitively simplest infinite structure. We have usable visual representations of instances of it that provide some awareness of the nature of the structure. But these representations are category specifications, not images. The fact that we can never have a visual experience encompassing the whole of an instance of this structure marks a qualitative difference between our grasp of this simplest infinite structure and our grasp of the finite structures discussed earlier, such as the two-generation binary tree or the power set algebra of a three-element set.

2.5 Conclusion

We can have cognitive grasp of some structures by means of visual representations. For small simple finite structures we can know them through sensory experience of instances of them, somewhat as we can know the butterfly shape from seeing butterflies. This is a kind of knowledge by acquaintance. Though we cannot have knowledge by acquaintance of any infinite structure, some simple infinite structures can be known by visual means, and not merely as the structure of models of this or that theory. The crucial representations in these

---

[20] This question was put to me by Stewart Shapiro.
[21] Apparently Georg Kreisel used to say that ordinals less than ϵ_0 are visualizable (Paolo Mancosu, personal communication), but it is not clear why he was confident that visualizability did not stop well before. There is an interesting earlier discussion of the matter by Oskar Becker in a letter to Weyl in 1926 and in Becker (1927) in his attempt to provide phenomenological foundations for the transfinite ordinals reported in Mancosu and Ryckman (2002).

cases are category specifications, which are not items of conscious experience. Nonetheless, through the images they give rise to they can give us a grasp of structure that is sometimes operative in practice. How useful this resource can be and how it links up with propositional knowledge represented in words or formulas is a matter for future investigation.

Bibliography

BECKER, Oskar (1927), 'Mathematische Existenz', *Jahrbuch für Philosophie und phänomenologische Forschung*, 8, 439–809.
GENTNER, Derdre (1983), 'Structure-mapping: A theoretical framework for analogy', *Cognitive Science*, 7(155–170). Reprinted in *Readings in Cognitive Science: A Perspective from Psychology and Artificial Intelligence*. A. Collins and E. Smith (eds.), (Palo Alto, CA: Kaufmann), 1988.
GIAQUINTO, Marcus (2007), *Visual Thinking in Mathematics: an Epistemological Study* (Oxford: OUP).
GÖDEL, Kurt (1972), 'On an extension of finitary mathematics which has not yet been used', in *Collected Works*, volume II (Oxford: OUP). This is a revised and expanded version translated into English of 'Über eine bisher noch nicht benützte Erweiterung desfiniten Standpunktes', *Dialectica*, 12, 280–287.
KOSSLYN, Stephen (1994), *Image and Brain* (Cambridge, MA: MIT Press).
MANCOSU, Paolo and RYCKMAN, Thomas (2002), 'The correspondence between O. Becker and H. Weyl', *Philosophia Mathematica*, 10, 130–202.
RESNIK, Michael (1997), *Mathematics as a Science of Patterns* (Oxford: Clarendon).
RUSSELL, Bertrand (1912), *The Problems of Philosophy* (Oxford: OUP).
SUTTON, Walter (1902), 'On the morphology of the chromosome group in *Brachystola magna*', *Biological Bulletin*, 4, 24–39.

3

Diagram-Based Geometric Practice

KENNETH MANDERS

Demonstrations in Euclid's *Elements*, from Proposition I.1 on, use their diagram essentially to introduce items such as that notorious intersection point of the two circles, for which Euclidean demonstration has no alternative justificational resources.

In the 19th century this style of reasoning received critical attention from mathematicians needing to articulate various alternative geometries and their interrelationships. Twentieth century philosophers of mathematics have tended to dismiss it altogether as a means of justification, often citing modern logic as setting not only a more appropriate standard for mathematical justification, but the only acceptable one. The article in this volume by Giaquinto cites some sources on these matters.

One student, apparently the beneficiary of enthusiastic instruction in the virtues of the modern logical account, recently compared the study of Euclidean demonstration as mathematical justifications to the study of the Flat Earth. While myself not in a position to judge the contributions of the flat earth tradition to modern plate tectonics, I believe this expresses at least two complaints that many professional philosophers of mathematics would endorse: that diagram-based reasoning in Euclid is unreliable and justificationally inadequate, and that the study of a tradition of argument so obsolete cannot benefit the philosophy of mathematics. Let me take these in turn.

Assessments based on diagrams are held to be unreliable on several grounds. Drawn diagrams are *imperfect* in that, say, lines are not perfectly straight; regardless, human assessments of straightness or equality of line segments would be imperfect. Moreover, geometrical figures are individual, or at least *atypical* compared to the generality of geometrical conclusions. Next, there are *different forms* of geometry, which differ in their conclusions, and so a single diagram-based form of reasoning cannot serve them all; indeed there are forms (such as plane coordinate geometry restricted to rational coordinates) in which the two circles would not have an intersection point. Finally, there are geometricals such as space-filling curves which utterly defeat diagram-based reasoning. All these observations are correct.

The existence of different forms of geometry is, of course, fatal to the claim that Euclidean geometry is unique as a conception of space. Once this is abandoned, however, it becomes clear (a) that the mere existence of diagrammatically intractable geometricals, such as space-filling curves or general Riemannian geometries, fails to count against the justificational adequacy of traditional diagram-based reasoning that does not purport to deal with them; and (b) that the mere existence of different forms of geometrical reasoning does not impugn the justificational adequacy of any one of them.

In particular, it is no less reliably the case that the circles intersect in the real-number coordinate plane because they do not intersect in rational coordinates. Indeed, because diagrams appear not to provide resources for judging whether the coordinates of the intersection points they show would be number-theoretically special in this or that way, the stable policy is to attribute the intersection points the diagrams show; and all traditional diagram-based geometries do so, including the famous non-Euclidean alternatives.

Imperfection and atypicality objections have force only to the extent that these characteristics of diagram assessments have the power to affect conclusions of Euclidean-style argument. On this point, I detect ignorance and empty (arrogant, even panicked) dismissiveness on the part of the critics of diagram-based demonstration. Logical investigation would instead be the appropriate response, as routinely pursued in other reasoning contexts of serious philosophical interest (modal logic, or demonstratives, for example). It is hard to escape the impression that philosophers of mathematics have had something at stake in dismissing properly geometrical reasoning methods *out of hand*.

Our main paper below, *The Euclidean Diagram* (ED), starts to lay out logical theory for Euclidean diagram-based reasoning. Already the simplest observation on what the texts do infer from diagrams and what they do not suffices to show the intersection of the two circles is completely safe, and severely limits the scope of the imperfection objection. The plot does thicken, more remains to be done, and the atypicality objection may point to new phenomena in logical theory. But that brings me to our student's other thought: that Euclidean-style geometrical reasoning is hopelessly obsolete.

Whether this is so as a practical way of gaining new geometrical knowlege in the 21st century, I will not contest. That someone might, out of purely historical interest, characterize an ancient reasoning practice warts and all, the student would not contest. But there are many reasons for a philosopher today to analyze diagram-based geometrical reasoning, starting with the challenge of explaining its epistemic success in the face of so many apparent challenges.

Though some philosophers haughtily deny there could be mathematics, properly speaking, before modern logic, mathematicians generally, including ones at the forefront of radical 20th-century reconceptualizations of geometry, recognize Euclid as rigorous mathematical reasoning to nontrivial correct conclusions (witness recent books of Hartshorne (2000) and Artmann (1999)). Maybe they just like having a long history?

No, modern mathematics *subsumes* Euclid's geometrical conclusions in real analytic geometry, real analysis, and functional analysis. In research profile, only a corner of modern geometry; but in terms of the footprint of mathematics in the modern world, that is most of modern mathematics.

Euclid (see Vitrac (2004)), and Apollonius and Archimedes, are virtually without error: their every result has a counterpart in modern mathematics, even if subsumed in patterns of claims and proofs recognized much later. But there was ample scope for error: some claims even of the early books of Euclid are subtle, and especially the many equality claims are versatile in indefinitely extendable combinations of great power. Think of the equalities of angles, sum-of-angles, and Pythagorean theorems in Book I, and in Book III the equality of angles on a chord from the points on a circle and the equality of rectangles (products) of secants through a point to a circle.

Moreover, Euclid's *Elements* already sets patterns that are extrapolated by more abstract 20th-century mathematics. The Pythagorean theorem lives on in Euclidean and even Riemannian metric spaces; even Euclidean results 'every dog knows' set the framework of more abstract modern mathematical thought, as with the triangle equality I.20 in the definition of metric spaces.

Surely one of the most basic intellectual responsibilities of justification-centered philosophy of mathematics is to account for the justificatory success of diagram-based geometry. Nor should we fool ourselves that modern logical reconstructions by Hilbert (1899/1902) and Tarski (1959), however great their interest in other respects, give such an account: ancient geometers achieved their lasting, subtle and powerful results precisely by the means that philosophers dismiss so high-handedly today, without the benefit of modern logic and Hilbert's refined control of coordinate domains. (Mumma (2006) gives a logical reconstruction that is more sensitive to diagram use in demonstration.)

Still, our student might protest, even if it discharges a pressing intellectual responsibility, giving an account of the success of ancient mathematical demonstration might be mere grunt-work and not teach us anything new philosophically. Here too he would be wrong.

Euclidean diagram use forces us to confront mathematical demonstrative practice, in a much richer form than is implicit in the notions of mathematical theory and formal proof on which so much recent work in philosophy of mathematics is based; and to confront rigorous demonstrative use of non-propositional representation. The philosophical opportunities are extrordinary.

In the remainder of this introductory chapter, we outline a theory of diagram-based inference in Euclidean plane geometry based on the exact/co-exact distinction; and then survey further issues about diagram use in geometrical demonstration: how particular diagrams justify general conclusions, and the role of our theory in understanding diagram use in geometry beyond Euclidean plane geometry. For a view of what can be understood about diagram use from careful analysis of ancient texts, see Netz (1999).

Our topic parallels that of Giaquinto in this volume; but the two contributions address very different issues. Justificatory diagram use requires strongly shared standards of inference; Giaquinto looks at the benefits of visualization where such agreement is unavailable. The difference is stark. Geometrical demonstration got the 'famously agonistic Greeks' (Lloyd, 1996, pp. 21–23), and the later Arabic, Latin, and modern worlds, to agree on a body of detailed mathematical claims and arguments. In contrast, every time a mathematically informed audience responds to a repertoire of mathematical picture-proofs, they disagree on what the pictures show; even though all agree that the pictures show something important.

3.1 Basic rules of Euclidean diagram-based plane geometry

We give only the briefest sketch, anticipating 'The Euclidean Diagram'.

1. Demonstration *step*. Euclidean demonstrations read as discrete sequences of claims, each licensed by prior claims and the 'current' diagram. Beyond traditional text–text licensing, demonstration steps may draw from the diagram (diagram-based premises), and they may take responsability for additional diagram elements (diagram-based conclusions). The current diagram consists of those items previously so introduced, together with the regions, segments, and distinguished points arising by their interaction.

When in I.1, for example, we have said 'let AB be the given straight segment', we have not thereby admitted further features or diagram elements such as a circle with center A and radius AB; only a later demonstration step admits it. That a diagram for the entire demonstration may already be present detracts from this no more than that subsequent claims of the demonstration text are already on the page detracts from the fact that at a given step in the proof, only claims up to that point are available as premises. The linear structure of text allows one to track how far the process of taking responsibility has progressed; including how much of the diagram is current.

The general form of a single demonstration step license is thus:

$$\frac{\text{claims in prior text,} \quad \text{attributions to current diagram}}{\text{new claims in text,} \quad \text{new elements in diagram}}$$

Not all elements need be present.

2. Diagram-based attribution: *exact vs. co-exact* claims.

Recent critics of traditional geometrical demonstration see potential for error, precluding justification, when we claim based solely on what is in the diagram, as with the intersection point of circles in I.1; the logical tradition then claims a *gap* in a traditional proof. This, however, ignores how the ancient texts limit diagram-based attribution.

The only claims based on diagram appearance in a demonstration recognize conditions that are insensitive to the effects of a range of variation in diagram entries: lines and circles that are not perfectly straight or circular, and cannot be taken to be without thickness. As we distort the 'circles' in I.1, their intersection point C may shift but it does not disappear. Such conditions I call *co-exact*. They include: part–whole relations of regions, segments bounding regions, and lower-dimensional counterparts. Call the totality of these conditions in the current diagram its *appearance*.

In contrast, many (most?) conditions considered in Euclidean geometry would fail immediately upon almost any diagram variation: notably, equalities of non-identicals and proportionalities. Such conditions I call *exact*; they are never claimed based on what the diagram looks like. (Some naive diagram-unreliability criticisms presuppose that they are.)

3. *Diagram entries* in elementary geometry are line segments and circles, by Postulates I.1–3 directly or via prior constructions. Whenever a diagram entry is made, the text records the exact character (straight, circular) of the element entered. There is thus no need to later judge this from the diagram.

Diagram entries must be adequate both with respect to their co-exact and exact character. They are (at a minimum) continuous non–self-intersecting curves, a line segment connecting its endpoints, a circle closed. Participants can apply these co-exact criteria immediately and decisively (diagram size permitting, see 'sensitivity avoidance' below). Egregiously un-straight or un-circular entries can be immediately rejected; but on these exact criteria, entries do not appear to admit of immediate decisive acceptance. What matters here is whether inaccuracies can cause spurious co-exact diagram consequences; accounts of demonstrative justification could invoke sensitivity avoidance options, but also deferred rejection in the form of challenges to subsequent co-exact attribution (a logical novelty).

It is striking that diagram-entries, here typically one-dimensional, control zero-, and two-dimensional diagram entries (distinguished points, regions). This mutual relatedness of diagram elements appears to underlie the facility for shifts in emphasis and groupings of diagram elements that Macbeth (2007) thematizes in diagrammatic reasoning.

4. Case-branching. How additional entries in the diagram change its appearance (attributable co-exact character) may depend on previously non-attributable (metric) features. Example: every triangle, taken by itself, has the same appearance; but whether a perpendicular dropped from the vertex point to the base falls within the base or outside it (co-exact) depends on the shape of the triangle. Thus the same construction, applied to same-appearance diagrams (in our technical sense) may give different-appearance results.

A demonstration may attribute a feature to its diagram (say, that the perpendicular from the vertex falls within the base) that would not arise had we made the prescribed entries in a different initial diagram for the proposition. Generality then requires case-branching: separate diagram and continuation of demonstration for each combination of attributions that could so arise from an instance of the initial configuration (see below). How Euclidean practice ensures this is thus critical to its ability to justify general claims.

There are two theories on how a demonstration is to determine, given the inaccuracy of drawn diagrams, which diagram appearances may actually arise when it adds an entry: *enumeration/exclusion* or *diagram control*. Miller's (2001, 2007) work exemplifies enumeration/exclusion theory: for any diagram entry, he enumerates all conceivable topological arrangements; the demonstration must explicitly argue to exclude unrealizable arrangements before it demonstrates its conclusion for the remaining ones. This gives vastly more cases than we encounter in the texts, limiting its utility as an explication of ancient geometrical method. Mumma's version (2006) implements a stricter conceivability, giving significantly fewer cases.

Diagram control theory (see Manders (2007)) invokes our ability, using geometrical constructions, to produce reasonably accurate physical diagrams, and so limit the diagram appearance outcomes to be considered by physical diagram production rather than discursive argument. Conversations with specialists suggest this is the basic tool of ancient practice, with reductio argument for the exclusion of putative alternatives as backup. Harari (2003) discusses the 20th-century alternative view that constructions serve as existence proofs.

Some treatments (Miller (2001); Manders (2007), arguably) take each different-appearance diagram after an additional entry to require separate case treatment. Using Euclid I.2 as example, Mumma (2006) shows that limiting case branching to when a different-appearance feature is actually attributed, while sufficient for generality, vastly reduces the number of separate cases to about what we find in Euclid and Proclus.

5. Sensitivity of co-exact conditions.

As long as the 'circles' in I.1 are continuous closed curves, no amount of distortion can eliminate intersection points. Most co-exact conditions, however, will be affected by sufficiently large distortions violating exact conditions: a 'straight' line segment, say, that loops out far enough, will spuriously intersect any other line in the diagram. Normal drawing practices obviously avoid such egregious spurious co-exact conditions; nonetheless, this raises a challenge to the justificatory ability of diagram-based attribution: might we not encounter 'sensitive' diagram situations, where normal drawing practices would not suffice to avoid a spurious co-exact occurrence, which a demonstration would then be entitled to attribute?

Traditional practice aims to avoid such sensitive situations. There seem to be two mechanisms.

(a) By breaking up a development into a sequence of separate propositions, Euclid keeps each diagram simple: omitting construction lines from prior lemmas avoids point- and region-forming interactions with those absent lines.

(b) No matter how simple a diagram, if it is too small we cannot draw it accurately enough to rely on its looks. Even a good-sized diagram might have too much going on in too tight a corner—in a seminar paper, Matthew Weiner trenchantly called this a 'smudge'. I propose that a diagram in a demonstration is required to provide a *clear case* for its co-exact attributions, on pain of rejection. For example, to show what happens when an exact condition fails (say in I.6, in a reductio proof of the exact condition), one must use a diagram in which it fails in an exaggerated way. In actual geometrical reasoning, providing demonstrative grounds for a given co-exact attribution might require re-drawing the diagram. We will, however, need to consider the consequences of this proposal for the epistemology of generality in geometrical demonstration.

In the face of such complications, keep in mind that ancient diagram-based reasoning did work! The complications are only in *our* understanding of why it works.

3.2 Geometric generality

A traditional quandary about geometrical demonstration, famously discussed by Locke, Berkeley, and Kant, is how one particular drawn diagram can justify a general claim.

1. Some 20th century commentaries on geometrical generality may miss the mark because our 'strict-universal' standards for universal quantification, by including singular (in the mathematical sense) instances, disagree with traditional mathematical usage.

A traditional geometrical demonstration only claims to establish its proposition for non-trivial instances of its initial diagram, in which items such as angles and triangles contain proper regions (respecting the co-exact component of these notions, explained above). The demonstration of I.1, for example, does not purport to construct an equilateral triangle on a 'side' consisting of a single point. Individuating claims in this way is rational: as the example suggests, direct argument for (analogs of) claims in such limiting cases tends to be much easier than the demonstration at hand.

Perhaps because later algebraic and calculus methods establish (most) limiting cases automatically, non-triviality understandings on primitive geometric notions eventually vanished. From this later point of view, singular situations became 'exceptions'; traditional general claims and arguments have the *force* of 'admitting' exceptions, i.e. not applying to certain singular instances.

Strict-universal usage is surprisingly recent: 'The subject treated... illustrates well one of the most striking tendencies of modern algebraic and analytic work, namely, the tendency not to be satisfied with results that are merely true "in general", i.e. with more or less numerous exceptions, but to strive for theorems which are *always* true.' 'The great importance of this tendency will be apparent if we remember that when we apply a theorem, it is usually to a special case. If we merely know that it is true "in general", we must first consider whether the special case... is not one of the exceptional cases in which the theorem fails.' (Bôcher (1901)) For the phenomenon of theorems 'with exception', see also Sorensen (2005).

2. Whether particular drawn diagrams suffice to justify general geometrical claims (in the sense just recovered) depends on how demonstrations use diagrams. Particularity is not an incurable infectious agent.

Beth (1956) suggested, and Hintikka developed, the idea that the setting-out of the diagram in ancient demonstration be understood by analogy to the setting-aside of letters in universal generalization (UG) rules in modern logic, say as opening a scope for UG in a natural deduction system. But does analogy ensure that general claims are justified? The key point is how proceedings in UG scopes are *un-responsive,* at the level of inference rules, to distinctive features of instances.

3. Modern logic employs what one might call representation-enforced (for short, representational) unresponsiveness: the UG letter, placeholder for an instance, lacks features on which differential responses to non-shared features of instances might be based. This ensures that formal inferences cannot depend on such distinctive features, and hence apply to all instances uniformly. Algebraic methods in geometry since Descartes also exploit representational unresponsiveness: a letter, originally standing for a Euclidean line segment, lacks features to preclude it holding the place of a negative or even complex quantity, allowing one to display uniformities lacking uniform traditional geometric proof.

Drawn geometric diagrams, however, are not mere placeholders; they display endless distinctive features to which geometrically pertinent responses might be made. It is therefore sometimes suggested, assimilating Euclidean reasoning generality to the representational unresponsiveness model, that proper Euclidean proof requires *generic* diagrams, ones only displaying features that the demonstration indeed attributes.

Such a conception does not allow uniform treatment of diagrams: those with, for example, irrelevant right angles or equalities of sides could not be used in demonstrations. But if there are generic diagrams, this objection would not undercut the ability of demonstrations using them to justify their claims also for instances with exceptional (but non-singular) diagrams that do have all features the demonstration attributes; even if they have further features as well. (That Locke somehow lacks this flexibility would threaten to render his idea-of-a-triangle paradoxically equilateral.)

Nonetheless, among further objections to generic-diagram conceptions of demonstration (medieval manuscripts notoriously use non-generic diagrams, which may or may not reflect earlier practice; how generic must a generic diagram be?), the emergence of multiple non-exceptional cases seems fatal: because each case demonstration requires attributions that are false in the other cases, such proofs have *no* generic diagram spanning their UG scope.

4. Representation-enforced unresponsiveness to distinctive features of instances, however, is not the only appropriate way to justify general conclusions! Even if the (placeholder for the) instance has distinctive features to which humans might respond with claims of the very sort being justified, standards of reasoning may sucessfully prohibit such differential response.

This is fundamental to the generality- and understanding-generating quality of a great range of reasoning (including the use of examples in ethics). Even some formal systems of quantifier logic enforce uniformity across instances in this way: prohibiting UG-letter-containing premises from outside the UG scope.

Euclidean demonstration, I propose, attains uniformity of reasoning for its instances by licensing attribution based on what the diagram looks like only for co-exact conditions clearly displayed. The scope of generality in a geometrical demonstration, i.e. the range of instances for which the claim is to be justified, is precisely conceptualized in advance (contra Norman (2006)) by the setting-out statement (ekthesis) in the demonstration text that opens its UG scope, say, 'let AB be the given finite straight line' (hence, any line segment, I.1) or 'let ABC be a triangle having the angle ABC equal to the angle ACB' (I.6).

This respects Kant's conception (cf. Shabell (2003), Goodwin (2003)) that intuitions (diagrams) are particular, and connected to general claims via schematization (conceptualization via the diagram construction conditions). That diagram-based (co-exact) claims are stable under diagram distortion, hence independent of any particular empirical realization, might then motivate the necessity or apriority of geometrical intuition

5. A demonstration considering one particular diagram justifies its general conclusion just to the extent that each diagram-based attribution it makes would be appropriate to any instance, i.e. initial configuration satisfying the setting-out statement, were the constructions in the demonstration carried out on it. Precisely when this constraint is violated, case branching, i.e. suitably exhaustive separate case treatment, becomes necessary (but also sufficient). Given Proclus' emphasis on case distinctions, they should not have gone unnoticed; but historically, the generality debate largely ignored them, perhaps because it focused on examples such as I.1 and I.32 that require none.

Because an ongoing demonstration displays only the diagram it is developing, it remains to be fully clarified what resources ancient geometric practice had to locate diagram cases requiring alternative demonstration, or to establish that a demonstration has covered all such cases. Taken in conjunction with the aspect of diagram sensitivity unresolved above, that unlimitedly diverse cases might be hiding in tiny corners of one's diagram, this might seem particularly threatening to cogency of demonstration.

Tarski's (1959) mutually inverse interpretations between his Elementary Geometry and the theory of Real-closed Fields, however, endow the totality of (ideal) diagrams for demonstration of a particular claim with the topology of definable sets over real-closed fields. The finite cell decompositions of such sets imply both that the number of relevant cases is always finite (hence, keeping diagrams simple gets a grip on the problem) and that these cases may be found by systematically exploring the 'boundaries' of the region in which one's diagram lies. Mumma (2007) is currently expanding his analysis of case distinctions in Euclidean geometry by developing this.

6. Lest relaxing the currently favored representation-enforced-unresponsiveness means to generality, by giving a role to response prohibition, seem radical: contrast both with a third mathematically crucial strategy (though not in traditional geometry) for justifying a general claim: reduction to a special case.

A striking geometrical example is Poncelet's use of 'maximally-Euclidean' configurations to prove projectively invariant claims. Faced with an ellipse and some lines, he specializes to the case of a circle with some parallel lines; allowing him to efficiently exploit various kinds of equalities and congruent triangles to get the desired result, in the form it takes in this special situation (Poncelet, 1822) [Figs. 57–67]. This argument is supplemented by the reduction: an analysis that identifies the projectively invariant form of the claim, together with a demonstration that any instance of its projectively invariant hypothesis may be connected by projections to the special configuration considered.

Reduction to a special case is distinctive in that its treatment of the special case expressly relies on special features not shared with typical instances of the general claim under consideration (otherwise why bother with reduction), and fails for precisely for this reason to justify it. Instead, further argument, the reduction, justifies the general claim based on the special case. In their relaxing of representational unresponsiveness, such generality strategies are thus more radical than what we find in Euclid.

3.3 Tasks for the future

The analysis of diagram use in ED is specific to plane geometry in the style of Euclid I, III. Focus on one definite, cogent reasoning practice is its strength: it leads to substantive, sustained inquiry that invites us to go beyond our ingrained ideas. The philosopher's task is to articulate, using such a special case, notions, or principles of general interest, a distinctive usage for discussing rigorous diagram-based reasoning practices.

This conceptual framework then puts us in a position to articulate differences that matter between the base case and the role of diagrams in any other comparable area of geometry. Doing so would also test the broader import of our theoretical notions: what range of related practices do they allow us to understand?

1. Natural candidates include: roles of diagrams in Euclid Book II, or in Apollonius' *De ratione sectionis*, or Euclid's solid geometry; in ancient work

on conics (Apollonius), spherical geometry (Menelaos), and mathematical astronomy. Attractive targets also include later non-analytic work in geometry (Steiner!) and its foundations (Leibniz, Legendre); and early projective authors who aim to clarify Euclidean geometry rather than play a different game (Desargues, Poncelet). Diagrams in early modern physics (Galileo, Huygens, Newton) provide the additional ability to study the geometry of tangency to as yet unknown curves, hence to tackle problems now handled by first-order differential equations. What non-diagrammatic means (sign conventions) make it possible for Lagrange's *Méchanique Analytique* to dispense with diagrams altogether in theoretical mechanics?

I myself have used ED to contrast with Descartes' analytic geometry and later versions. One striking difference: analytic geometries have expressive resources allowing certain cogent inferences based solely on exact geometric information (using polynomial equations) separately from co-exact information (polynomial inequalities); Euclidean diagram-based reasoning requires both in tandem. These expressive means include invariants of a determinate geometric problem, the coefficients of its (irreducible) polynomial equation, which, up to selection from a finite set using inequalities, suffice for its solution. Moreover, algebraic manipulation, perhaps together with initial geometric constructions, suffices to derive the equation of a problem.

2. A special challenge is the role of diagram use in investigations of the *parallel postulate* (PP). It licenses inferences that would be based on the behavior of the diagram 'infinitely far' out, or, as is easily seen equivalent by Book I arguments not using PP, in the 'infinitely small': precisely circumstances under which our ability to control or attribute features in physical diagrams must give out.

Neither the mere consistency of alternatives to PP (hyperbolic geometry) with the remaining tenets of Euclidean geometry, nor even the 'Euclidean models' of such alternatives (hyperbolic geometry as the restriction of the Euclidean plane to the interior of an ellipse) fully settle the original question: whether Euclidean diagram-use has resources to obtain the results for which our Euclid texts invoke PP. For at least in their modern forms, consistency results are based only on propositionally articulated aspects of Euclidean practice, and the 'Euclidean models' explicitly modify the way diagram elements (even line segments) are handled. Both are obvious distortions of ancient practice.

The methods of ED in contrast, may be applied to earlier, still diagram-based investigations of PP (Saccheri, Lambert). Notably, Saccheri (1733) tries to exclude alternatives to PP by reductio analysis; but he clearly treats even line segments differently from what one would encounter in a non-reductio Euclidean demonstration. Is his treatment within the limits for a proper Euclidean reductio? Such questions will force us to refine our conception of ancient diagram-handling practice, eventually perhaps to the point where we can settle the original question about PP.

3. The role of diagram use in *ancient geometric analysis* is highly non-trivial, and 20th-century discussions have shed little light on this. Compared to what we find in the *Elements,* ancient analytic practice (Pappus' *Collectio* IV, VII) seems much less affected by case distinctions; probably due to matters of logical order (Behboud (1994)) as well as to the structuring of analytic reasoning by the collections of case-free what-determines-what lemmas such as Euclid's *Data.* Indeed, one as yet undescribed factor in geometrical analysis is the sense of what-determines-what that one only obtains by actually constructing diagrams. (Study of geometrical analysis must include the rich Islamic tradition, cf. Bergren and van Brummelen (2000).)

4. The nature of geometricality. Descartes' method in geometry is guided by algebraic form rather than by a geometrical diagram, and its algebraic manipulations are not intelligibly related to the original problem diagram—perhaps because a problem has the same equation as infinitely many others, whose diagrams need not be at all similar to the one intended. This strikes many authors, notably Leibniz, as a lack of geometricality; motivating two centuries of ultimately fruitful work to provide an 'analysis situs', some complementary theory of spatial arrangements that would respect geometrical form independent of the merely metric properties to which Descartes' method reduces geometry.

For example, Poncelet (1822) aims to regain diagrammatic intelligibility of geometrical reasoning while selectively maintaining uniform treatment of variant diagrams achieved by algebraic methods. Via a system of reinterpretations guided by algebra and continuous variation of diagrams, invoking complex ('imaginary') and ideal objects ('at infinity'), he grounds a novel diagram reading practice and an accompanying algebra of generalized line segments that together eliminate case-branching even more than Descartes. (This is a clear case of enhanced uniformity of conception and argument that exploits standards of reasoning rather than choice of representation to suppress differential response to certain different-looking diagrams.) At the same time, Poncelet's diagram maintains his grasp on a similarity class of Euclidean-individuated diagrams much narrower than those sharing a common Cartesian problem equation.

ED and most other recent discussions cited address, in the spirit of logical foundations, primarily justificational roles of the geometrical diagram. But diagrams in Poncelet have lost some cogency roles that ED attributes them in Euclidean argument. Because of this, further contrasting Descartes' ungeometrical geometry with Poncelet's (and Steiner's) diagram use may help bring out diagram contributions to geometricality that we have so far missed in looking at Euclid.

The Euclidean Diagram Due to its already somewhat wide distribution, and its role in guiding subsequent work, I have chosen to give the text in its original 1995 form that was distributed.

Bibliography

ARTMANN, Benno (1999), *Euclid: the Creation of Mathematics* (Berlin: Springer-Verlag).
BALL, W. W.Rouse (1892), *Mathematical Recreations and Essays* (New York: MacMillan).
BEHBOUD, Ali (1994), 'Greek Geometrical Analysis', *Centaurus*, 37, 52–86.
BERGREN, J. L. and VAN BRUMMELEN, Glen (2000), 'The Role and Development of Geometric Analysis and Synthesis in Ancient Greece and Medieval Islam.', in Patrick Suppes, Julius Moravscik, and Henry Mendell (eds.), *Ancient and Medieval Traditions in the Exact Sciences* (Stanford: CSLI).
BERKELEY, George (1710), *A Treatise Concerning the Principles of Human Knowledge*.
BETH, Evert W. (1956), 'Über Locke's "Allgemeines Dreieck"', *Kant-Studien*, 48, 361–380.
BÔCHER, Maxime (1901), 'The Theory of Linear Dependence', *Annals of Mathematics (second series)*, 2, 81–86.
EUCLID (anc.), *The Thirteen Books of Euclid's Elements*.
GOODWIN, William (2003), *Kant's Philosophy of Geometry*, Ph.D. dissertation, Berkeley.
HAHN, Hans (1980), 'The Crisis in Intuition', in *Empiricism, Logic and Mathematics: Philosophical Papers* (Dordrecht: Reidel), 73–102.
HARARI, Orna (2003), 'The Concept of Existence and the Role of Constructions in Euclid's Elements' in *Arch. Hist. Exact Sci.*, 57, 1–23.
HARTSHORNE, Robin (2000), *Geometry: Euclid and Beyond* (Berlin: Springer-Verlag).
HILBERT, David (1899/1902), *Foundations of Geometry* (La Salle: Open Court).
HINTIKKA, Jaako (1967), '*Kant's mathematical method*', *The Monist*, 51, 352–375.
LLOYD, G.E.R. (1996), *Adversaries and Authorities* (Cambridge: Cambridge University Press).
MACBETH, Danielle (2007), 'Diagrammatic Reasoning in Euclid's Elements', *manuscript*.
MANDERS, Kenneth (1995), 'Diagram Contents and Representational Granularity', in J. M. Seligman and D. Westerstahl (eds.), *Logic, Language and Computation* (Stanford: CSLI).
—— (2007), 'The Euclidean Diagram', *this volume*.
MILLER, Nathaniel (2001), *A Diagrammatic Formal System for Euclidean Geometry*, Ph.D. thesis, Cornell. <http://hopper.unco.edu/faculty/personal/miller/diagrams>
—— (2007), *Euclid and his Twentieth Century Rivals. Diagrams in the Logic of Euclidean Geometry* (Stanford: CSLI).
MUMMA, John (2006), *Intuition Formalized*, Ph.D. thesis, CMU.
—— (2007), Proofs, Pictures and Euclid, *manuscript*.

NETZ, Revel (1999), *The Shaping of Deduction in Greek Mathematics* (Cambridge: Cambrige University Press).
NORMAN, Jesse (2006), *After Euclid* (Stanford: CSLI).
PONCELET, Jean Victor (1822), *Traité des Propriétés Projectives des Figures*, 2nd edn (1865).
____ (1862/1864), *Applications d'Analyse et de Géométrie* (Paris).
PROCLUS (anc.), *A Commentary on the First Book of Euclid's Elements*, G. R. Morrow (trans.), 1970, (Princeton: Princeton University Press).
SACCHERI, Gerolamo (1733), *Eublides Ab Omni Naevo Vidicatus...*, George Bruce Halsted (ed. and trans.), 1920, (London: Open Court). Chelsea Reprint, 1986.
SHABELL, Lisa (2003), *Mathematics in Kant's Critical Philosophy*, Studies in Philosophy: Outstanding Dissertations Series, Robert Nozick (ed.) (London: Routledge).
SORENSEN, Henrik Kragh (2005), 'Exception and Counterexamples: Understanding Abel's Comment on Cauchy's Theorem', *Historia Mathematica*, 32, 453–480.
TARSKI, Alfred (1959), 'What is Elementary Geometry?', in L. Henkin, P. Suppes, and A. Tarski (eds.), *The Axiomatic Method*, 16–29 (Amsterdam: North-Holland).
VITRAC, Bernard (2004), 'A Propos des Démonstrations Alternatives et Autres Substitutions de Preuves dans les Eléments d'Euclide', *Arch. Hist. Exact Sci.*, 59, 1–44.

4

The Euclidean Diagram (1995)

KENNETH MANDERS

[Geometrical] figures must also be regarded as characters, for the circle described on paper is not a true circle and need not be; it is enough that we take it for a circle.

(Loemker, *Leibniz: Philosophical Papers and Letters,* p. 84.)

In Euclidean geometry, a diagram has standing to license inference, just as do relationships recognized in the text. It is now commonly held that this is a defect of rigor. But the extraordinary career of Euclidean practice justifies a fuller consideration. It was a stable and fruitful tool of investigation across diverse cultural contexts for over two thousand years. During that time, it generally struck thoughtful and knowledgeable people as the most rigorous of human ways of knowing, even in the face of centuries of internal criticism in antiquity.

We here undertake to reconstruct the role of diagrams in the inferential standards of Euclidean practice, seeking a more accurate view of the strengths and weaknesses of traditional geometrical argument. It is interesting to reclaim our philosophical grip on traditional geometrical reasoning especially because such reasoning coordinates two means of expression with very different characteristics: diagrams and ordinary text-based argument. Philosophers of our age seem to have ignored such representational contrasts, which nonetheless appear to be of great interest.

If it is to give a non-trivial grip on life, in its particular way, an intellectual practice must give us—all too finite and human beings—a game we can play; and play well, together, and to our profit. To succeed in this, intellectual practices harness our abilities to *engage* their artifacts, as I will call it: to produce, preserve, and respond to artifacts in controlled ways—to play games,

Research supported by NSF Grants DIR-90-23955 and SBER-94-12895, and a Howard Foundation Fellowship for 1994/95.

I suppose. Indeed, they must harness our abilities to do so in controlled ways responsively to our living, and to live in controlled but decisive ways responsively to our artifact games—so these games are for real, I suppose. Be they spoken dialogue, judicial argument, or geometrical diagrams, the detour through artifacts is the trick.

From the scant evidence of early Greek geometrical practice, I suppose that, to a much greater extent than since Euclid's time, 'seeing in the diagram' must have been the primary form of geometrical thought and reasoning. Whether or not this is so, we can seek out at every step not only the familiar pressures which promote the ultimate propositionally explicit foundation of geometry in the late 19th century, and the less familiar ones relieved by Descartes' geometry and by projective geometries, but also the pressures on a more fully diagram-based practice which would be relieved by the distribution of labor between diagram and text sequence we find in Euclid and beyond. As we become aware of this, the coordinated use of these two artifact types in Euclidean geometrical argument allows us a glimpse of what makes the detour through artifacts tick.

There is a long tradition of commentary on Euclid. Such works are animated by the authors' normative visions on geometric reasoning. For it is in the light of such normative visions that this gap appears; that that alternative proof seems superior for avoiding superposition; that the development of such a chapter seems roundabout; that the position of so a result seems puzzling. Only with a normative vision on geometric argument do the chains of argument become animated, more than just the chains of argument they are.

We here try, just so, to bring Euclidean geometry as a whole to life, as a practice among other geometrical practices historical or imagined, by confronting it with a normative vision on what a mathematical practice can be. This will sometimes allow us to see that it is reasonable that the practice be arranged as it is.

While we feel that there is an intuitively recognizable unity to 'traditional diagram-based geometry' from Plato's time to Hilbert's, a unity which our account of Euclidean reasoning tries to capture, it is of course plain, and confrontation with our account should make it plainer, that the tradition is anything but monolithic. We do not assume that geometers ever held the kind of view of their activities proposed here, or would have to do so to do geometry properly.

The conception of the role of diagrams in traditional geometrical thought given here should be understood as incomplete, probably on philosophically central points. While I go beyond inferential reconstruction in traditional logical terms, the roles to which we in the 20th century are sensitive and know to articulate are broadly but constrainedly 'inferential'. It is plain that we are missing a lot. Present philosophical resources for understanding the understanding seem insufficient to explain why intercourse with diagrams remains essential to geometrical thought. While their role is much less clear-cut in 20th-century geometrical theories, one look at the diagrams Hilbert enters next to his axioms of order should convince us of this—what are they doing there, and throughout the rest of his *Foundations of Geometry*?

4.1 Euclidean diagrams: artifacts of control or semantics?

At its most basic, a mathematical practice is a structure for cooperative effort in *control* of self and life. In geometry, this takes many forms, starting with the acceptance of postulates, and the unqualified assent to stipulations—and as it appears, for now, to conclusions—required of participants. Successes of control may be seen in the way we can expect the world to behave according to the geometer's conclusions; the way one geometer centuries later can pick up where another left off; the way geometers can afford not to accept contradiction. When the process fails to meet the expectations of control to which the practice gives rise, I speak of *disarray*, or occasionally, *impotence*. Such occurrences are disruptive of mathematical practices; they tend to reduce the benefits to participants and to deter participation. At best, they motivate adjusting artifact use, modifying the practice to give similar benefits with less risk of disarray.

The notion that the intellectual might be seen as some kind of game has recently been at the center of contention. Can't games *just* be made up? Kids make up games all the time; grown-ups get elected and paid to do it. If intellectual practices are games, what about the image of rigorous hard-working Science giving us a solid grip on Reality?

Theoretical geometry, which after all can be practiced in many forms and for many purposes, is hardly a context in which to address many such concerns, and facing any head-on would anyhow detract from our investigation. The impression, however, that making up a game involves excessive freedom, reflects a lack of grasp how hard it is to get a good game going, how hard it is to satisfy the varied and stringent constraints on an effective intellectual practice, especially one that must handle a fairly subtle and informative theory (such as Euclidean geometry) with minimal risk of disarray. Along the way, we shall become acquainted with such constraints and the forces shaping the practice to which they give rise.

4.1.1 Diagram and control

As does control itself, diagnosis and remedy of failures of control in geometry can take many forms. The grip on living a practice can give arises in a delicate interplay between the way its artifacts lie in the game, and the endowments and limitations humans bring to it. Two examples (a–b) of control through attaining uniformity, in quite different guises, may illustrate this.

(a) We take participants to be responsible for attaining uniformity of response in reading off features in diagrams. But the production and reading of diagrams cannot humanly be controlled so as to obtain a uniform response to diagrams directly on questions such as equality of lines. If a practice licenses a type of response to artifacts, then in order to attain uniformity of response, it must provide means by which participants can resolve their differences in responses to those physical artifacts they might reasonably produce. In traditional geometry, the division of responsibility between diagram and text, in which the text tracks equality information, is a way to meet this challenge. Subsequent geometrical representations provide alternatives.

If participants are to be able to respond in a stable and stably shared fashion, a practice must limit the repertoire of basic responsive roles which it requires. Geometers need to recognize a triangle in a diagram; and about thirty other such items. This repertoire of responses is drastically less diverse than are the physical objects which would qualify as diagrams (even more so, if we allow diagram imaginings).

(b) Under the banner of 'generality', practices exploit repertoire restriction also in artifact production. Why make distinct artifacts to secure the same response, after all? Why not support uniformity of response through uniformity of artifact? Geometrical diagrams are drastically less diverse than are the circumstances in life with which geometrical practice may let us cope; in that its generality (in Aristotle's sense?) consists. And in any particular geometrical demonstration, the diagrams displayed are drastically less diverse than are the circumstances in life with which that argument can help us cope; in that its generality (in Frege's sense) consists. We want to get the control or coping ability we need, through argument or otherwise, with the scantest artifact-and-response repertoire that will do.

The advances with respect to Euclidean geometry in Descartes' geometry, and later in projective geometries, are practice modifications to exploit more uniform artifact strategies, which implausibly somehow retain argumentative grip. The uniformity is achieved by suitably reducing the diversity either of artifacts or of responses to them. On the other hand, unsuitable limitation on the diversity of artifact-and-response repertoire leaves a practice blind and irrelevant, unable to provide its participants a non-trivial grip on life: whatever participants have to go on in the game must support distinctions where different actions are required.

If the only response ever required to p_1 and p_2 is q, we could disregard the difference between p_1 and p_2, always write p_1, or always write p. But participants typically can't assess 'only response ever required', nor—here our schematic p_1, p_2 notation breaks down—can they typically assess what range of alternatives might ultimately be relevant. In its diagram use, Euclidean geometry 'writes p_1' until it becomes clear some 'p'' could require a different response. A diagram adequate at one stage of a proof may upon development be 'un-representative'; this forces a case distinction in which multiple diagrams are considered separately. The peculiar success of various later geometries rests not only on their surprising reduction of diversity in artifact and response, but also on the way the correlative collapse of ranges of alternatives, quite as a by-product, often eliminates imponderables that participants in diagram-based geometry cannot count on being able to assess.

The way its artifacts lie in the game is not innocent to the control its players can achieve.

4.1.2 *Diagrams and semantic role*

Artifacts in a practice that gives us a grip on life are sometimes thought of in semantic terms—say, as representing something in life. There is, of course, an age-old debate on how geometrical diagrams are to be treated in this regard.

Long-standing philosophical difficulties, on the nature of geometric objects and our knowledge of them, arise from the assumption that the geometrical text is in an ordinary sense true of the diagram or a 'perfect counterpart'. These difficulties aside, a genuinely semantic relationship between the geometrical diagram and text is incompatible with the successful use of diagrams in proof by contradiction: reductio contexts serve precisely to assemble a body of assertions which patently could not together be true; hence no genuine geometrical situation could in a serious sense be pictured in which they were.[1] This simple-minded objection has nothing to do specifically with geometry: proofs by contradiction never admit of semantics in which each entry in the proof sequence is true (in any sense which entails their joint compatibility).

To help be a bit more precise: I take it that among logical formalisms, natural deduction systems reconstruct most directly the actual patterns of occurrence

---

[1] David Sherry has directed this argument at Berkeley's conception of the role of diagrams, and draws a similar conclusion (Sherry, 1993, p. 214). But Sherry makes only incidental proposals on how diagrams support inferences (p. 217) and expects to invoke Pasch-style axioms (n. 5).

of sentences in rigorous informal argument. Notably, they model *restricted* reasoning contexts—hypothetical, reductio, and quantificational contexts—by assigning distinct roles to sentences depending on their position in those contexts. These distinct roles have the effect of keeping straight the hypotheses under which an assertion is made, and conversely, what prior entries are in force at any given stage of an argument. (This in contrast to systems such as Frege's and Hilbert's, which treat all entries in the inferential sequence equally as assertions to which the reasoner is committed, differentiated as to inferential role only by their displayed structure and their order of occurrence. This requires some re-casting, perfectly straightforward, of patterns of occurrence in informal reasoning, such as Frege's explicit carrying along all hypotheses in force as antecedents of each sentence.) Although we do not take all inferences in the geometrical text to be purely logical, modeling patterns of sentence occurrence in restricted reasoning contexts seems the most illuminating formal approach to geometric argument.

With reference to this understanding, explicated in natural deduction systems, of being in force in a context, the claim becomes: proofs by contradiction never admit of semantics in which each claim in force in the reductio context is true (in any sense which entails their joint compatibility). In particular, neither an imperfect diagram nor a perfect counterpart can picture or represent, in any sense which entails their joint compatibility, all claims in force within a reductio context.[2]

It does not follow that there could not be a picturing-like relationship between the diagram and *some* claims in force within a reductio context. Rather, these facts put a different pressure on semantic conceptions of the relationship between diagram and text than the usual notice of imperfections (judged by the inferences licensed in the text) in the diagram, or (as in Leibniz) our ability to live with those imperfections. If diagram imperfections only were in play, one might well hold that the function of diagrams could fruitfully be approached by first elaborating a notion of perfect geometricals of which

---

[2] Juxtaposition of Euclidean diagram usage with natural deduction formalism suggests another remarkable conclusion. Euclid, by and large, lives by 'one proof, one diagram'. But reductio context or not, any broadly semantic conception of the relation between diagram and claims in force within a context would suggest instead: 'one reasoning context, one diagram'. Here, however, the upshot is not a restriction on the sense in which the relation between text and diagram could be a semantic one. Rather, the appropriate conclusion seems to be the narrowly logical one that Euclidean argument prefers 'one proof, one genuinely geometric reasoning context': except for purely logical unpacking of propositions (which in standard natural deduction formalism would require multiple nested contexts) there is preferably only one reasoning context per demonstration in which a diagram plays a special role. We have to allow disjunctive case analysis with distinct diagrams (see later), but that too appears not preferred.

the text is literally true, then treating diagrams actually drawn in geometrical demonstrations as approximations to perfect ones; finally deriving from all this an understanding of the bearing of the imperfect diagram on inferences in the text. But no detour through ontology and semantics which treats of truth in a diagram in a sense which entails joint compatibility of all claims in force in the reductio context can speak to the difficulty with the role of diagrams in reductio arguments, which are pervasive in Euclid.[3]

Thus one is forced back to a direct attack on the way diagrams are used in reductio argument; the problem of the relationship between diagram and geometric inference here turns out to be one of standards of inference not reducible in a straightforward way to an interplay of ontology, truth, and approximate representation. But once this is admitted, there seems to be no reason why direct inferential analysis of diagram-based geometrical reasoning should not be the approach of choice to characterizing geometrical reasoning overall, with or without reductio. If this order of analysis proves fruitful, ontological and semantic considerations will seem decidedly less central to the philosophical project of appreciating geometry as a means of understanding. For in their then remaining role of making the standards of geometrical reasoning seem appropriate, ontological-and-semantic pictures will have to compete with other types of considerations which we will find have potential to shape a reasoning practice. The failure, since Plato's time, of diagram ontology strategies to sponsor a convincing account of Greek inferential practice with diagrams, perhaps does not bode well for their prospects.

I therefore take it that traditional geometric demonstration has a verbal part, which for contrast I will call the *discursive* text; and a graphical part, the *diagram*. The discursive text consists of a reason-giving ordered progression of assertions, each with the surface form of an ascription of a feature to the diagram (*attributions*). A lettering scheme facilitates cross-references to the diagram. A step in this progression is licensed by attributions either already in force in the discursive text or made directly based on the diagram as part of the step, or

---

[3] Jerry Seligman has suggested that one might avoid the argument via some kind of 'compositional semantics' of diagrams. To avoid the argument, however, geometrically incompatible sets of sentences would have to be 'compositionally true' in the same diagram. That is to say (and the point has nothing to do with diagrams or even geometry), a 'semantics' that fills the bill here thereby lacks *minimal soundness,* the minimum requirement for notions of (weak) truth, that a set of sentences all 'true' in the same situation, cannot trivially entail a contradiction. Because of its failure of mininal soundness, such a 'semantical' conception of diagrams could not account for a role of diagrams in geometric inference (the original point of the detour); and it would deprive ontology of its traditional philosophical point, taking truth really seriously. A corollary: the problem of reductio argument forces semantics/truth considerations apart from semantics/meaning considerations, which might indeed effectively be approached by compositional semantics, presenting *separate* situations in which the components of a contradiction are true.

both. A step consists in an attribution in the discursive text, or a construction in the diagram, or both. For example, the application of Postulate 3 in the argument of Euclid I.1: 'With center A and distance AB let the circle BCD be described' (Euclid) consists in the entry of a certain curve in the diagram, the establishment of anaphoric ties between discursive text and diagram, and the stipulative attribution in the discursive text that the curve is a circle with given center.

4.2 Diagram-based attribution

Diagram-based attributions—conclusions licensed in part by what is *seen* in the diagram rather than already in force in the discursive text—are of special interest, given that our official model of reasoning is motivated by linguistic rather than diagrammatic representation.

Work in the program of locating 'implicit assumptions' has amply shown that genuinely diagram-based moves occur throughout traditional geometrical argument (Mueller, 1981, p. 5, and throughout). With obvious exceptions, almost every step in traditional geometrical argument finds its license partly in the arrangement of the diagram. From a modern foundational point of view (Hilbert (1899/1902), Tarski (1959)), it is clear that continuity considerations enforced by the diagram, in close conjunction with exact relations, contribute much of the existential import to geometry. Moreover, the very objects of traditional geometry arise in the diagram: we enter a diagonal in a rectangle, and presto, two new triangles pop up. Though it has proved remarkably difficult to explain why this is so important, without diagram-based inferences, organized rather as they are through the actual use of diagrams, spatial thought is handicapped beyond recognition. Even as far as proofs are concerned, the Euclidean discursive text, unlike 20th-century formal logic, generally lacks alternate resources for obtaining the diagram-based conclusions which the Euclidean diagram provides. Judging from the historical record of the reputation of geometrical reasoning as a model of rigor, it does so without causing undue disarray.

Nonetheless, it has been commonplace for at least the last century to castigate traditional geometry for 'gaps in arguments' and 'implicit premises' due to 'reading off from the figure.' By the end of the 19th century, the view had taken root that drawing conclusions from the diagram was a defect of traditional geometrical practice. The additional axioms introduced in this period (Pasch, Hilbert), notably concerning betweenness and continuity, aimed to remedy this by strengthening the discursive text. Felix Klein comments: 'The significance of these axioms of betweenness must not be underestimated... if we wish to develop geometry as a really logical science, which, *after* the axioms are selected, no longer needs to have recourse to intuition and to figures for the deduction of its conclusions... Euclid, who did not have these axioms, always had to consider different cases with the aid of figures. Since he placed so little importance on correct geometric drawing, there is real danger that a pupil of Euclid may, because of a falsely drawn figure, come to a *false* conclusion.'[4]

Twentieth-century philosophers have tended to follow mathematicians in their assessments of traditional geometry. Ayer, for example: 'diagrams... provide us with a particular application of the geometry, and so assist us to perceive the more general truth that the axioms of the geometry involve certain consequences.... most of us need the help of an *example*... [T]he appeal to intuition... is also a source of danger to the geometer. He is tempted to make assumptions which are accidentally *true of* the particular figure he is taking as an illustration.... It has, indeed, been shown that Euclid himself was *guilty of this,* and consequently that the presence of the figure is essential to some of his proofs.'[5]

Thus, the first task of an account of geometrical reasoning in which diagrams are treated as genuinely inferentially engaged 'textual' elements is to augment the well-established logical construal of the discursive text by a reconstruction of the standards for reading and producing diagrams.[6]

Extant ancient texts give little explicit discussion of these standards, on which we anyhow cannot expect full unanimity over the centuries of Greek mathematical practice. Some light is shed on them by occasional records of dispute on geometrical propriety, and by records of proposed cases or objections.[7]

The key to reconstructing standards for producing and reading diagrams is the realization that diagram and text contribute differently, so as to make up for each other's weaknesses. Some things may be read off directly from

---

[4] Klein (1939, p. 201); quoted from Mueller (1981, p. 5).

[5] (Ayer, 1936, p. 83). (My emphasis. I thank Ali Behboud for pointing out this passage to me.) Philosophers' concern about diagram use in geometry go back at least as far as Plato's *Cratylus:* '...geometric diagrams, which often have a slight and invisible flaw in the first part of the process, and are consistently mistaken in the long deductions that follow' (Jowett transl., at 436D).

[6] Wilbur Knorr has pointed out to me that the standards for diagrams in solid geometry, and in its applications in mathematical astronomy, must be quite different indeed from those in force in plane geometry. Only plane geometry is considered here.

[7] The most informative remaining source is Proclus' *Commentary on the First Book of Euclid.* For discussion, see especially Heath (*Euclid*, vol. 1) and Mueller (1981). There may be some bias in focussing on the foundationally oriented commentary tradition represented by Proclus; see the comments on mathematical impact of the Alexandrian foundational school, in Knorr (1980, esp. pp. 174–177).

a diagram, some may be inferred only from prior entries in the discursive text. More careful consideration reveals that the inferential contribution of the diagram depends also on standards of diagram production, in perhaps not entirely expected ways.

Directly attributable features are ones with (a) certain explicit perceptual cues that are (b) fairly stable across a range of variation of diagram ('co-exact' rather than 'exact') and (c) not readily eliminable by *diagram discipline,* proper exercise of skill in producing (still imperfect) diagrams required by the practice.

For later reference, we define the *appearance* or *topology* of a diagram to comprise the inclusions and contiguities of regions, segments, and points in the diagram (strictly speaking, only in so far as not eliminable by diagram discipline). These will turn out to determine its attributable features.

Already in antiquity, diagram-based moves were open to challenges, as we see from *objections* recorded by Proclus and others, discussed below; but the moves were defended and retained their license. By the middle and late 19th century, one does indeed want to consider alternative geometries such as those with restricted coordinate domains characterized by algebraic conditions; one does indeed want to consider the properties of 'curves' such as space-filling or continuous nowhere-differentiable ones. Traditional diagram usage does indeed provide inadequate representational support for the reasoning required in these cases.[8] But such challenges did not arise within traditional geometry; and the unquestionable need to shift to other representational means to meet them should not (as it has for most of this century) cloud the question of how traditional geometry met *its* challenges in such exemplarily stable fashion.

4.2.1 Explicit perceptual cues

Greek geometry requires us to respond to diagrams as made up primarily of regions, not the curves which bound them. Euclid defines, 'A figure is that which is contained by any boundary or boundaries.' The Greek word for figure here, *schema,* refers to a shaped region; in contrast, we usually attend to the delimiting curve and treat as incidental such regions as appear. In analytic geometry, a circle is a curve satisfying an equation; in Euclid, it is 'a plane figure contained by one line such that...'[9]

---

[8] This, and not the outright invalidity of diagram-based reasoning ('spatial intuition') would be the proper conclusion from similar considerations adduced by Hans Hahn (1980).

[9] Euclid I definitions 14–15; see further Heath *Euclid* I, 182–3. Proclus brings out this sense of *schema* very clearly: 'the circle and the ellipse ... are not only lines, but also productive of figures. ... if they are thought of as producing the sorts of figures mentioned, ...' (103). He attributes to Posidonius a contrary view: 'it seems he is looking at the outer enclosing boundary, while Euclid is looking at the whole of the object.' (143.14)

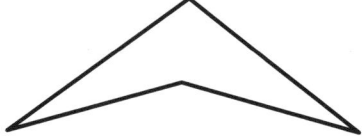

Fig. 4.1.

Traditional geometric practice is surprisingly literal in what it recognizes in the diagram beyond items and relationships attributed stipulatively in the discursive text. Proclus points out that '...the triangle as such never has an exterior angle' (*In Euclidem*, 305). One must explicitly say, 'if one of the sides is produced', and so indicate in the diagram, before referring to an exterior angle in a demonstration. Even so, Proclus counts a proof (Euclid I.32) as inferior, with 'the middle term used here only a sign,' instead of a 'demonstration based on the cause,' because it involves extending one of the sides to form an exterior angle (206.16–207.3).

Another example: Proclus accepts a *four-sided* triangle (!) (Fig. 4.1), because it '...was the practice of Greek geometers not to recognize as an "angle" any angle not less than two right angles' (Morrow).

Evidently, the outer angle does not count here; presumably because angles are regions delimited in a certain way, and two lines at two right angles or more do not do so.[10]

The primary recognitions sanctioned by traditional practice include suitably bounded *regions* or extended objects (line segments, triangles, angles), associated containment relationships (part-whole, point-in-region, line-in-region), and intersections of two (but not more) curves. For example, when, as in the proof of Euclid I.6, we draw a line from one vertex of a triangle to (the interior of) the opposing side, we are to recognize that the triangle has been partitioned into two triangles, parts of the original, sharing a common side; that the angle at the vertex in question is divided into a sum of two angles, and that the line forms a pair of angles where it meets the opposing side.

Standards for treatment of equality and inequality (greater and less) help further bring out the explicitness requirement in diagram-based attribution. Many things are said to be equal in geometry; and this is a key to the power of geometrical inference. But judgements of equality, if to be made from a

---

[10] Morrow, fn. to Proclus 329.7; see also Heath *Euclid* I p. 264. At 301–302, Proclus introduces a three-grade distinction which can plausibly be aligned with explicitness: between the familiar (i) knowledge about already existing things (that the base angles of the isoceles are equal), and (iii) making something (bisecting an angle), he suggests (ii) a 'finding of things', 'to bring what is sought into view and exhibit it before the eyes' (to find the center of a given circle, or the greatest common measure between two given commensurable magnitudes).

diagram, rather than inferred from prior entries in the text, require not just equality but instead its most explicit possible form: *coincidence* that arises in an extremely direct way from the constraints under which the (imperfect) diagram is produced; such as the common side in the triangles which arise in the proof of I.6. This restriction on diagram-based attribution is not so surprising, as equality among less directly related diagram elements cannot be controlled by the standards under which diagrams are produced; but the restriction applies just so to inequality (which appears to pass our 'co-exactness' stability test below).

A striking example is Euclid I.20, 'any two sides of the triangle taken together are greater than the third.' The sometimes maligned requirement that this diagrammatically evident fact be proved no doubt reflects a relatively advanced theoretical foundational development in Greek geometry; but it is that tradition so developed which we are to characterize. Geometrical reasoning frequently obtains inequalities directly from the diagram, but (corresponding to the restriction for equality) only when an object in the diagram is a *proper part* of another, rather than from any kind of indirect comparison. Equalities and inequalities (greater, less) among objects distributed throughout the diagram are of great importance, and inference principles for both are available within the discursive text. In order to allow these notions to function also in diagram-based geometrical argument, the *Elements* makes it its first order of business to secure their transfer around the diagram so that direct comparisons may be made (notably, I.2, I.3, I.4): both (i) so that inequalities already available in the discursive text may be realized as proper inclusions in the diagram (I.6), and (ii) so that inequalities in the diagram may be related to explicit inclusions there which then may be read off (I.16).

4.2.2 Exact and co-exact attribution

Geometric attribution standards may be understood further by distinguishing what will be called *exact* and *co-exact* attributes. Typical alleged 'fallacies of diagram use' rest on taking it for granted that an—to the eye apparently realized but false—exact condition would be read off from a diagram; but the practice never allows an exact condition to be read off from the diagram. Typical 'gaps in Euclid' involve reading off some explicit co-exact feature from a diagram; and this is permissable. It will not be hard to discover a rationale for this: the distinction of exact/co-exact attributes is directly related to what resources the practice has to control the production of diagrams and provide for resolution of disagreement among judgements based on their appearances. The practice has resources to limit the risk of disagreement on (explicit) co-exact attributions from a diagram; but it lacks such resources for exact attributions, and therefore could not allow them without dissolving into a disarray of irresolvably conflicting judgements.

The distinction, stated loosely, is this. Consider all (imperfect) diagrams which might reasonably be drawn from given specifications; some of which may be regarded as related by deformation or *variation*. Co-exact attributes are those conditions which are unaffected by some range of every continuous variation of a specified diagram; paradigmatically, that one region includes another (which is unaffected no matter how the boundaries are to some extent shifted and deformed), or the existence of intersection points such as those required in Euclid I.1 (which is unaffected no matter how the circles are to some extent deformed). Exact attributes are those which, for at least some continuous variation of the diagram, obtain only in isolated cases; paradigmatically, straightness of lines or equality of angles (neither of which survive any except exceptional types of deformation, no matter how small). The distinction is perfectly apparent to anyone who has made enough diagrams to sense what one can and cannot control in a reasonably simple diagram.[11]

Many attributions have both co-exact and exact components. A triangle, for example, is a non-empty region bounded by three visible curves (all that is co-exact) which are in fact straight line segments (exact). Once realized, this does not give any difficulty.

Exact attributes can, at least since since Descartes' time, be expressed by algebraic equations. In traditional geometry, many were expressed or defined by equalities or proportionalities. Prominent examples include equality of lines (segments), angles, or other magnitudes, congruence of triangles or other figures, proportionality of lines; that an angle is *right* (not that it is an angle), that four points are con-cyclic (lie on a common circle); the geometric character of lines or curves, or the regions they delimit: a *circle*, an *ellipse*; that lines are *straight* (derivatively also, rectilinear angles, triangles, and so on) or *parallel*; that three lines or curves, while not initially stipulated to do so, intersect in a common point (rather than pairwise in three distinct points); that a line is tangent to a curve (rather than intersecting it in two or more distinct points close together).[12]

---

[11] It is a matter of mere patience to provide an explicit modern topological account of diagrams, (in terms of smooth embeddings of linkages of lines and circles into the Real plane) and their variation (in terms of homotopic families of such embeddings), which does not suppose them made with perfect accuracy (because a smooth embedding of a straight line need not be straight).

[12] 'Expressing' here carries no philosophical weight, as their equational expressability is not what makes attributes exact. But it is perhaps prudent to note that this 'expressing' involves considerable complications. Notably, the notion of straight line in the Euclidean diagram sense is presupposed: it is an unexplained primitive in Descartes' geometry, and must be independently available in order to relate coordinate geometry to Euclidean diagrams. The algebraic properties of the coordinate equation

Exact attributes (indeed, by definition) are unstable under perturbation of a diagram. This renders them well beyond our control in drawing diagrams and judging them by sight. Thus, diagramming practice by itself provides inadequate facilities to resolve discrepancies among responses by participants as they draw and judge diagrams. It is therefore unsurprising that exact attribution is licensed only by prior entries in the discursive text; and may never be 'read off' from the diagram. We observe this throughout ancient geometrical texts. That the curves introduced in the diagram in the course of the proof of Euclid I.1 are circles, for example, is licensed by Postulate 3 in the discursive text, and is recorded in the discursive text to license subsequent exact attributions, such as equality of radii (by Definition 15, again in the discursive text).

Exact attributions license a variety of extremely powerful inferences, which are central and pervasive in traditional geometrical discourse. Besides substitution inferences

$$a \simeq b, \Phi(b) \quad \Rightarrow \quad \Phi(a)$$

which notably include transitivity inferences, we find equalities licensed by definitions (circle has equal radii), and congruence attributions, and licensing congruence attributions. We frequently find proportionality attributions licensed by the various rules for manipulating proportionalities, similarity of figures, and 4-point con-cyclicity (by Euclid III 35–37). Without such inferences (whether in this or some other form), spatial reasoning is handicapped beyond recognition.[13]

Co-exact attributes express recognition of regions (and their lower-dimensional counterparts, segments, and points) and their inclusions and contiguities in the diagram. We might say they express the topology of the diagram. Prominent examples include ascription of non-empty delimited planar regions: triangles, squares, ... (but not that sides are straight); circles (but not that they are circular rather than elliptical or just irregular); angles (must be less than two right angles, to delimit a region). In lower dimension, non-empty segments (but not the character of the curve of which they are segments); points as two-place (but not three-place or tangent) intersections of curves; *non-parallel* lines; non-tangencies. Further, contiguity and inclusion relations among these: point lies within region or segment; side *opposed* to vertex; line *divides region* into two parts; one segment, angle, or triangle is *part of* another; *alternate* angles, a point *lying between* two others, and so on.

---

of the straight line no doubt does constrain its own applications; but the equation still captures only a certain *inference potential* of the Euclidean diagram notion of straight line, not that notion itself.

[13] As in Fuzzy Logic, compare Hjelmslev (1923).

Co-exact attributions either arise by suitable entries in the discursive text (the setting-out of a claim, the application of a prior result or a postulate, such as that licensing entry of a circle in the proof of I.1); or *are licensed directly by the diagram;* for example, an intersection point of the two circles in Euclid I.1. This poses no immediate threat of disarray, because co-exact attributes (again, by definition) are 'locally invariant' under variation of the diagram: they are shared by a range of perturbed diagrams.

The allowable range of variation, not specified by our definition of co-exact attributes, differs from diagram to diagram. Some features, such as an intersection point in I.1 or in Pasch's axiom (a line crossing into a triangle must, if suitably extended, cross exactly one other side or a vertex), arise from fundamental topological properties of the Real plane and are extremely stable. On the other hand, in complicated configurations the topology of the diagram and the ordering of salient points on a line can be quite sensitive to perturbation of the diagram. Such sensitivity poses a genuine threat of disarray. We will return to this. In many cases, this challenge may be met in traditional geometry by simplifying the diagram, by splitting off a preliminary result.

In so far as exact attributes are expressable by equalities and co-exact ones by strict inequalities, one would expect that a diagram inappropriate with regards to co-exact feature could not lead to an inappropriate exact conclusion: strict inequalities don't imply non-trivial equalities. Both are unfortunately false. Certain equational relationships are co-exact, and appropriately read off from topological relationships in the diagram: inclusions in a diagram license the assertion that the whole (region, angle, line) equals the sum of its disjoint parts.

Notably, when a line segment is seen to coincide with its parts put end to end, it may be written as the sum of those parts; say (in Fig. 4.2), $AB = AP + BP$. (In traditional geometry, until the 19th century, line segments are non-directed; one writes BP and PB *indifferently*.) This is co-exact if P is constrained by stipulation to lie within AB: no matter where P lies in AB, there will be a range of variation of P within AB (by stipulation, its only degree of freedom), and $AB = AP + BP$ will of course continue to hold throughout this range.

4.2.3 *Diagram-based attribution tested: are all triangles isoceles?*

The preceding theory readily eliminates those 'fallacies' of diagram-based geometry that depend on reading an exact condition off from the diagram. Much more serious challenges, however, arise from co-exact equational conditions: the equality of the whole and the sum of its parts. Because the resulting attributions are equational, they can lead to very strong conclusions;

THE EUCLIDEAN DIAGRAM (1995) 95

Fig. 4.2.

because they are (properly) based on diagram appearance, there is a premium on controlling this.

An example is the striking argument that all triangles are isoceles, popular in discussions of 'geometrical fallacies' (Maxwell (1961), Dubnov (1963)). Let the given triangle be $\triangle ABC$ (Fig. 4.3). Let the bisector of angle (BAC) and the perpendicular bisector (at D) of the side BC intersect at O. Draw the perpendiculars OQ and OR to the remaining sides AC and AB; connect OB and OC. We obtain the following three congruences: (a) $\triangle ODB \simeq \triangle ODC$, by side-angle-side; hence $OB = OC$. (b) $\triangle AOR$ congruent to $\triangle AOQ$, by (bisected) angle-(common) side-(complemetary) angle; hence $OR = OQ$. But then (c) $\triangle BOR$ congruent to $\triangle BOQ$, because they are right triangles with equal side and hypotenuse; hence $RB = QC$. So far, so good.

It is, however, possible to draw a diagram with AO not quite bisecting the angle, and DO not quite bisecting the side (Fig. 4.4), so as to display R and Q either (i) *both* on AB respectively AC, or (ii) *both* beyond AB

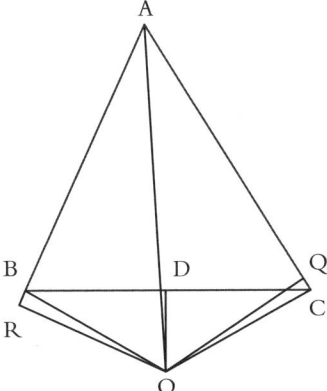

Fig. 4.3.

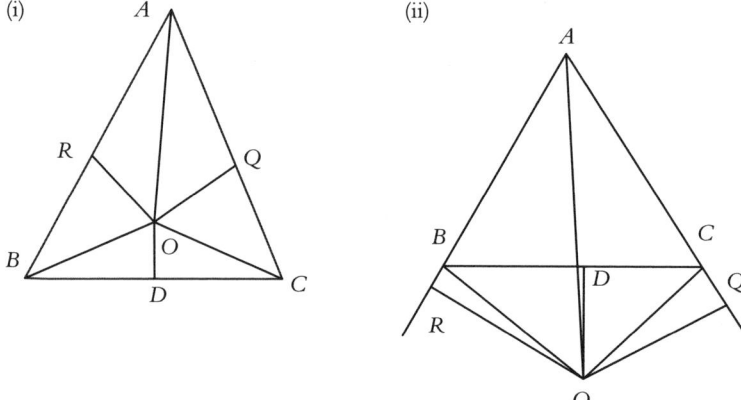

Fig. 4.4.

respectively AC. Either display is explicit and co-exact. Then one threatens to conclude (i) $AB = AR + RB = AQ + QC = AC$ or (ii) $AB = AR - RB = AQ - QC = AC$, so the triangle is isoceles!

The resources of traditional geometry to deal with arguments of this nature consist in control of the diagram; backed up by probing of diagram behavior. For triangles that are not close-to-isoceles, all but the initially given appearance would be eliminable by refinement of the diagram subject to 'diagram discipline' discussed below; so that one could block attribution of the other appearances that lead to the undesired conclusion. But the ability of geometric argument to exploit this is restricted by the unlimited sensitivity, for close-to-isoceles triangles, of the appearance of the diagram constructed.

4.3 Controlling diagram appearance

Diagram appearance, then, in the technical sense—the inclusions and contiguities of regions, segments, and points—plays a decisive role in geometrical demonstration. The basic general resource of traditional geometrical practice in controlling the appearance of the diagram is *diagram discipline*, standards for the proper production and refinement of diagrams.

These standards are implemented by exploiting the considerable (though nonetheless limited) practical determinateness of geometrical constructions, and our (also nonetheless limited) ability to detect certain types of defects in candidate diagrams. The control that may be obtained in this way is further limited by the (sometimes extreme) sensitivity of diagram appearance to metric relationships, and the occurrence of multiple cases. Traditional geometrical practice is on guard against failures of diagram appearance control, by a process of *probing* diagram behavior.

4.3.1 Diagram discipline

Diagram-based attribution requires a third element, that the feature attributed appear in *appropriately produced* diagrams, and indeed, is not readily eliminable from them upon refinement. We appear to lack ancient prescriptions for the proper production of geometric diagrams. Nor should we be too ready to suppose that those found in contemporary or 19th century teaching materials reflect ancient practice.[14] Older texts, and diagrams in a few surviving though less ancient sources, do make clear that at least the use of letters to indicate points or lines is ancient.[15]

The following basic and incomplete proposals may nonetheless be made. Because of its reliance on diagram-based attribution, traditional geometric reasoning is essentially bound up with physical skills of diagram production, for which there are standards of practice but not explicit and complete rules. Diagrams are constructed line by line (this includes curves), with points taken and lines drawn as stipulated in the discursive text. In the *Elements* these constructions are those enumerated in the postulates, but elsewhere further means of construction (*neuses*, for example) are available.

We discuss exact and co-exact attributes separately. In entering a diagram element according to a construction postulate, one readily produces an unequivocally readable presentation of its stipulated connectivity ('let *CA* ... be joined'); and on this type of co-exact attribute I take the practice to require success. One might, however, worry whether this always holds of the initial

---

[14] (a) I don't know of any use before the 19th century of the familiar 'hash marks' to indicate equality of lines or angles. This practice seems too effective to have been readily lost once established; and is thus presumably a recent innovation. (b) Jones remarks of the extant manuscripts of Pappus' *Collectio*: 'The most apparent, and paradoxical, convention is a pronounced preference for symmetry and regularization in diagrams, introducing equalities where quantities are not required to be equal in the proposition, parallel lines that are not required, right angles for arbitrary angles, and so forth. Modern practice discourages the introduction of this kind of atypicality in geometrical figures.' (N. Jones, *Pappus of Alexandria, Collectio VII*, vol. 1, p. 76.) It is not clear to me whether this tendency towards recognizably atypical diagrams reflects geometrical practice in Pappus' (let alone Euclid's or Plato's) time, or later scribal practice detached from doing geometry; but we will anyhow find the modern 'atypicality discouragement' standard inadequate.

[15] Texts: Older sources, viz. Hippocrates (as reported by Eudemus) and Aristotle write expressions such as 'that [point] at which κ [stands]', where Euclid would write simply 'the [point] κ'. (See Tropfke (1923) vol. 4 pp. 14–15.) Ancient examples: letters are visible in the diagrams for I.9 in a papyrus from Herculaneum (P. Herc. 1061) the text of which is dated about 100BC, and of I.39 in an Egyptian papyrus (P. Fay. 9) of the second century AD. (See Fowler (1987) Plates 1, 3 esp. pp. 208–209, pp. 212–213, and references given there.)

diagram of a proposition; for arbitrarily complex patterns of connectivity might be required; and certain patterns of connectivity would be incompatible with exact attributes such as that the lines be straight. Only certain co-exact requirements, moreover, can be directly controlled: only lines can be drawn directly; planar regions (as well as certain distinguished points and segments on them) just pop up as a result of this.

Whether a region is, for example, three- or four-sided, may depend on the initial data of the construction in a manner sufficiently inscrutable to thwart control. One purpose of Euclid's treating numerous construction problems in Book I may therefore be to underwrite the availablity of diagrams which propositions might require us to consider. It is probably prudent to regard the availability of diagrams satisfying complicated co-exact conditions as potentially a sore point in traditional practice; and potentially contributing to the pressure to get inferential weight away from such diagrams. Such pressure, however, is only exerted to the extent that one has the expressive means, and intellectual priorities, to bring complex diagram specifications in play in the first place; one should not simply extrapolate the urgency of such problems back from the 19th century to ancient mathematics.

On the other hand, *exact* stipulations concerning diagram elements, such as that lines be straight or circles perfect, cannot be fully attained even for the simplest of configurations. The needs of Euclidean practice are higher here than one might imagine; rather than unequivocal readability of exact attributes (which is superfluous in so far as the discursive text records the stipulations), what will turn out to be needed is metric accuracy sufficient to render unequivocally readable co-exact properties associated with further constructions which might need be applied to the diagram: both the co-exact pre-conditions of applicability of those constructions, and the appearance of the diagram resulting by their application.

I take it that presentation of exact stipulations in the diagram is therefore subject to *quality control:* defects are recognizable, and when they appear severe, or pertinent to co-exact attributions made from the diagram, complaints are in order. In some situations a diagram must be re-drawn, or rejected as inadequate. (i) If a line said to be straight is so crooked in your diagram that I can noticeably improve it, and it looks like switching to the improved line might affect the topology of regions, you are required to re-draw; (ii) equally so if I can point to outrageous features of the circle or circle segment in your diagram, or (iii) clear failures of alleged equalities or (iv) of parallelism, which appear of some moment for what you are doing.

These quality control standards for exact attributes concern diagram elements individually (straight lines), or in pairs (equalities, parallelism of lines) and perhaps occasionally threes and fours (proportionalities, con-cyclicity). In each case, quality control operations (both the detection of defects which require questioning the suitability of the diagram and corrective control when defects arise) involve relationships between diagram elements specifically identified in the given stipulation.

When diagram discipline is in force with respect to an exact claim (we will see that in reductio arguments this does not always take place), we say that 'the diagram is *subject to*' the exact attribution; say, that a line AB is straight. The term 'subject to' marks off that although the diagram under these circumstances need not (and in general will not) satisfy the condition in unequivocally readable fashion, it is to be held to certain standards in relation to its function in argument, standards which might require diagram replacement at any point. The explicit focus of the standards is indicated by the attribution in the 'to' position of 'subject to'. Any explicit co-exact condition which is properly attributable in a diagram, i.e. not eliminable by refinement with respect to the exact conditions to which the diagram is subject, will be said to be *indicated in* that diagram. On the other hand, at least the justificational burdens which moderns think to target by quality control standards on the diagram as a whole or on unspecified subsets of diagram elements, notably requiring the diagram to avoid special properties (atypicality), would have been carried by other features of ancient practice:

(i) Because exact attributions may anyhow not be read off from the diagram, it has no inferential import if some special relations of that type might seem (spuriously) realized in a diagram. This is just as well, as there are no limits to the equalities which might be judged to emerge in the construction of a complex diagram; and except in the very simplest diagrams it is therefore implausible that participants could effectively control their work so as to satisfy such an atypicality requirement, or agree whether they had succeeded in doing so.

(ii) Diagrams may be atypical, relative to what is called for, in their co-exact properties as well. If such atypical properties are explicit in the diagram (see above), traditional practice cannot afford to simply block us from reading them off, as the discursive text has inadequate resources to acquire by other means the co-exact attributions it needs. Something else is therefore required to control the effects of atypicality at the global diagram topology level; this function is fulfilled by practices of case and objection proposal (see below). Here too, the modern idea of atypicality avoidance would not work: it has the wrong structure. For the atypicality avoidance standard presumes that there is such a thing as a 'generic' diagram, and as long as you re-draw until you have one, you will then be OK. This is fine for genericity with respect to exact attributes (as we can learn from the ubiquity of the notion of *generic point* in algebraic geometry earlier in this century). With respect to co-exact attributes, however, no single diagram need be generic. For non-trivial configurations, the effect naively envisaged as atypicality avoidance must instead be achieved through redrawing practice together with *disjunctive* case examination.[16]

4.3.2 Constructions and diagram control

Constructions contribute to diagram appearance control by their determinateness: they allow us to form a particular one of the many diagrams with curves with the desired connectivity.

Recent commentators have extensively debated whether Euclid's construction postulates and propositions have existential import, but virtually without mention of their uniqueness import or determinateness. Mueller (1981) denies that Euclid shows 'any concern with the question whether a constructed object is unique'—in connection with I.30 (p. 19), and a straight line connecting two points (p. 32); not, though, because Euclid would doubt uniqueness, but because it is clear what uniqueness results from the basic constructions, and this need not be explicitly addressed.[17]

It seems plain that geometric constructions purport to give unique outcomes, or outcomes unique up to finitely many variants in any given case. For example, the line connecting two given points is plainly understood to be unique (see also Proclus on an objection to I.4). The equilateral triangle construction of I.1 is plainly understood to give two possible triangles, congruent but perhaps differently located with respect to the data, if (as Proclus 225.16 appears to regard as relevant) space allows. Occasional Euclidean propositions, such as I.7, could, moreover, be read as addressing uniqueness questions for constructions. We propose that traditional practice had, in principle if to varying extent depending on the case at hand, access to two quite distinct resources in regards uniqueness of the outcome of constructions subject to diagram discipline.

---

[16] There are other levels of function of diagrams in traditional geometry, on which we have not touched. Besides providing 'inferential' licenses, diagrams serve to organize and motivate geometrical inquiry. These considerations have not yet found their proper philosophical casting; and it goes beyond our present purpose to attempt that task here. In the context of such non-justificational functions, diagrams license a wider range of 'attribution', including what we have called exact conditions. There is a corresponding need for stricter diagram discipline, as well as different criteria for case distinctions in atypicality control. In particular, diagram discipline in this sense can be expected to extend to suitably selected *inequalities*.

[17] The index to Knorr (1986), for example, contains a series of headings under 'existence'; notably, 'problems as existence proof' gives 10 entries, but there is nothing on 'uniqueness'. Mueller mentions no work on the topic since Heath.

The primary resource is a non-propositional acquaintance with control in executing constructions in a given diagram. The practice requires one to grasp that the attempt to connect two points by a straight line, for example, has no legitimately competing alternate outcomes. In this attenuated and inarticulate sense, uniqueness is a part of what must be undertaken and subscribed to in Postulate 1. Little does it matter that Euclid might have asserted uniqueness, making it a propositional content of Postulate 1, by using some appropriate verbal form such as a definite article (as Mueller objects, p. 32), but did not do so. Just so, one is required to grasp that the control one exercises in attempting to make the circles in the construction of I.1 is such that we should take precisely two intersection points to arise as seen from the diagrams we produce. At this level, there is no need of I.7 or III.10 (as Heath envisages, p. 261) to establish the uniqueness of the triangle in I.1. Such a requirement of non-propositional grasp is of course compatible with the lack of explicitly articulated concern with uniqueness which Mueller observes in Euclid.[18]

In spite of this non-propositional grasp, challenges arise to what 'just happens' in making diagrams. Perhaps the realm of non-propositional grasp is a fool's paradise? As constructions pile up, adequate grasp of what control may be exerted through proper diagram discipline may be unattainable for participants. For example, the two possibilities in I.1 could each lead to multiple and dissimilar possibilities in further constructions. What possibilities arise in this way will typically depend on the initial data. It helps that, as one can now establish by metatheoretical means, a determinate construction has at most finitely many outcomes; each corresponding to an in principle independently intelligible geometrical condition on initial data.

But also, the conditions on initial data under which various possibilities arise need not be open to grasp in carrying out the construction in a particular diagram: in the all-triangles-are-isoceles argument, what clearly happens for clearly non-isoceles triangles might inscrutably cease to occur for certain close-to-isoceles triangles; perhaps with some unimagined threshold. We may therefore need to survey the possible outcomes at each stage. Under such circumstances, a secondary resource in the practice comes to the fore: propositionally explicit claims and demonstrations concerning the outcome of

---

[18] The expectation of control in constructions is illustrated, for example, by a remark of Newton: he compares the traditional geometrical *neusis* construction of the conchoid with a regulation Cartesian construction of solutions to its cubic locus equation (pointwise by intersecting a varying circle and fixed hyperbola), and comments '...this [Cartesian] Solution is too compounded to serve for any particular Uses. It is a bare Speculation, and geometrical Speculations have just as much Elegancy as Simplicity, and deserve just so much praise as they can promise Use.' (*Universal Arithmetic* 1728 pp. 229–230.)

constructions. Examples would be uniqueness results such as I.7, or arguments using circumcircle and Simson-line to determine the appearance of the all-triangles-are-isoceles diagram (Dubnov or Maxwell, pp. 24–25). Arguments of this sort shift part of the burden of grasp of construction outcomes, from what 'just happens' in making diagrams with proper discipline (purely diagram-based) to a mixture of diagram- and discursive text-based moves.

Of course, geometrical constructions are in general only *quasi*-determinate: adding elements to a diagram by a prescribed construction may yield finitely many diagrams from any given initial diagram, rather than just one; as in the two circle intersections in I.1. Instead of proving uniqueness, the task in settling a theoretical question (such as whether all triangles are isoceles) becomes to determine all diagrams that have standing as the result of the constructions indicated; and verify that they have the properties claimed of the construction. Moreover, it is understood that metrically different initial diagrams with the same appearance may lead, via the same construction, to different finite collections of result-diagrams. (We will return to this.)

In the example at hand, however, the question involves properties of the initial diagram only: whether the given triangle is isoceles. It is then intolerable, whether or not the auxiliary constructions are unique in the ordinary sense, as they are here, that after their application *to one and the same initial diagram,* one result-diagram leads to a proof of the claim and another to its disproof. Because the claim considered involves attributions to the initial diagram only, conflicting conclusions (that the triangle is both isoceles and non-isoceles) from distinct outcomes of the constructions indicated, applied to the same initial diagram, would—except in a reductio context—put singular pressure indeed on any sense that geometric property attribution behaves like a truth predicate (or assertability): holds indifferently of the wider context of the objects of the claim.

Until somehow relieved, such a situation would constitute an extreme form of disarray; quite distinct from that which one would encounter if unable to resolve which of two diagrams would be the outcome of one way of carrying out the construction, and the two potential outcomes led to conflicting conclusions on the main claim: then one would just not have a proof of either conclusion, perhaps a disconcerting situation but a thoroughly familiar one. Indeed, if what looked to us all along as properties of a diagram were instead implicitly relational qualities with respect to the wider context of the objects of the claim, this would undercut one of the key contributions of the discursive text in geometrical practice: it mediates application relations between geometrical arguments, indifferent both to how the configuration to which the lemma applies is embedded in a diagram in the application, and to how the configuration to which the lemma applies is embedded in a diagram enriched by constructions in the course of demonstration of the lemma.

In view of the long record of entrenchment of the sense that geometric property attribution behaves like truth, as well as the central role of context-free application relationships among the chain of propositions, it does not seem that this particular type of deep conflict has been widely felt to be in force for any notable interval in the documented history of the practice. This may motivate what seems to be the standard of traditional geometrical practice for dealing with the possibility of multiple outcomes of specified constructions on one given initial diagram: it suffices to treat one outcome (provided its standing as such is not in question), leaving it to the process of probing (see later) whether something funny comes up when a different outcome diagram is considered. Perhaps this is best seen as a gamble, which serves to reduce quasi-uniqueness to effective uniqueness; together with an implicit acceptance of vulnerability to the underlying risk.

The determinateness import of constructions is plainly not exhausted by their (quasi-) uniqueness (and existence) properties: the information that a certain construction, if properly carried out, will give a unique result need not help in establishing what the metric properties or even appearance of the result-diagram will be. Although extreme-sensitivity cases such as arbitrarily close-to-isoceles triangles may press constructional determinateness past its limits, most diagram appearance issues are adequately dealt with by employing constructions, subject to diagram discipline.

4.3.3 *Stipulative diagram control*

A practice can resolve a *limited* number of artifact control problems by stipulation. Perhaps the Parallel Postulate should be understood to arise in that way. As we vary the angle between two intersecting lines, we get a transition of behavior across the range of angles in which the intersection lies too far away to draw. The intersection now lies to one side, ever farther out; but if we approach the same range of angles (for which the intersection lies too far away to draw) from the other side, the intersection lies to the other side, ever farther out. Because of this transition of behavior, something must be said about the appearance of diagrams in this range; but precisely when the intersection lies too far away to draw, we have lost our diagramming artifact support: appearance control by diagram has been lost.

With the benefit of hindsight, we can list some stipulations which could be made in this situation, compatibly with enough of traditional geometrical practice to render the proposals competitors: (i) there is a single intersectionless, equal-angles position (Euclidean parallel postulate); (ii) there is a tiny interval of intersectionless positions, symmetric around the equal-angles intermediate position (hyperbolic geometry); (iii) at the equal-angles position, there is an intersection in each direction, as there is in any other position (elliptic geometry); and (iv) as in every position, there is a unique intersection in the intermediate position, and it lies simultaneously in both directions (Real projective geometry). The history of the subject, however, brings out well that such stipulative resolutions raise questions: does the practice control the appearance independently of the stipulation, in some as yet unseen way, so as to render the stipulation either superfluous or incompatible?

4.4 Case distinctions

Traditional geometrical reasoning distinguishes many cases. Poncelet describes the situation perceptively, if with discontent:

> The diagram is drawn, and never lost from view. One always reasons about real magnitudes; every conclusion must be pictured... One stops as soon as [in varying the diagram] objects cease to have definite, absolute, physical existence. Rigor is even taken to the point of not accepting conclusions of reasoning on one general arrangement of a diagram, for another equally general and perfectly analogous arrangement. This restrained geometry forces one to go through the whole sequence of elementary reasoning all over, as soon as a line or a point has passed from the right to the left of another, etc. (Poncelet, 1822, pp. xii–xiii)

For example, *De Sectione Ratione* of Apollonius Apollonius concerns the problem:

> Given two lines, a point on each (Q, Q'), and a point P not on either line: to locate a line through P that (with the given points) cuts off from the given lines segments QR, $Q'R'$ in a given ratio.

Starting with the major division as to whether the two given lines intersect, Apollonius treats some eighty-seven (!) cases, with occasional further subdivisions. Although such complete treatments appear to be rare (Euclid, for example, tends to leave many cases to the reader), Apollonius' careful treatment appears to reflect the highest standard of geometrical reasoning as far as rigor (disarray avoidance) is concerned. The standards of traditional diagram-based geometrical reasoning force case distinctions: 'diagrams individuate claims and proofs'! This we must see. We recognize a 'case distinction' when capacity for uniform proceeding (in artifact and response) in one aspect or stage within a practice falls short of such capacity elsewhere, in a finite-to-one fashion: as when a geometrical assertion or argument admits of finitely many cases.

The capacity for uniformity in diagram-based reasoning is limited by the dependence of geometrical demonstration on what we have called its appearance or *topology*—inclusions and contiguities of regions, segments, and points in the diagram. A geometrical proof typically responds to the appearance of its diagram. If, in varying a diagram, 'a line or a point has passed from the right to the left of another', as Poncelet puts it, we would need to reconsider whether some region or inclusion invoked in the proof might have disappeared. The all-triangles-are-isoceles argument shows that such considerations as whether a point lies on one or the other side of a given line are crucial; if differential responsiveness to such conditions were allowed to relax, disarray is a virtual certainty.

Geometric text is typically more flexible, giving the appearance of greater uniformity of presentation than can be attained in a diagram-based demonstration. In the case of Apollonius, the resources of geometric text allow uniformity of formulation ultimately 87-fold beyond what geometric argument requires.

As one would expect from a practice which engages its artifacts in that way, topologically distinct diagrams are treated in separate argument; whereas—maximizing uniformity in artifact use—separate argument is inappropriate for diagrams with the same appearance, however otherwise dissimilar. We will refer to this as the 'one appearance–one diagram' principle. In this way, the artifact strategy of the practice, combining diagram and discursive text, attains a 'local optimum' in making more manageable the diversity of spatial forms: by reducing it while retaining an important inferential grip.

Euclidean geometry often prefers to avoid making case distinctions at the outset of a proof: the resources of geometric text allow one instead to substitute a formulation in which the appearance of diagrams is more fully described. Thus, a claim $(p \to q)$, where diagrammatically, p is one of p_1 or p_2, would tend to be replaced by separate discursive claims $(p_1 \to q)$ and $(p_2 \to q)$. Indeed, under some circumstances, separate claims are required: construction problems often have preconditions of possibility (*diorismoi*), which must be stated as assumptions of a claim; these preconditions are often different for different diagram topologies. The claim must then be made separately for such different diagram types.

In other circumstances, case distinctions arise only *within* geometrical demonstration and so are not readily avoidable. Arguments by reductio often require a case distinction because of trichotomy: when two quantities can be compared, the first is either greater, equal, or less than the second. If, for reductio, one of these possibilities is denied, the other two must be considered, which typically requires a case distinction. (Reductio argument is considered more fully below.)

Far greater difficulties arise because—quite aside from ambiguities as to the appearance of the diagram formed under proper discipline—there is another reason that the *appearance* of a diagram at one stage in its construction does not determine its appearance after further constructions are applied in the course of an argument, even if those further constructions are fully constrained. The topological data of initial appearance simply does not determine what topology arises from construction of additional elements. Only more detailed *metrical* data—such as might now be given by algebraic inequalities—can determine this. For example, whether a point in the diagram will lie inside a newly entered circle or not would depend on whether its distance from the center exceeds the radius. Such metric data is unavailable (not attributable) in a Euclidean diagram.

As a consequence, topologically similar initial diagrams may become dissimilar through constructions in the course of a proof. When this occurs, a proof must, from that point on, deal separately with the dissimilar diagrams which have arisen. We might call this *case branching*. It occurs, here as elsewhere, because capacity for uniform proceeding after the construction falls short (in a finite way) of such capacity before it. (It seems a merely expository quirk that written proof typically gives variant diagrams from the outset; the format makes it inconvenient to display the stages in which a diagram is built up and case distinctions arise.)

If the practice is to avoid disarray in applying propositions, proofs must be applicable to any diagram which satisfies the conditions stated in the text of a proposition. In light of 'one appearance–one diagram', this requires that precisely those appearances which might possibly arise in an argument be presented in the argument, one diagram each.[19]

Any diagram arises from arbitrarily chosen, approximately metrically determinate, initial data (say, the ratios of distances of the vertices of a triangle) by constructions—be it the initial diagram of a proposition, or a subsequent elaboration by auxiliary constructions. We have encountered the limitations of diagram control in obtaining the appearance of a single diagram, constructed using a single system of arbitrary choices. Here, an additional and independent limitation on case branching control comes to the fore: even if there were

---

[19] This is perhaps too strong: once it is clear how an argument will go, one could in principle dispense with differences in appearance that were plainly immaterial to it. Authors anyhow seem to differ in the completeness of their case enumerations.

never ambiguity as to the appearance of the diagram formed under proper discipline from given data, adequacy of case distinction still requires control over complete families of diagrams.

But how are participants to see whether a given selection of variant diagrams (after a construction complicates the diagram, or even at the outset of argument) is exhaustive of the possibilities which require separate argument? If it is not exhaustive, how to locate further alternatives? Traditional practice lacks procedure either to certify completeness of case distinctions, or to generate variants. Except in special situations (negation of equalities) there are no discursive arguments to this effect; and while one would expect to look to the diagram for suggestions in locating variant appearances, there is no hint of clear-cut diagram-based procedures to do so.

Traditional geometrical practice largely lacks artifact support that would allow participants to survey diagram appearances once and for all, or even to recognize when they had done so. First, the possibilities for free choices, whether at the beginning or along the way in building up a diagram, are unsurveyable. Second, (even without ambiguity as to the appearance of diagrams properly constructed from given data) the relationship between those free metric choices and ultimate diagram appearance is unsurveyable. Diagrams come one at a time; having made one set of choices, we may hope to tell the consequences; but we thereby usually gain little grasp of the range of possible alternative outcomes from other choices.

These limitations of artifact support in traditional geometry may be brought out by contrast with a quite limited but historically significant exception: *locus* descriptions of curves. A good case is the well known result (extending III.21 and converse) that the locus of points O that make a fixed given angle POQ with two given points P and Q is a circle through P and Q. In a weak sense, this gives a survey of all possible ways, given P, Q, and the angle, of locating O to satisfy the condition. It allows us to read off a little about (say) how triangles OPQ of this type would interact and form regions with further diagram elements. We can tell what the possibilities are for O to lie with respect to other regions in the same diagram, by seeing which such regions contain points of the diagram; we can find all such triangles OPQ with side parallel to a given line by intersecting the circle with the lines through P or Q parallel to the given one... But we still can't read off all such triangles OPQ; we can draw them one at a time, but if we draw more than one or two, the diagram becomes too cluttered to be of any use (we lose appearance control).

The relationship between case distinctions and discrepancies in capacity for uniform proceeding is, however, quite general. One may also come to recognize 'cases' in traditional geometry when variant geometrical practices reconstrue traditional geometrical claims, but can proceed more uniformly with them. The traditional geometrical text often allows more uniform formulation of claims than do diagrams; but here the scope for uniform proceeding is limited by the need for diagram-based inferences. On the other hand, Cartesian geometrical representation by equations, and others inspired by it, such as Poncelet's, allow for such uniform treatment of results treated separately in traditional geometry that we regard them as analogous or as cases of 'the same problem'. Often, ancient texts already show, at least by their expository arrangement, that analogies are sensed among such distinctly treated questions; often, they do not. Notably, whereas traditional geometry handles limiting or degenerate cases such as tangencies or coincidences separately (Heath *Euclid* II p. 75), algebraic representation usually includes them in the treatment of the generic case; the importance of continuity in this sense was stressed by Leibniz and Poncelet.

For example, when two lines through a point O meet a circle, say, one in A and B and the other in C and D, then the rectangles (products) OA by OB and OC by OD are equal (*cf.* Euclid III 35–37, and Heath's summary of related subcases and claims, (Euclid, vol. 2, pp. 71–77)). Of the various distinct diagrams which can arise, Euclid III.35 concerns specifically the one where O lies inside the circle, and III.36 the case in which O is outside but one given line is tangent to the circle. This requires separate statement as a proposition because the traditional discursive text cannot be read to include the square on a single segment OP under the product OA by OB. Ultimately, this restriction on discursive representation seems again forced by topological distinction behavior in reading diagrams, given the need for a workable cross-reference system between discursive text and diagram: OP cannot at all be visually located in a diagram showing distinct non-tangent OA and OB, and in the opposite case one would at least have to allow double-labeling (Fig. 4.5).

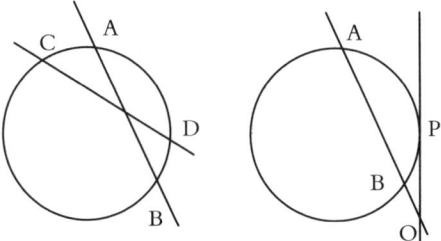

Fig. 4.5.

Thus, practice with diagrams, and not the geometrical text, controls the individuation of claim and proof in traditional geometry; in contrast with Cartesian and projective geometrical practices, which exploit different representation in order to individuate geometrical claims and proofs less finely.

4.5 Diagrams in reductio proof: torturing the diagram?

Although traditional diagram-based geometrical practice has built-in quality-control standards, both on the production/reading of diagrams and through the case/objection proposal mechanism (discussed below), it lacks an external vantage point from which to impose a standard (truth?) by which diagram-based reasoning can, in a unified and independent way, be evaluated for stepwise cogency. The verdict of a diagram can be objected to; and if so, localized responses can be made; but by and large, these in turn must accept the verdict of diagrams.

It is therefore strange to encounter arguments (in Euclid) in which the diagram is simply impossible, or clearly violates some condition invoked in the argument; for it is unclear what force the verdict of such diagrams could have.[20] The matter is worth careful examination, for after all, Euclidean practice shows no signs of having broken down in disarray; and we can hardly dismiss its use of such diagrams. We find them also used to refute objections; that will require separate analysis.

4.5.1 Example: Euclid I.6

In the *Elements*, we first encounter *reductio* in I.6. Given a triangle ABC with the angle ABC equal to the angle ACB, it is to be shown that the sides AB and AC are equal. For reductio, we deny this conclusion in the presence of the premises. If AB and AC were not equal, one would exceed the other, say AB exceeds AC (by symmetry, one case alone suffices). The inequality is converted into the combination of (i) an equality between AC and BD constructed on AB (the equality is registered in the discursive text, perhaps manifestly violated in the diagram) and (ii) the topologically explicit diagrammatic presentation of the inclusion of BD in AB. Draw the diagram, starting from an isosceles triangle.

---

[20] I thank David Israel for pressing me on this point.

Let's spell out the conditions. We write 'LineSeg(*ADB*)' to indicate a line segment (in the antiquated sense of line found in Euclid, which does not entail that the 'line' is straight, rather than, say, a circle, ellipse, or other simple non-self-intersecting curve), with endpoints *A* and *B*, and an intermediate point *D*. When the dust has settled on the assumption for reductio, we have the co-exact conditions indicated in the diagram:

(d-i) LineSeg(*ADB*), (d-ii) LineSeg(*AC*), (d-iii) LineSeg(*BC*),
(d-iv) LineSeg(*DC*) ,(d-v) NonNull(*AD*);

and the exact conditions indicated in the discursive text:

(t-i) Straight(*ADB*), (t-ii) Straight(*AC*), (t-iii) Straight(*BC*),
(t-iv) Straight(*DC*), (t-v) angle(*ABC*) = angle(*ACB*), (t-vi) $DB = AC$.

Of these exact conditions, the diagram we find in Euclid is subject to (t-i)–(t-v), but, reflecting the unsatisfiability of the assumptions in force for reductio, not (t-vi). Any other diagram, however, indicating precisely (d-i)–(d-v) subject to (t-i)–(t-iv)—that is, with straight segments *ADB*, *BC*, *DC*, *AC*—would be topologically equivalent; any such diagram would do just as well to display all geometrical objects and their inclusions required for the proof. The exact conditions here do contribute to the topology, to some extent: if *AC* were curved, say, it might intersect *CD*. Of all the constraints on the diagram, at most (d-v) and the straight-line combinations (of (d-i)–(d-iv) and (t-i)–(t-iv) respectively) are relevant to the conclusions which the argument reads off from the diagram as it proceeds:

(d-vi) Region(*DBC*), with precisely three angles: Angle(*BCD*), Angle (*CDB*), Angle(*DBC*)–hence, because the sides are straight, Triangle(*DBC*);
(d-vii) region(*DBC*) IsPartOf region(*ABC*)—hence, because they are triangles, triangle(*DBC*) IsPartOf triangle(*ABC*) = triangle(*ACB*).

From (d-vi) and (t-i)–(t-vi), the text applies the side-angle-side congruence criterion I.4, giving:

(t-vii) triangle(*DBC*) = triangle(*ACB*),

which is incompatible with (d-vii)—parts are *proper* parts.

That the diagram is not subject to (t-vi) is thus not terribly significant. The equalities (t-v)–(t-vii) are irrelevant to the inferential *use* of the diagram here: they would in any case directly license moves only because they are indicated in the discursive text. Indeed, the diagram for the reductio proof of Euclid I.7 appears subject to neither segment equality in question there, nor to either angle equality obtained in the course of the proof.

A proof is unaffected by any diagrammatic incompatibility between the premises and the denial of the conclusion (diagrammatic incompatibility: no diagram indicates all the co-exact conditions and is subject to all exact ones) in so far as responsability for move licensing is divided between diagram and discursive text so as to leave it indifferent whether the diagram is subjected to all exact conditions in force. This allows one to use a perfectly ordinary Euclidean diagram, in co-exact conditions if not in every respect metrically appropriate to the discursive text.

Whether or not the diagram was subjected to specific equalities indicated in the discursive text—indeed, any metric property of the diagram—could affect which topological conclusions it would indicate; notably, if the proof employed constructions more metrically sensitive than our straight-line joining of *DC* in I.6. In so far as a diagram is not subject to all exact conditions in the discursive text, there would seem to be no general assurance for the standing of co-exact conclusions which emerge after constructions; although the case/objection proposal mechanism would in any case provide means for criticism of such conclusions.

Because the exact conditions to which the diagram is subjected are a subset of those indicated in the discursive text, it is easier than one would expect from those conditions taken together (rather than harder because of their ultimate incompatibility) to produce variant diagrams subjected to the same exact conditions but indicating different co-exact ones. One would then seek contradiction for those cases in turn, employing all exact conditions available in the discursive text together with such co-exact conditions as might be indicated by the diagram. (To keep variants from proliferating in this process, one would subject the diagram to as many at a time as appears possible of the exact conditions indicated in the discursive text.)

There is thus no special reason to doubt the cogency of diagram-based reasoning in reductio proof: it does not presuppose that everything asserted in the discursive text be 'true' of the diagram, not even in the attenuated sense provided by the notion of diagram discipline.

4.5.2 The hypothesis for reductio

So far, we have taken something else for granted: for straightforward proceeding according to the account given above, the data of a reductio context must form a (conjunctive) system of conditions (unsuccessfully) put forward concerning a single diagram. It would of course also be possible to consider several variant diagrams disjunctively, though this does not seem to have been the preference. In any case, we can give sufficient conditions for this success in launching a reductio argument.

Suppose the statement to be shown set out in a diagram. Logically, this involves considering an instance; as we hereafter suppose done, in order to avoid quantifiers. Let the statement have the form $(C_1 \& \ldots \& C_n) \to D$. We assume that the Cs and Ds are what we call *diagram conditions*: either exact conditions that would be asserted in the discursive text, or explicit co-exact conditions, that could be indicated in or read off from a diagram. The class of diagram conditions is not closed under negation: there is no condition equivalent to non-circularity, for example.

To proceed by reductio, we must consider $C_1 \& \ldots \& C_n \& \neg D$. It remains to convert $\neg D$ into a diagram condition. Typically, D asserts an equality, or is straightforwardly interderivable with one; thus, its negation can be converted by trichotomy into a disjunction of two strict inequalities (x is greater than y). With luck, as in I.6 (but not I.25), symmetry considerations allow this to be reduced to one. A strict inequality, in turn, may be reduced to a proper-part relationship in the diagram by using an auxiliary equality:

x is greater than $y \Leftrightarrow$ for some proper part z of x, $z = y$.

In principle, one could hope to deal with claims of the form, for any m and n,

$$(C_1 \& \ldots \& C_n) \to (D_1 \vee \ldots \vee D_m).$$

When denied as hypothesis for reductio and set out in a diagram, these give the form

$$C_1 \& \ldots \& C_n \& \neg D_1 \& \ldots \& \neg D_m.$$

In general, though, each of the $\neg D_j$ reduces to a disjunction of diagram conditions, and for increasing m putting this as a disjunction of conjunctions gives an exponential explosion of interacting case distinctions. It is therefore unlikely that we would encounter, in a single proposition established by reductio, m greater than one.

I.7 shows a special case, where in effect $m = 0$; the statement is then equivalent to $\neg(C_1 \& \ldots \& C_n)$. Such assertions are infrequent; perhaps because recognizably negative claims are frowned upon, perhaps because negations of diagram conditions often cannot be converted into diagram conditions, and hence the resulting proposition could only be applied in further reductio arguments. The hypothesis for reductio is then $C_1 \& \ldots \& C_n$, which is immediately of the form to put forward of a diagram. III.13, that two circles do not have two distinct tangents, and Proclus' corollary from I.17, that there is only one perpendicular on one side of a line at a given point (313), are also of this type.

THE EUCLIDEAN DIAGRAM (1995)   113

What would not work well is a reductio argument starting with an inarticulate denial of a diagram: 'Things are not like *this*'; in logical terms, a diagram-based reductio proof of $(D_1 \& \ldots \& D_m)$, starting with $\neg(D_1 \& \ldots \& D_m)$. (Of course, there might be such a reductio argument entirely within the resources of the discursive text.) In diagram-based reasoning, reductio gets its grip only when it leads determinately to a diagram.

4.5.3 Further constraints on the diagram for reductio: I.27

Our text of Euclid does not fully conform to the account given so far. I.27 considers the straight line *EF* falling on two straight lines *AB*, *CD* such that the alternate angles *AEF*, *EFD* are equal (Fig. 4.6); and then assumes for reductio, and without loss of generality, that *AB* and *CD* produced meet in the direction of *B*, *D*, at *G*. The diagram in the text would be read as subjected to the equality of alternate angles with *EF*, which of course precludes straight lines from meeting a short distance away at *G*. Thus, the diagram gives up on subjecting *AB* and *CD* produced to being straight, in order to indicate their intersection at *G*. (So far, so good.)

We moderns might indicate this intersection by producing either or both *AB* and *CD* free-form monotonically curved toward their intersection *G*. This would allow one to attribute the triangle *GEF* referred to later in Euclid's argument (the requisite straightness of *GE* and *GF* is indicated in the text even though the diagram is not subjected to it). Admittedly, ancient practice does not appear to have utilized free-form curves; but instead apparently sought to satisfy the requirements of argument by a diagram somehow made up from the ordinary types of diagram elements. This desideratum can be satisfied for any of the proposals just made, however, by substituting a circular arc for free-form curves. In his diagram for an objection to I.4 (239), Proclus

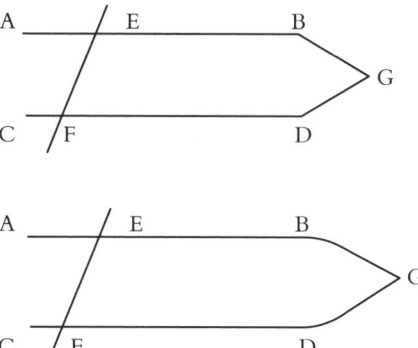

Fig. 4.6.

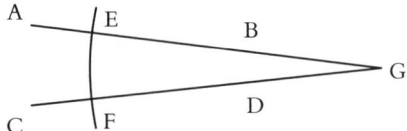

Fig. 4.7.

subjects the diagram to a circular arc where the discursive text puts forward a straight line.

Another approach would smoothly curve the segment *EF* so that *AB* and *CD* produced intersect at *G* even though the alternate angles are subject to equality and the lines are produced subject to being straight (Fig. 4.7); yet another would have *EF*, *EG*, and *FG* straight but give up on equality of the alternate angles in the diagram. Proclus (313) in effect does this in a related application of I.17.

Our text of Euclid does none of these. Instead, the line segments *AB*, *CD*, and *EF* are all subjected to straightness, and to equality of alternate angles with *EF*. An intersection at *G* is indicated by inserting straight segments *BG* and *DG* at an angle to segments *AB*, *CD* respectively. Evidently, as a result, we would say that the diagram puts forward angles *EBG* and *FDC*. This example suggests several lessons, which appear confirmed in other instances.

(1) It appears preferred to treat equivalently in the diagram items which play equivalent roles in the discursive text; even when the diagram cannot be subjected to all conditions in the discursive text. Thus, for example, both *EG* and *FG* are broken, even though the intersection could be indicated with only one broken and one straight (Fig. 4.8).

(2) Given that a choice must be made among exact conditions, the procedure here suggests a priority to subject the diagram to those indicated first in the discursive text. Similarly, the diagram for I.6 is subject to all exact conditions in force except for the one ($AC = DB$) arising from the hypothesis for reductio, and the subsequently drawn conclusion that the triangles *DBC* and *ACB* are congruent. The examples do not allow us to discriminate whether the priority arises simply from the sequence of the discursive text, or whether it accords

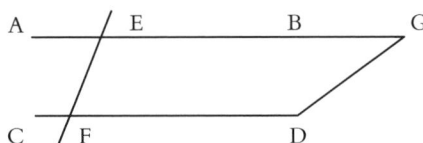

Fig. 4.8.

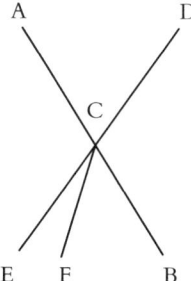

Fig. 4.9.

equal priority to all conditions in force from before the scope of the reductio argument, over those which arise within that scope.

Again, consider Proclus' diagram for the converse of I.15, that if, upon a straight line *AB*, we assume two straight lines *CD* and *CE* on opposite sides [at the same point *C*] and making the vertical angles *ACD* and *BCE* equal, these lines lie on a straight line *DCE* with each other (299–300). The diagram is subjected to all these straightnesses (including that of the intended conclusion) and the angle equality. For reductio, Proclus assumes *DCE* not straight, and enters *CF* at an angle to *DE* while stipulating in the discursive text that *DCF* is straight (Fig. 4.9).

This may strike us as strange, for it would come to us more naturally to make *DCF* straight in the diagram, enter *CE* off at an angle, and hash-mark as equal the (visibly unequal) vertical angles *ACD* and *BCE*. But Proclus first gives a diagram in which *the data are set out,* subjected to the given exact relations; this forces *DCE* to be straight (as we are out to show). Only then is the hypothesis for reductio entered, properly unpacked. These priorities for subjecting the diagram to certain exact conditions rather than others thus become understandable if we assume that the diagram must be suitable for sequential display as the demonstration proceeds. For then we must set out the claim first, before we announce and diagram the reductio strategy.

4.5.4 A complication in I.27: attribution blocking?

According to our account so far, one is licensed to attribute angles *EBG* and *FDC* to the diagram of I.27, but the *Elements* refrains from doing so. Indeed, that attribution would preclude treating *EFG* as a triangle. That would block the Euclidean argument: it rests on an application to *EBGDF* of I.16, which holds for triangles but fails for quadrangles (squares, for example) and beyond. For the argument to proceed unimpeded, something must be in force here, not just to sanction refraining from attributing the spurious angles, but indeed, to block (or disarm) the obvious objection: 'But *EBGDF* is a pentagon, not a triangle!'

This goes to the heart of reductio demonstration. For according to our account, reductio employs an altogether ordinary Euclidean diagram, in an only slightly unusual way: one refrains from subjecting the diagram to one or more exact conditions in force in the discursive text. In the case at hand, diagram discipline is clearly suspended with respect to keeping *EG* and *FG* straight. But according to our account, the diagram functions in the ordinary fashion; in particular, we must endorse such licenses to attribute co-exact conditions based on the appearance of the diagram as allow the argument to proceed to contradiction. The *mere* incompatibility of some such condition with some attribution in force in the discursive text cannot suffice to block it, or diagram-based contributions to reductio arguments would essentially all be blocked.

In our initial example I.6, the *Elements* makes the diagram-based attribution (d-vi) that *DBC* is a triangle, and applies the side-angle-side criterion I.4 to conclude that *DBC* equals *ACB*. Only then does it conclude from the diagram that *DBC* is a (proper) part of *ACB*; if this attribution were blocked, no contradiction would be available.

There are several resources available here to deal with this issue. First, one could accept the pentagon objection to this particular diagramming strategy; as an alternative would seem available. Even accepting the decision to subject the diagram to all exact conditions except that *EG* and *FG* be straight, we can connect *BG* and *DG* by producing *EB* and *FD* by circle segments tangent at *B* and *D* respectively. Then no spurious angles can be attributed. Similarly for the 'line' *DCF* in Proclus' converse of I.15 (above).

On the other hand, one can imagine that a suitably selective rule on defeasibility of attribution licenses would allow the present argument to go forward without undercutting the very possibility of diagram-based attribution in reductio argument; a natural candidate emerges. Genuinely diagram-utilizing arguments by reductio require a diagram; hence (if need be, after disjunctive case analysis) some construal of the hypothesis for reductio as a conjunctive collection of exact and co-exact conditions $D_1 \& \ldots \& D_m$. According to observation (2) just above, there is a lower priority for subjecting the diagram to exact conditions among these, compared to exact conditions coming into force at an earlier stage. It would thus be natural to single out—as having their attribution licenses undercut—co-exact conditions which taken by themselves directly contradict exact conditions from among the *D*s to which the diagram is not being subjected.

We here touch upon a general problem inherent in any use of (quasi-)semantic methods for reductio along the lines suggested, not just diagrams, but also most other forms of 'semantic representation' and 'models'. Suppose the artifact d (here, a diagram) is not subjected to a constraint e (here, exact). On the one hand, if e fails in any d satisfying the appropriate constraints, then it would be appropriate to consider this to conclude the reductio, in that our commitment to e (in the text) is incompatible with the failure of e in any d satisfying the constraints we impose on it. On the other hand, it is part of what we expect from (quasi-)semantic artifact use that e may fail in one artifact satisfying the constraints we impose but not in another; that is, that our production and reading of d-type artifacts proceeds less uniformly that our setting up of conditions (e-type artifacts).

In view of the former possibility, we would want to be able to read off $\neg e$ from d to conclude the reductio; but in view of the latter possibility, we cannot afford to do so on pain of disarray: our text inferences would then proceed more uniformly than corresponds (via production and reading procedure) to d-type artifacts. What is a practice to do in such a bind?

In the example of I.6, reading off of $\neg e$ is simply blocked, in that it is an inequality, and hence not explicit in our technical sense. Most negations of exact conditions are in fact not diagram conditions, and this blocks their attribution; evidently, geometrical reductio typically just has to do without the ability to close off by attributing $\neg e$ even if a d-type object satisfying the constraints we impose on it cannot be subjected to e. Undercutting attribution licenses should not cause worry of unfounded conclusions or disarray in diagram-based reductio argument: it only weakens the argumentative resources at hand.

This, however, has nothing to do with the source of the problem with quasi-semantic artifact use in reductio proof, which we already identified: 'quasi-semantic' artifact use provides less support for uniform proceeding than ordinary text. The appropriate solution then, as always when capacity for uniform proceeding in one aspect or stage within a practice falls short of such capacity elsewhere, in a finite-to-one or otherwise surveyable fashion, is case distinction. In reductio, then, we should not aim to simply block the attribution $\neg e$ outright. Rather, if there are different diagrams with potential to affect the argument, we should consider them as cases, and if we fail to do so, an objection is in order, presenting some diagram of a type not yet considered. There is nothing special about reductio, or geometry, in this.

If this general policy is in force, however, then the present difficulty cannot be avoided simply by diagramming the lines smoothly to avoid ascribing a pentangle; for the pentangle diagram could still be brought as an objection. There must, therefore, be a way of dealing with that diagram, either by disqualifying the diagram outright (*pace* Euclid) or by blocking its pentagram problem.

Choosing the first direction, we might insist that broken lines count as made up of separate segments, and hence here violate the requirement that co-exact conditions assumed for reductio must be be indicated in the diagram. In support of this, we might note that the diagram attributions that the pentangle objection exploits, which are evidently co-exact, are precisely that the region has five distinct sides. We might, then, interpret what we see in Euclid as a refusal to read the diagram in regards the separateness of the sides, not via an additional mechanism for blocking attributions in reductio proof, but simply in order to disregard a point in which the diagram is known to be defective. In all, this reading is now seen as neutral on the choices we just envisioned.

This leaves the question why one would not make a correct diagram, with lines curved, in the first place. There might be several reasons for this. When the diagram cannot be subjected to a straightness condition indicated in the discursive text for the production of a line, there appears to be a preference to produce the line at an angle (but straight). Perhaps there is a general preference for straightness when straight lines are put forward in the text, perhaps because straightness discipline gives the most powerful control, perhaps because straight-line completion is least likely to produce potentially spurious regions and hence give rise to atypicality in the diagram which would provoke a case analysis. We are not in a position to pursue these questions at this stage.

4.5.5 *Summary: principal standards for diagrams in Euclidean reductio proof*

1. The relationship of the diagram to conditions in force upon entering the reductio context is unaffected.
2. The diagram indicates all co-exact conditions from the hypothesis for reductio.
3. The diagram may be subject to some exact conditions from the hypothesis for reductio—but not all.

4.6 Roles in geometrical practice: cases, objection, probing

Current philosophical conceptions of mathematics are predominantly agentless: geometry too is cast as a mathematical *theory,* a body of statements articulated by starting-points and proofs. Of course, geometry is done by humans, and their practice contains a variety of distinguishable roles—author, teacher, student. But these seem beside the philosophical point: they have to do solely with the interface between mathematical theory and social institutions, or even humans in the flesh; as such they are of no greater philosophical pertinence to mathematics proper than the role of college president, or of mother.

At a minimum, though, bringing out the distinctive character of mathematics would seem to require that we say something about the kinds of stands that mathematical practice requires agents to take in order to qualify as properly mathematical; what we, with some oversimplification, above called 'unqualified assent'. We miss something about number theory, for example, if we think that after I assert that n is prime and you doubt or even deny it, we can properly just move on happily; in the way that we properly can and perhaps should, when I admire the *Water Lilies* in the museum but you respond with indifference. The body of requirements on properly mathematical agents which one might analyze out from various such contrast cases are (part of) what in agentless terms is usually captured by traditional conceptions of mathematical certainty or mathematical truth. Such conceptions license, it seems, but a single stance or role within traditional geometry (except perhaps for differences in stance corresponding to different stages of the science, such as before and after a fundamental theorem has been established). The ideally competent geometer grasps the content of geometrical practice up to his time and takes responsibility for it personally: being able and willing to affirm geometrical claims and present geometrical constructions on his own responsibility, and respond to challenge and question by appropriate performances, including recognizably geometric research.

Traditional geometry so conceived in effect recognizes only one role, which we might call that of *protagonist;* and except possibly for the advance of the science, no differences among protagonists have any standing. Our usual agentless conception of mathematical theory is, while philosophically less informative, nonetheless neatly compatible with such a conception of geometrical roles. According to this theoretical conception of mathematics, any discordant response to the protagonist is out of order, or expresses incompetence. That the student is not ready to shoulder responsibility for Pythagoras' theorem after seeing it for the first time just shows he is no geometer yet. One who questions the straightness of a side of a square in the sand has simply missed the point of the practice. Proclus, however, acknowledges standing for roles beyond that of the protagonist, of an arguably theoretical nature. Admitting these additional roles perhaps modifies the character of the protagonist's role so as to render it incompatible both with contemporary agentless conceptions of mathematics, and with our initial suggestion that geometrical practice requires 'unqualified assent'.

4.6.1 *Proposing a case or an objection*

According to Proclus, 'the proposer of a case ... has to show that the proposition is true of it...' (212). The proposer of a case exercises standing to respond to the protagonist: 'you have not dealt with a diagram like *this*.... Here is how it goes...' A 'case' shows how a claim made by the protagonist applies to a variant diagram topologically inequivalent to those considered by the protagonist, be it the initial diagram of a claim or one arising from a construction (all illustrated in Proclus' case analysis of I.3, pp. 228–232): 'a case proves the same thing in another [diagram] (alloos)' (289). It is perhaps preferable that the proposed proof be analogous to the proof being responded to: Proclus praises Euclid for a construction which can be modified to fit a great variety of cases (222). There may be some room for debate as to whether a proposed case meeting these standards is to the point.

It may be unclear why the role of proposer of a case merits serious philosophical consideration. Case arguments are often either similar to the ones given, or concern degenerate situations in which a simpler argument is available. Shouldn't their treatment be governed by strictly expository or pedagogical considerations? To write out every case would be boring and unnecessarily long; better to let students exercise by working out the variants. So it's sensible to leave the burden of proof on the proposer of a case.

It would be wrong to dismiss case-proposing in that way. As we argued, traditional geometrical practice has inadequate facilities for case-branching control; so that case distinction management remains in principle open-ended. The critical attitude toward the protagonist's diagram choices required in traditional practice is not supported by any clear-cut or complete procedure, and therefore leaves geometric inference open-ended in a way which we moderns, spoiled by complete systems of iron-clad inference licenses, hardly expect.

In the 20th century, Tarski (refining Pasch and Hilbert) was able to deploy complete first-order inferential machinery with which one can, if not locate cases, at least conclusively establish that given enumerations of cases are complete. But Tarski uses representational (artifact) resources that Greek geometric practice lacked: formal quantifier-predicate logic with complete proof systems and associated foundational methodology.

In a geometrical practice without such means at its disposal, the critical attitude toward the protagonist's diagram choices displayed by the proposer of a case has an intrinsically theoretical, even inferential, function; it is an element of the intellectual strategy by which the practice, in the long run, achieves unanimity on 'general' claims from the manipulation of a small number of physical diagrams. The generality so attained is neither full uniformity of treatment (distinct cases are argued with varying degrees of analogy), nor universality in the modern model-theoretic sense with respect to specified collections of 'individuals'; it is that acceptable diagram variants not covered by the argument don't come up any more.

Case-proposing criticism works in traditional geometry because the open-endedness of case-branching control is in fact modest; for the scope of case-criticism is nontrivially constrained. In a way that is not less real for being difficult to formalize, the diagram of an argument gives access to a 'space' to be probed for variants; for variants may generally be located by distortion of metric features of a diagram. The tradition thanks its stability in part to the quite limited repertoire of variants requiring separate treatment that come up in this way. In today's Real Algebraic Geometry, this might now be made precise. It follows from this work that, even by traditional geometrical standards, only a finite repertoire of alternative diagrams need be considered for any argument in 'elementary' geometry. As we will see later, moreover, traditional geometry had resources to keep diagrams relatively simple; and in this way to control the numbers of cases, so they were, by and large, manageable rather than merely finite.

According to Proclus, '... an "objection" *(enstasis)* prevents an argument from proceeding on its way by opposing either the construction or the demonstration' (212). Proclus speaks so often of objection that the Teubner edition does not undertake to index all occurrences. Among objections that he records: the triangle constructed in I.1 need not be equilateral if two distinct lines can coincide in a segment and then branch apart; the diagram of I.7 (triangles with the same base, and the same sides upstanding at the same ends of the base, have the same vertex) admits a variant not covered in Euclid.

Proclus takes making objection to be a responsive role distinct from proposing a case: 'A case and an objection are not identical; the case proves the same thing in another [context], but an objection is adduced to show an absurdity (eis atopian epagei to enistamenon) in the proof objected to.' (289) He lays responsibility for dealing with case and objection on different shoulders: '... unlike the proposer of a case, who has to show that the proposition is true of it, he who makes an objection does not need to prove anything; rather it is necessary [for his opponent] to refute the objection and show that he who uses it is in error...' (212).

It is puzzling, though, that the distinction should go very deep in this way. For if a case is viable, in the sense that its initial diagram satisfies all stipulations in force, but no proof of the claim is readily available, it surely becomes material for an objection. Moreover, an omitted but treatable case in a reductio argument would look to Proclus as an objection; this happens, for example, in I.7 (262–263). Unsurprisingly, then, Proclus continues after setting out the distinction, 'By not discriminating between these, commentators have introduced them all together and have not made it clear whether they are asking us to diagram (graphein) cases or objections' (289). It was evidently not a commonplace for two roles to be distinguished at all. Perhaps participants have standing indifferently to propose cases or objections which they discern. An uncovered case too is potentially a defect of argument. That responsibility for a straightforward case, once recognized, remain with the proposer, could be justified by pedagogy alone: often enough, generation and treatment of variant diagrams is just routine. Thus, whatever other serious reasons there might be to distinguish case and objection, differences in allocation of the task of responding need not have any theoretical significance.

Our evidence on cases as well as objections is skewed because of the way in which ancient geometrical texts come to us through a long process of revision and recasting.[21] Even in the commentary literature, a successful case-proposing would tend to be recorded only if there was some special interest in the variant diagram or proof. An objection which in fact succeeded in blocking an argument might result in the deletion or modification of the argument objected to. Notably, if a variant diagram were discovered in which a construction could not be carried out, the result would most likely be revision of the text to exclude the variant by an additional hypothesis in the construction theorem. Proclus (330) regards the condition of I.22, that the two sides taken together exceed the third, as having such an objection-refuting role. On the other hand, successful response to an objection might be incorporated into the text as a subsidiary argument. As the order of propositions, though evidently not all details of their formulation, appears to have been fixed in Euclid's time, that particular mode of inquiry must predate him.

4.6.2 Probing

In addition to the role of protagonist, we propose to recognize one single role as theoretically central in traditional geometric practice, that of providing critical scrutiny, or as we will call it, *probing*. Participants may, of course, recognize various distinctive styles of critical scrutiny, including those of proposing case

---

[21] cf. W. Knorr (1989).

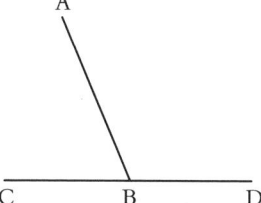

Fig. 4.10.

and making objection, and may make distinctive arrangements for carrying them on, such as the division of labor that Proclus records; but we see no overriding significance to the intellectual function of the practice in such divisions. In particular, we just saw how case and objection are easy to confuse, sometimes indistinguishable, sometimes distinguished only by whether the outcome is comfortable to the protagonist; and in any case tend both to be submerged in revision of the text.

Proclus records ample instances of critical scrutiny of *The Elements*. Many point to the need to state a proposition just right. The 'Euclidean' text, which Proclus often praises for its precise formulation, is probably the product of centuries of refinement of formulation after Euclid as well as before.

For example, consider the statement of I.14: If, with any straight line *AB*, and at a point *B* on it, two straight lines *BC* and *BD not lying on the same side of AB* make the adjacent angles *ABC* and *ABD* taken together equal to two right angles, the two straight lines will be in a straight line with another (Fig. 4.10). (Adapted from Heath, Euclid I, p. 276.)

Proclus (297–298) gives a counterexample of Porphyry to show that the condition 'not lying on the same side' cannot be omitted: if we do so, an initial diagram with different appearance will satisfy the remaining conditions, but not the conclusion.

To our sensitivity, this makes absolutely clear that the added condition is needed. It replaces otherwise implicit contributions of the diagram to the statement of a proposition, by a fuller verbal description of the diagram. But why must one do so? Although this is not the way Euclid's text works, we can imagine treating the initial diagram as part of the statement text of a proposition. One could then strip the discursive statement of I.14 down to: *Let three line segments as in the diagram have two adjacent angles taken together equal to two right angles. Then the two straight lines making the two angles with the third will be in a straight line with another.* This too seems to work. If the diagram is treated as part of the statement, we seem to have the same effect as from the verbally more complete Euclidean version. Even in Proclus, a case is occasionally described in part: 'as in the diagram'. Both the diagram and the discursive text appear to have the resources to record such co-exact contents.

It is in fact not immediately clear what decisive virtue favors fully discursive formulation. Proclus' commentary is a teaching text; and it is no doubt useful to sensitize the student to the relationship bewoen diagram and text. But beyond the pedagogical role of attending to such refinements, one can still ask whether continued pursuit of explicit verbal statement beyond the level attained in Euclid's time is a genuine contribution to the effectiveness of geometrical practice; or just the product of a scholastic philosophical commentary tradition, out of touch with what matters in geometry, and in single-minded pursuit of discursively complete argument.

We shall return to explain why a diagram does not function well as part of the statement of a proposition. But on the surface, leaving unambiguously readable contiguity specifications implicit in the initial diagram of a proposition would seem not to pose undue risk of disarray. For an initial diagram can be designed to show a 'clear case', avoiding diagram sensitivities that might cause an appearance control problem to limit our ability to read off its contiguities in the diagram. On the other hand, traditional practice is plainly limited in its artifact support of what we have called appearance control (limited by our lack of metric control of diagrams individually) and case branching control (limited instead by our lack of intellectual grip on the totality of diagrams satisfying conditions laid down in the text, quite independently of limitations of metric control). Because traditional diagram-based geometrical reasoning is somewhat open-ended in both these ways, it requires a sustained critical exploration of diagram use.

We probe case branching control in an argument by feeling around for diagram variants which satisfy the conditions laid out in the text. Attempting a more principled distinction between case and objection than that given by Proclus, we might propose to call a variant recognized as arising from probing case branching control a *case;* and one recognized as arising from probing appearance control, an *objection*. The variants, for example, in the all-triangles-are-isoceles argument arise from probing appearance control, not case branching control. For they purport to depart from the same initial metric choices by fully determinate construction; any inability on our part to tell which is appropriate is a breakdown of appearance control.

By this standard, Dubnov mischaracterizes his response to the challenge: what is called for is not that we 'consider all cases' (p. 24) but that we, given one of the diagrams, probe so as to come up with an account of its appropriate appearance. This might, however, take the form of surveying the variants viable by some metrically relaxed standard, and then providing further reductio ad absurdam arguments showing which variants are not viable as outcomes of the determinate constructions (Dubnov, p. 24). But it might also take the form of a direct argument showing that a particular variant is the appropriate diagram (Maxwell, pp. 24–25).

Again, however, we seem to have gained little by our sharper distinction between case and objection.

While probing is directly evident in the commentary tradition, it is of great interest to consider what in Euclid might best be understood as already a response to probing, even if only conjecturally. III.2 is a striking case. III.2 argues that a chord of a circle lies inside it, a matter read off from the diagram in the proof of III.1. From a logical point of view, this order of presentation is puzzling. Mueller (p. 179) notes that the theory of circles in Book III makes many 'implicit assumptions', 'facts of spatial intuition', sometimes taking them for granted and sometimes, as in III.2, arguing for them. Though this still offends logical sensitivity, its status as a refutation of an objection to III.1 could account for its following rather than preceding that result.

In view of other unprobed diagram-based steps in Book III, it is perhaps somewhat puzzling why this point of appearance control would be probed. Unless we worry about situations in which the end-points of the chord on the circle lie so close together that the separation of chord midpoint and circle becomes questionable, there seems no particular metric sensitivity or risk of disarray addressed by demonstrating III.2—but the same may be said for many propositions in Book III, which might readily be read off directly from the diagram.

Alternatively, III.2 might be a probing of a matter needed in the demonstration of its only application III.13, that two circles cannot have two distinct points of outer tangency. The diagram in our text shows two circles intersecting twice, in accord with the theory of diagrams for reductio given here: the diagram is subjected to all but the tangency conditions, which are exact conditions from the hypothesis for reductio. From this diagram, one could read off directly that there are common interior points, contradicting external tangency without invoking III.2. Given the opposite curvatures of the circles, the diagram presents no particular appearance control sensitivity unless the two points of tangency are particularly close together, so it is not completely clear why this observation would be an urgent candidate for probing. The diagram could, however, be made differently, still in accord with our expectations for diagrams for reductio: with either or both circles distorted to keep them disjoint with two points of tangency. In that case, an argument would be needed; and not only the statement of III.2 but a reductio argument with precisely the diagram we find there would be about what was called for. This could help explain why III.2 is argued by reductio rather than directly, as a Book I style fact about lines in triangles as proposed by De Morgan (Heath Euclid II 9–10). Even if part of Book III derives from an earlier work on circles composed without the benefit of a well-developed theory of triangles, recognizing III.2 as a probing of diagram behavior for the benefit of III.13 would help make its diagram and argument seem appropriate.

This example should suffice to show how the notion of probing developed here can provide an additional resource in interpretation and historical reconstruction of Euclid. As a further sample, consider I.7: triangles with the same base, and the same sides upstanding at the same ends of the base, have the same vertex. Imagine constructing a triangle with sides of given lengths on a given base by circling the side lengths around the ends of the base, just as in I.1; the vertex lies where the circles intersect. But this could be subject to a diagram control challenge: maybe the circles intersect in many nearby points rather than in just one point. This would be similar to Protagoras' challenge, that a circle and its tangent intersect in a segment rather than a point. I.7 could be read as responding to this challenge: if there were more than one triangle in this way, we are led to a contradiction. It is especially effective in that it does so without using any properties of the circle, which are after all in question.

The greatest sensitivity of the two-circle construction is when the two circles are nearly tangent. This occurs as the vertex angle of the triangle approaches 0 or 180°. In the first situation, multiple intersections of the circles would lead to multiple triangles related as in the diagram of I.7; in the second, to multiple triangles as in Proclus' variant diagram, in which the one triangle completely contains the other (262–63). Taking the curvature of the circles into account, the threat of misreading whether there are one or more intersection points is greater for small vertex angles (where the circles curve together, in the same direction) than for large vertex angles (where the circles curve apart, in the opposite direction).

While there is no direct evidence that I.7 was originally motivated by this particular challenge, that would explain why Euclid deals with one diagram appearance variant rather than the other added in Proclus: Euclid's diagram draws the consequence from the reading of the two-compass diagram that most compellingly competes with the standard one. By Proclus' time, the original motivation would have been lost. The statement of I.7 would then be seen as including the second variant diagram, as writers at least since Proclus have done.

I.7 shows another way in which distinguishing types of probing has limited value: whether a given variant diagram probes appearance control or case branching control is a relational matter. Given the statement of I.7, Proclus' alternative is a different-appearance diagram arising from different metric choices (but not limitations on control in making the diagram from those metric choices) satisfying the verbal statement of hypothesis for reductio. Hence it is a case arising in the course of the argument, by our test of different degrees of uniformity in text and diagram. On the other hand, both diagrams in I.7 can be regarded as arising from appearance control probing; not in I.7 but in earlier constructions such as that of I.1.

4.6.3 Probing and knowledge

Traditional geometrical practice has no access, except through diagramming and proof, to ulterior standards of geometrical truth, by which it might back up its standards for handling text and diagram in demonstration. Not far under the surface of Proclus' frequent praise of the subtlety of Euclid's formulations lie buried hosts of objections based on variant readings and earlier texts, objections which are only avoided through what must long have seemed open-ended refinement of expressive means and exploration of diagrammatic possibilities. Using diagrams is a much more effective strategy in gaining an intellectual grip on space than we have been able to show here; but it does come at the price of a certain open-endedness in geometrical reasoning: although 'exact' conclusions are not read off from the diagram, diagram use also remains in need of probing, due to the limitations of diagram control.

Because of this, the long-range stability of traditional geometry—across centuries, participants, and the extended body of accumulated geometrical argument—cannot be understood without taking into account the constructive critical attitude towards geometrical argument expressed through probing. In our philosophical account of geometrical knowledge, probing must therefore share center stage with reasons.

But if probing is conceptually central to traditional geometric practice, we can no longer treat that practice as requiring unqualified assent to geometric claims. Instead, geometrical practice must cultivate dissent in order to avoid disarray! This dissent takes the form of vigorous critical scrutiny of geometrical argument. The normative structure of this dissent, however, differs from that of dissent in other practices, in which dissenting parties may stay clear of each other while they seek power (dissent in the body politic) or in which they may sometimes agree to disagree (dissent in the body philosophic). The geometer's obligation is to assert (the protagonist), to probe, *and* to seek a resolution of differences; suited to stand up in that very process continued by other participants bound only by that same obligation. Taken all together, this is a structure of unqualified responsibility, with the widest possible scope, for geometrical claims. This, rather than unqualified assent, is to be shared across space, time, and participants. Probing is an integral part of that responsibility, its element of criticism.

The normative structure of this dissent, with its insistence on uniformly acceptable resolution of differences, explains how the practice is compatible with an agentless conception of mathematics even though dissent implies more than one epistemic role. For, given that the practice was reasonably successful over the long term in providing artifact support for uniform responses, the distinct roles we recognized do not give rise to sustained distinct-commitment stances of the sort which would force us to admit multiple epistemic agents in our long-term perspective. We were able to construe I.7, each of its cases in fact, as a response to probing the outcome of certain compass constructions in extremes of sensitivity; where a uniqueness question arises gratifyingly similar to Protagoras' challenge on the intersection of circle and tangent. We might notice, however, that such appearance control probing tends to be urgent primarily in extremes, in atypical diagrams. Our discussion of III.2 confirmed this: sensitivity problems in telling whether the interior of a chord lies inside the circle arise only in extreme situations. In particular, neither natural target of probing from which III.2 might arise, in the arguments for III.1 and III.13, involves points near the end of a chord. Thus, only extremely short and hence atypical chords raise an appearance control concern in those arguments.

We must therefore cast around for what drives the practice to develop explicit proofs for the many propositions that conclude some explicit co-exact condition that could be read off from all but quite atypical diagrams. A practice might conceivably hold disarray at bay by simply not considering such atypical diagrams.[22] Or rather than not considering such diagrams, one might hope to assess them in light of 'neighboring' diagrams—ones related by continuous variation of the diagram which encounters no recognizable change of diagram appearance—which are clear; leaving the assimilations to be probed. For example, one could presume that very short chords of circles give the same appearance as longer ones, just as one presumes that diagrams too large to draw give the same appearance as smaller ones; all the while accepting the obligation to probe whether some boundary is crossed on the way to some extreme case, at which the appearance would change; and to probe for appearance-changing degeneracy of extreme cases themselves.

---

[22] An analogous attempt was made by the Italian School in algebraic geometry at the turn of the century; which made tremendous progress by limiting its attention mainly to 'generic cases'; and while this is now discredited, a similar strategy underlies the focus on 'stable' singularities in current singularity theory (Arnold (1992) Ch. 8.).

Either way, such a strategy might nonetheless lead to disarray in some unforeseen way; say, due to disagreements as to what diagrams are legitimate, or due to extreme-sensitivity diagrams arising in the course of argument by construction from otherwise apparently typical diagrams, in such a way that their occurrence is unpredictable or their appearance inscrutable.

Evidently, such was not the response of Greek geometry to sensitivity extremes. Perhaps this was due to disproportionate emphasis on generality and perfection to verbal standards in a strand of (Eleatic?) Greek thought seminal to geometry? Or perhaps it was found that disarray was not to be avoided in such a way. For example, the parallel postulate resolves the nature of an appearance transition; there is no saying that the appearance 'persists to the limit'.

At this point, however, it should strike us as puzzling that so much in geometry should be driven by extreme cases not explicitly referred to. The proof of III.2 plainly addresses the behavior of ordinary chords in circles, not extremely short ones (for which its use of the diagram would remain obscure). Neither sensitivity extreme of two-circle intersection construction—nor for that matter, any putative instance of non-uniqueness of such intersections—supposedly addressed by the full generality of I.7 would in fact, if diagrammed, be adequate to support the proof of I.7. For that argument has the same sensitivities: two triangles on the same base with vertex angles so close that one might be tempted to regard them as on the common intersection segment of two circles would not clearly show the regions and intersections which a satisfactory diagram of I.7 must provide. Evidently, both the force of the argument of I.7 near its sensitivity extremes, and the force of its contribution to appearance and case branching control with respect to the two-circle intersection construction, remain based on some policy on sensitivity extremes such as we tentatively rejected just above.

It should strike us that I.7 forcefully addresses the two-circle intersection construction primarily away from its sensitivity extremes; where it is already just as clear what happens in the two-circle intersection diagram. The paradigmatic example of clarity before proof is I.20: two sides of a triangle taken together exceed the third, as 'every dog knows'. Evidently, proof of diagrammatically clear propositions, including explicit co-exact results which could be read off, is typical in diagram-based geometry.

We came to this impasse by focussing on how proofs, eventually supplemented by probing, serve *justificatory* purposes in ancient geometry. We followed out threats that the practice might come out with some improper assertion, or fail to resolve disputes among participants. Knowing from history that disarray in geometry somehow never stayed far out of control, we sought to attribute rules to the practice which would account for this. Having recognized several rings of defense against error and indecisiveness, we find that remaining threats to geometry fail to motivate further discernable measures; indeed, fail to motivate the major actions of the practice.

This embarrassment results from too severe a restriction of philosophical focus, by which we have deprived ourselves of an explanatory perspective on geometrical diagram and proof. We have looked at inferential roles, demonstration, and probing alike, exclusively as defense against error, unreliability, or disagreement. This leaves us scarcely able to motivate why Euclid proves what his diagram shows!

Drawing on our broader perspective on intellectual practices as means of control, however, we can overcome this restriction of focus. For justification by no means exhausts the control intellectual practices aspire to. To admit that something *just happens* when we construct a chord in a circle, for example, that a particular region just *pops up* inside another one, is to admit defeat along a broad front. In their disdain for 'empirical geometry' or 'mechanical curves', participants in the geometrical tradition express revulsion for such impotence. This sensation that confronts us when something 'just happens' in a diagram may be contrasted with (a) the diverse measures that if taken may enhance our control; (b) the diverse questions, to which the ability to answer is part of being in control; and also (c) the diverse virtues we recognize when we are more in control.

Starting with measures: guided by the diagram, we can (as in III.2 as a response to the chord location inside a circle) probe variant diagrams by way of case and objection, and seek claims and arguments to clarify the original diagram.

For another example, we might have gone on our way to I.20, starting from circling two segments around the two endpoints of another segment and noticing a triangle pop up when we draw lines from the intersection to the two endpoints. This brute occurrence could then be probed by trying to get it to fail. If we stretch the base segment but not the others, we can see what happens to the intersection point; this could lead to the diorism of I.20. We could, however, apply the challenge to the construction of a triangle in this way in the demonstration of I.1, reported by Proclus: the straight lines from the circle intersection to the ends of the base might have a common segment, and so not give a triangle with the given segments we circled around as upstanding sides. Or we might give the diagram of I.7 and object to the implicit assumption of diagram control in the uniqueness of the outcome of the construction. Plainly, the brute diagram occurrence we started with does not call forth a unique challenge; and although the initial diagram (and, perhaps, grasp of its construction) provides some control over what is an appropriate challenge, this control is singularly inarticulate.

The claims and arguments which might be appropriate to the situation probed may be felt to address a variety of questions, which might or might not have appeared inarticulately bound up in the original sensation of impotence: when the diagram showed what it did, (i) could that have been a breakdown of appearance control? (ii) What might have been relevant about the free metric choices: can we identify features or ranges of these choices that demonstrably lead to the outcome appearance? (iii) What other appearances might arise from different metric choices? (iv) What is the range of control possibilities and outcomes?

When we have achieved a favorable resolution, acquired a measure of enhanced intellectual control over the original diagram occurrence, we may choose to characterize the enhancement as the acquisition of one or more of several broad virtues: we may say we have justified a claim about the outcome, or that we have explained why the construction gave the outcome, or that we have articulated claims, conditions, reasons, arguments...

The responsibility for geometrical claims that comes with being a geometer is thus much broader than a concern for reliability. Probing, too, should now be seen as far more than an element of criticism in an overall justificational strategy. Probing is the form of action within geometrical practice through which participants undertake their responsibility to improve its overall intellectual standing. Presumably, such a conception is not special to geometry, but instead common to a range of intellectual practices.

We tend to fail to recognize probing at work, fail to apply the concept—as it were, miss its unity—in two distinct ways. We can be distracted by the diverseness or special character of probing *actions* in particular situations; as when we are tempted to distinguish case and objection by who can conveniently respond. Or we can be distracted by the diverseness or special character of evaluations that drive probing or praise its accomplishments in particular situations; as when we take formal proofs in Tarski's system to settle justificational issues while asserting that they, taken by themselves, are geometrically unintelligible.

We can learn much, perhaps, by making such discriminations among action types and among evaluation types, in contexts where those distinctions are viable. It is important, however, to insist on the unity of probing over these two multiplicities because they are incompatible: varieties of probing actions do not match up with varieties of intellectual virtues that arise from them. Case proposal does not go with justification while objection goes with explanation, or vice versa. Probing the two-circle intersection construction by articulating the properties of the resulting linear configuration is as an action not usefully separable into parts addressing justificational and non-justificational concerns. It furthers overall intellectual control, and only as such is it properly understood. If we restrict our perspective on what virtues probing serves, we get into blind alleys in which the role of probing in the practice becomes unintelligible.

Acknowledgments. I would like to thank Fabrizio Cariani for converting my text into LaTex and producing the diagrams.

Bibliography

APOLLONIUS OF PERGA, *On cutting off a ratio [De sectione rationis.]* E. M. Macierowski (trans.), Robert H. Schmidt (ed.) (Fairfield, Conn.: Golden Hind Press).
ARNOLD, Vladimir (1992), *Catastrophe Theory*, 3rd rev. and expanded edn (Berlin, New York: Springer).
AYER, Alfred, J. (1936), *Language Truth and Logic* (London: Dover).
DUBNOV, Y. (1963), *Mistakes in Geometric Proofs* A. Henn and O. Titelbaum, (trans.) (Boston: Heath).
EUCLID, *Elements*, published as *Euclid's Elements: all Thirteen Books Complete in One Volume* T. Heath (trans.), D. Densmore (ed.) (New York: Dover Books, 1959).
FOWLER, David (1987), *The Mathematics of Plato's Academy* (Oxford: Oxford University Press).
HAHN, Hans (1980), 'The Crisis in Intuition', in *Empiricism, Logic and Mathematics: Philosophical Papers* (Dordrecht: Reidel), 73–102.
HILBERT, David (1899/1902), *Foundations of Geometry* (La Salle: Open Court).
HJELMSLEV, Johannes (1923) *Die natürliche Geometrie*. Mathematischen Seminars der Hamburgischen Universität.
KLEIN, Felix (1939), *Elementary Mathematics From an Advanced Standpoint* E. R. Hedrick and C. A. Noble (trans. of 3rd German edn) (New York: Dover).
KNORR, Wilbur (1975), *The Evolution of the Euclidean Elements: a Study of the Theory of Incommensurable Magnitudes and its Significance for Early Greek Geometry* (Dordrecht, Boston: Reidel).
—— (1980), 'On the Early History of Axiomatics: The Interaction of Mathematics and philosophy in Greek antiquity', *Proceedings of the 1978 Pisa Conference* (Dordrecht, Boston: Reidel).
—— (1986), *The Ancient Tradition of Geometric Problems* (Boston: Birkhäuser).
—— (1989) *Textual Studies in Ancient and Medieval Geometry* (Boston: Birkhäuser).
LOEMKER, Leroy (1970), *Leibniz: Philosophical Papers and Letters. A selection.* 2nd edn (Dordrecht: Reidel).
MAXWELL, E. (1961), *Fallacies in Mathematics* (Cambridge: Cambridge University Press).

MUELLER, Ian (1981), *Philosophy of Mathematics and Deductive Structure in Euclid's Elements* (Cambridge, MA: MIT Press).

NEWTON, Isaac (1728), *Universal Arithmetic.*

PAPPUS OF ALEXANDRIA, *Book 7 of the Collection,* edited with translation and commentary by Alexander Jones, 1986 (New York: Springer).

PONCELET, Jean Victor (1822), *Traité des Propriétés Projectives des Figures.* (Paris: Bachelier); expanded/corr. 2nd edn, 1865 (Paris: Gauthier Villars).

SHERRY, David (1993), 'Don't Take Me Half the Way: On Berkeley on Mathematical Reasoning', *Studies in History and Philosophy of Science,* 24, 207–225.

TARSKI, Alfred (1959), 'What is Elementary Geometry?', in L. Henkin, P. Suppes, and A. Tarski (eds.), *The Axiomatic Method,* 16–29 (Amsterdam: North-Holland).

TROPFKE, J. (1923), *Geschichte der Elementar-mathematik,* Band. 4, *Ebene Geometrie.* 2nd edn (Berlin/Leipzig: W. De Gruyter).

5

Mathematical Explanation: Why it Matters

PAOLO MANCOSU

The topic of mathematical explanation has recently been the subject of much interest. Although I will point out below that attention to mathematical explanation goes back to the Greeks, the recent revival in the analytic literature is a welcome addition to the philosophy of mathematics. In this introduction I have set myself two goals. The first is that of giving a survey of the literature on mathematical explanation and how the different contributions in this area are connected. Secondly, I would like to show that mathematical explanation is a topic that has far-reaching ramifications for many areas of philosophy, including, in addition to philosophy of mathematics, epistemology, metaphysics, and philosophy of science.

Let us begin by clarifying two possible meanings of mathematical explanation. In the first sense 'mathematical explanation' refers to explanations in the natural or social sciences where various mathematical facts play an essential role in the explanation provided. The second sense is that of explanation within mathematics itself.

5.1 Mathematical explanations of scientific facts

The first sense is well illustrated by the following example taken from a recent article by Peter Lipton:

> There also appear to be physical explanations that are non-causal. Suppose that a bunch of sticks are thrown into the air with a lot of spin so that they twirl and tumble as they fall. We freeze the scene as the sticks are in free fall and find that appreciably more of them are near the horizontal than near the vertical orientation. Why is this? The reason is that there are more ways for a stick to be

near the horizontal than near the vertical. To see this, consider a single stick with a fixed midpoint position. There are many ways this stick could be horizontal (spin it around in the horizontal plane), but only two ways it could be vertical (up or down). This asymmetry remains for positions near horizontal and vertical, as you can see if you think about the full shell traced out by the stick as it takes all possible orientations. This is a beautiful explanation for the physical distribution of the sticks, but what is doing the explaining are broadly geometrical facts that cannot be causes. (Lipton, 2004, pp. 9–10)

In this sense mathematical explanation is explanation in natural science carried out by essential appeal to mathematical facts. It is immediate from the quotation that one of the major philosophical challenges posed by mathematical explanations of physical phenomena is that they seem to be counterexamples to the causal theory of explanation. The existence of mathematical explanations of natural phenomena is widely recognized in the literature (Nerlich, 1979; Batterman, 2001; Colyvan, 2001). However, until recently very little attention has been devoted to them. Accounting for such explanations is not an easy task as it requires an account of how the geometrical facts 'represent' or 'model' the physical situation discussed. In short, articulating how mathematical explanations work in the sciences requires an account of how mathematics hooks on to reality, e.g. an account of the applicability of mathematics to reality (see Shapiro (2000), p. 35; cf. p. 217). And this opens the Pandora's box of models, idealization, etc.

In the analytic literature, a first attempt at giving an account of mathematical explanation of empirical phenomena was made by Steiner (1978b). Steiner's theory of mathematical explanations in the sciences relied on his theory of explanation in mathematics (see below) and was not supported by a detailed set of case studies. The central idea of Steiner's account is that a mathematical explanation of a physical fact is one in which when we remove the physics what we are left with is a mathematical explanation of a mathematical fact. He discussed one single example, the result according to which 'the displacement of a rigid body about a fixed point can always be achieved by rotating the body a certain angle about a fixed axis.' Steiner's discussion was a contribution to the larger worry of whether one could use the existence of such explanations to infer the existence of the entities mentioned in the mathematical component of the explanation. His reply was negative based on the claim that what needed explanation could not even be described without use of the mathematical language. Thus, the existence of mathematical explanations of empirical phenomena could not be used to infer the existence of mathematical entities, for this very existence was presupposed in the description of the fact to be explained. Indeed, he endorsed a line of argument originating from

Quine and Goodman according to which 'we cannot say what the world would be like without numbers, because describing any thinkable experience (except for utter emptiness) presupposes their existence'. (1978b, p. 20) Such reasoning, however, can be blocked by arguing that the true statement 'there are two cows in the field' does not commit the speaker to the existence of the number two, for the apparent reference to the number two can be explained away using a standard elimination of the number term by using existential quantifications. Of course, this does not mean that in more complicated physical statements the situation does not agree with the characterization given by Quine and Goodman.

Whether one agrees or not with Steiner on the many issues raised by his position, it is important to point out that his account had the merit of addressing the problem of when the mathematics plays an 'essentially' explanatory role in the explanation of a natural phenomenon and when it does not. The issue has resurfaced in Baker (2005), where Steiner (1978b) is however not mentioned. Baker proposes a new line on the indispensability argument in which mathematical explanations play a central role. There are several versions of the indispensability argument, but the general strategy runs as follows. Mathematics is indispensable for our best science. We ought to believe our best scientific theories and therefore we ought to accept the kind of entities our best theories quantify over. There are several ways to question the cogency of this line of argument but the key feature related to Baker's discussion is the following. Many versions of the argument rely on a holistic conception of scientific theories according to which ontological commitment is determined using all the existentially quantified sentences entailed by the theory. No particular attention is given to an analysis of how different components of the theory might be responsible for different posits and to the roles that different posits might play. Baker proposes a version of the indispensability argument which does not depend on holism. His contribution takes its start from a debate between Colyvan (2001, 2002) and Melia (2000, 2002) which saw both authors agreeing that the prospects for a successful platonist use of the indispensability argument rests on examples from scientific practice in which the postulation of mathematical objects results in an increase of those theoretical virtues which are provided by the postulation of theoretical entities. Both authors agree that among such theoretical virtues is explanatory power. Baker believes that such explanations exist but also argues that the cases presented by Colyvan (2001) fail to be genuine cases of mathematical explanations of physical phenomena. Most of his article is devoted to a specific case study from evolutionary biology and concerns the life-cycle of the so-called 'periodical' cicada. It turns out that three species of such cicadas 'share the same unusual life-cycle. In each species the nymphal stage remains in the soil for a lengthy period, then the adult cicada emerges after 13 years or 17 years depending on the geographical area. Even more strikingly, this emergence is synchronized among the members of a cicada species in any given area. The adults all emerge within the same few days, they mate, die a few weeks later and then the cycle repeats itself.' (2005, p. 229) Biologists have raised several questions concerning this life-cycle but one of them in particular concerns the question of why the life-cycle periods are prime. Baker proceeds then to a reconstruction of the explanation of such fact to conclude that:

> The explanation makes use of specific ecological facts, general biological laws, and number theoretic result. My claim is that the purely mathematical component [prime periods minimize intersection (compared to non-prime periods)] is both essential to the overall explanation and genuinely explanatory on its own right. In particular it explains *why* prime periods are evolutionary advantageous in this case. (Baker, 2005, p. 233)

It is of course not possible in this brief introduction to even summarize the reconstructed explanation and the additional arguments brought in support of the claim that this is a genuinely mathematical explanation. Let us rather summarize how such explanations give a new twist to the indispensability argument. The argument now runs as follows:

(a) There are genuinely mathematical explanations of empirical phenomena;
(b) We ought to be committed to the theoretical posits postulated by such explanations; thus,
(c) We ought to be committed to the entities postulated by the mathematics in question.

The argument has not gone unchallenged. Indeed, Leng (2005) tries to resist the conclusion by blocking premise (b). She accepts (a) but questions the claim that the role of mathematics in such explanations commits us to the real existence (as opposed to a fictional one) of the posits. This, she argues, will be granted when one realizes that both Colyvan and Baker infer illegitimately from the existence of the mathematical explanation that the statements grounding the explanation are true. She counters that mathematical explanations need not have true explanans and consequently the objects posited by such explanations need not exist.

Mathematical explanations of empirical facts have not been sufficiently studied. We badly need detailed case studies in order to understand better the variety of explanatory uses that mathematics can play in empirical contexts. The philosophical pay-offs might come from at least three different directions.

First, in the direction of a better understanding of the applicability of mathematics to the world. Indeed, understanding the 'unreasonable effectiveness' of mathematics in discovering and accounting for the laws of the physical world (Wigner, Steiner) can only be resolved if we understand how mathematics helps in scientific explanation. Second, the study of mathematical explanations of scientific facts will serve as a test for theories of scientific explanation, in particular those which assume that explanation is causal explanation. A promising start has been made by Batterman through an examination of what he calls asymptotic explanation (Batterman, 2001, Ch. 4). Such explanations 'illuminate structurally stable aspects of a phenomenon and its governing equations', (p. 59) using highly sophisticated mathematical manipulations. Third, philosophical benefits might also emerge in the metaphysical arena by improved exploitation of various forms of the indispensability argument. Whether any such argument is going to be successful remains to be seen but the discussion will yield philosophical benefits in forcing for instance the nominalist to take a stand on how he can account for the explanatoriness of mathematics in the empirical sciences.

5.2 From mathematical explanations of scientific facts to mathematical explanations of mathematical facts

Since we have been discussing indispensability arguments I will take my start from there. In an interesting note to her paper Leng says:

> Given the form of Baker and Colyvan's argument, one might wonder why it is mathematical explanations of physical phenomena that get priority. For if there are, as we have suggested, some genuine mathematical explanations [of mathematical facts] then these explanations must also have true explanans. The reason that this argument can't be used is that, in the context of an argument for realism about mathematics, it is question begging. For we also assume here that genuine explanations must have a true explanandum, and when the explanandum is mathematical, its truth will also be in question. (2005, p. 174)

This comment reflects the general use to which indispensability arguments have been put. The main goal is to provide an argument for platonism in mathematics but no attention is truly given to the different kind of mathematical entities we are postulating. From this point of view the existence of the natural numbers is on a par with the existence of a Mahlo cardinal or of a differentiable manifold. It is, however, reasonable to ask whether mathematical explanations can be used not as arguments for realism in mathematics *tout court* but rather as specific arguments for realism about certain mathematical entities.[1] I am interested in articulating a possible parallel between uses of the indispensability of mathematics in science as described by Baker and Colyvan and the case of mathematics. Perhaps the best argument one can get here is one foreshadowed in Feferman (1964). Discussing Gödel's claim that the postulation of the Cantorian sets was just as justified as that of physical bodies in order to obtain a satisfactory theory of sense perceptions, Feferman claimed that the development of mathematics strongly supported the following interpretation of the argument:

> Abstraction and generalization are constantly pursued as the means to reach really satisfactory explanations which account for scattered individual results. In particular, extensive developments in algebra and analysis seem necessary to give real insight into the behavior of the natural numbers. Thus we are able to realize certain results, whose instances can be finitistically checked, only by a detour via objects (such as ideals, analytic functions) which are much more 'abstract' than those with which we are finally concerned. The argument is less forceful when it is read as justifying some particular conceptions and assumptions, namely those of impredicative set theory, as formally necessary to infer the arithmetical data of mathematics. It is well known that a number of algebraic and analytic arguments can be systematically recast into a form which can be subsumed under (elementary) first order number theory. (Feferman, 1964, p. 3)[2]

Feferman seems to think that a persuasive form of the Gödelian argument goes as follows:

1. There are scattered results in one branch of mathematics (the data, this might be finitistically verifiable propositions concerning the natural numbers) which call for an explanation.
2. Such an explanation is obtained by appealing to more abstract entities (say ideals and analytic functions).
3. We thus have good reason to postulate such abstract entities and to believe in their existence.

---

[1] This is different from, although related to, the issue of objectivity discussed in Leng (2005), pp. 172–173. She discusses, following Waismann and Steiner, an example concerning how the explanation of a known fact about the real numbers can be used to 'to support the non-arbitrariness of our extension of the number system to complex numbers' (p. 172) or of our 'representations of complex numbers' (p. 173). But the issue here is not that of realism about the mathematical objects.

[2] Feferman's work showed that predicative analysis could not be formally sufficient to obtain all the arithmetical consequences of impredicative mathematics. However, he also claimed until recently that predicative mathematics was sufficient to prove all of the arithmetical consequences of mathematical interest. He now agrees that certain arithmetical results (such as the modified form of finite Kruskal's theorem (FKT^*); see Feferman (2004) and Hellman (2004))) which cannot be proved predicatively are of mathematical interest.

If this is correct, we have something in the vicinity of an indispensability argument. In 'Platonism' Dummett rejects a similar argument on account of the fact that 'real numbers and ordinals do not act on each other or on anything else; so there is nothing which is left unaccounted for if we suppose them not to be there' (Dummett, 1978, p. 204). But it is evident that such rejection is based on the questionable assumption that all explanations must be causal. The argument presented above would of course leave a committed predicativist unimpressed (for the explanation in question would be a derivation which uses tools not available to the predicativist). However, just as standard indispensability arguments address those who are realists about theoretical entities in science, so here the intended audience for the argument would consist of those who are realists about a certain realm of mathematical entities (say, the natural numbers) and in addition are not already committed to a foundational position (such as predicativism) which forbids entertaining the entities being postulated by the explanation. Reconstructed as such the argument is useful in providing rational grounds for the acceptance of the mathematical entities appealed to in the explanation. Its strength becomes evident when one considers that the argument goes through even if it turns out that the entities in question are in principle eliminable on account of the fact that the explained result is derivable within a narrower framework (as in the case where you have a theory T' which is a conservative extension of T). However, if this derivation results in a loss of explanatory power then we still have good reasons to believe in the entities in question. This of course leaves us with the question of when a derivation is explanatory. And this parallels the situation we discussed in the case of mathematical explanations of physical phenomena. I should point out, however, that I am not endorsing the indispensability argument for mathematics I have been considering but that I do find it of interest.

The original form of the indispensability argument relied on a form of confirmational holism. This left the argument open to the objection, raised forcefully by Maddy, that scientific practice proceeds otherwise or to the objection that other accounts of confirmation block the conclusion (Sober, 1993). In response, advocates like Colyvan and Baker have argued that explanatory considerations lead to platonism even if we drop confirmational holism. But, as I pointed out, nobody really has an account of mathematical explanations of scientific phenomena.

In addition to being of independent interest, the move to mathematical explanations of mathematical facts is justified also by the following two considerations. First, it is conceivable that whatever account we will end up giving of mathematical explanations of scientific phenomena, it won't be completely independent of mathematical explanation of mathematical facts (indeed for Steiner the former is explicated in terms of the latter). Secondly, the vicissitudes with holism recounted above have their analog in the recent developments in the philosophy of mathematics. Quine originally used the indispensability argument to argue that we should believe in sets because they do the best job in tracking all our commitments to abstract objects. For Quine the appeal to empirical science was essential. Maddy's realism drops the connection to empirical science and tries to obtain the same conclusion just by focusing on pure mathematics. In Chapter 4 of Maddy (1990) we find a lengthy discussion of theoretical virtues, including explanatory ones, that play a role in 'extrinsic' justifications for axiom choice in set theory. However, the problem of mathematical explanation was not singled out but rather dealt with at the same level of other theoretical virtues (verifiable consequences, powerful methods of solution, simplification and systematization, strong intertheoretic connections, etc.). And although Maddy herself gave up the attempt in favor of 'naturalism' (see Maddy (1997)), mathematical explanation can still play an important role in this debate. For those who believe that her realism can be revived perhaps the detour through indispensability arguments that appeal to mathematical explanations might provide a more persuasive type of argument than the other varieties of 'extrinsic' justifications mentioned in 1990. Moreover, those who are persuaded by the 'naturalist' approach of her latest book will as a matter of fact have to welcome investigations into mathematical explanation as they are part and parcel of the kind of work the methodologist in this area ought to carry out. So both these options call for an account of mathematical explanations of mathematical facts.

5.3 Mathematical explanations of mathematical facts

The history of the philosophy of mathematics shows that a major conceptual role has been played by the opposition between proofs that convince but do not explain and proofs that in addition to providing the required conviction that the result is true also show *why* it is true. Philosophically this tradition begins with Aristotle's distinction between *to oti* and *to dioti* proofs and has a rich history passing through, among others, the *Logic of Port Royal* written by Arnauld and Nicole, Bolzano, and Cournot (see Harari (2008), Kitcher (1975) and Mancosu (1996, 1999, 2000, 2001)). This philosophical opposition between types of proof also influenced mathematical practice and led many of its supporters often to criticize existing mathematical practice for its epistemological inadequacy (see for instance Guldin's program in the 17th century (Mancosu, 1996, 2000) and Bolzano's work in geometry and analysis (Kitcher, 1975)). Steiner's model of explanation, to be discussed below, although not relying on the Aristotelian opposition, aims at characterizing the distinction between explanatory and non-explanatory proofs.

The opposition between explanatory and non-explanatory proofs is not only a product of philosophical reflection but it confronts us as a datum from mathematical practice. A mathematician (or a community of mathematicians) might find a proof of a certain result absolutely convincing, but nonetheless he (they) might be unsatisfied with it for it does not provide an explanation of the fact in question. The great mathematician Mordell, to choose one example among many, mentions the phenomenon in the following passage:

> Even when a proof has been mastered, there may be a feeling of dissatisfaction with it, though it may be strictly logical and convincing; such as, for example, the proof of a proposition in Euclid. The reader may feel that something is missing. The argument may have been presented in such a way as to throw no light on the why and wherefore of the procedure or on the origin of the proof or why it succeeds. (Mordell, 1959, p. 11)

This sense of dissatisfaction will often lead to a search for a more satisfactory proof. Mathematicians appeal to this phenomenon often enough (see Hafner and Mancosu (2005) for extended quotations from mathematical sources; and of course, their joint paper in this volume) as to make the project of philosophically explicating this notion an important one for a philosophical account of mathematical practice. But explanations in mathematics do not only come in the form of proofs. In some cases explanations are sought in a major conceptual recasting of an entire discipline. In such situations the major conceptual recasting will also produce new proofs but the explanatoriness of the new proofs is derivative on the conceptual recasting. This leads to a more global (or holistic) picture of explanation than the one based on the opposition between explanatory and explanatory proofs (in Mancosu (2001) I describe in detail such a global case of explanatory activity from complex analysis; see also Kitcher (1984) and Tappenden (2005) for additional case studies). The point is that in the latter case explanatoriness is primarily a property of proofs, whereas in the former it is a property of the whole theory or framework and the proofs are judged explanatory on account of their being part of the framework. This captures well the difference between the two major accounts of mathematical explanation available at the moment, those of Steiner and Kitcher. Before discussing them, I hasten to add that other models of scientific explanation can be thought to extend to mathematical explanation.

For instance, Sandborg (1997, 1998) tests van Fraassen's account of explanation as answers to why-questions by using cases of mathematical explanation.

Steiner proposed his model of mathematical explanation in 1978. In developing his own account of explanatory proof in mathematics he discusses—and rejects—a number of initially plausible criteria for explanation, i.e. the (greater degree of) abstractness or generality of a proof, its visualizability, and its genetic aspect which would give rise to the discovery of the result. In contrast, Steiner takes up the idea 'that to explain the behaviour of an entity, one deduces the behaviour from the essence or nature of the entity' (Steiner, 1978*a*, p. 143). In order to avoid the notorious difficulties in defining the concepts of essence and essential (or necessary) property, which, moreover, do not seem to be useful in mathematical contexts anyway since all mathematical truths are regarded as necessary, Steiner introduces the concept of characterizing property. By this he means 'a property unique to a given entity or structure within a family or domain of such entities or structures', where the notion of family is taken as undefined. Hence what distinguishes an explanatory proof from a non-explanatory one is that only the former involves such a characterizing property. In Steiner's words: 'an explanatory proof makes reference to a characterizing property of an entity or structure mentioned in the theorem, such that from the proof it is evident that the result depends on the property'. Furthermore, an explanatory proof is generalizable in the following sense. Varying the relevant feature (and hence a certain characterizing property) in such a proof gives rise to an array of corresponding theorems, which are proved—and explained—by an array of 'deformations' of the original proof. Thus Steiner arrives at two criteria for explanatory proofs, i.e. dependence on a characterizing property and generalizability through varying of that property (Steiner, 1978*a*, pp. 144, 147).

Steiner's model was criticized by Resnik and Kushner (1987) who questioned the absolute distinction between explanatory and non-explanatory proofs and argued that such a distinction can only be context-dependent. They also provided counterexamples to the criteria defended by Steiner. In Hafner and Mancosu (2005) it is argued that Resnik and Kushner's criticisms are insufficient as a challenge to Steiner for they rely on ascribing explanatoriness to specific proofs based not on evaluations given by practicing mathematicians but rather relying on the intuitions of the authors. By contrast, Hafner and Mancosu build their case against Steiner using a case of explanation from real analysis, *recognized as such in mathematical practice*, which concerns the proof of Kummer's criterion for convergency. They argue that the explanatoriness of the proof of the result in question cannot be accounted for in Steiner's model and, more importantly, this is instrumental in giving a careful and detailed scrutiny of various conceptual components of the model. In addition, further discussion of Steiner's account, aimed at improving the account, is provided in Weber and Verhoeven (2002).

Kitcher's model will be described at length in the research paper following this introduction. Criticisms of unification theories of explanation as insufficient for mathematical explanation have also been raised forcefully in Tappenden (2005). In the next section, I would like to point out some aspects of Kitcher's position that bring us back to the issue of generalization and abstraction. This will also be instrumental in introducing aspects of Kitcher's thought relevant to the research paper to follow.

5.4 Kitcher on explanation and generalization

I will start with a striking quote concerning generalization and its relation to explanation within mathematics. It is taken from Cournot:

> Generalizations which are fruitful because they reveal in a single general principle the rationale of a great many particular truths, the connection and common origins of which had not previously been seen, are found in all the sciences, and particularly in mathematics. Such generalizations are the most important of all, and their discovery is the work of genius. There are also sterile generalizations which consist in extending to unimportant cases what inventive persons were satisfied to establish for important cases, leaving the rest to the easily discernible indications of analogy. In such cases, further steps toward abstraction and generalization do not mean an improvement in the *explanation* of the order of mathematical truths and their relations, for this is not the way the mind proceeds from a subordinate fact to one which goes beyond it and *explains* it. (Cournot, 1851, sect. 16, Engl. trans. 1956, p. 24, my emphasis)

The central opposition in this text is that between fruitful generalizations vs. sterile generalizations. What distinguishes the two of them is that the former are explanatory while the latter are not. Genius consists, according to Cournot, not in generalization *tout court* but in those generalizations that are able to reveal the explanatory order according to which mathematical truths are structured. A remarkably similar statement is found in an article by the mathematician S. Mandelbrojt who claims that 'La généralité est belle lorsqu'elle posséde un caractére explicatif' and 'l'abstraction est belle et grande lorsqu'elle est explicative' (Mandelbrojt, 1952, pp. 427–428). Also Mandelbrojt pointed out that generalization can be cheap and boring. The generalization is informative when it is explanatory. Such explanatory generalizations can be obtained by the right degree of abstraction and should show the object being studied in its 'natural setting'. Finally, note that both Cournot and Mandelbrojt take it as a matter of course that mathematical explanations exist. Of course, the above quotes can only be the beginning of the job. What is then the relationship between explanation and generalization in mathematics? Such a major question cannot be answered in this introduction but it is important to raise it since this web of relations between explanation, generality, and abstraction is a recurring one in all attempts to talk about explanation. Here I propose to indicate how the problem recurs in Kitcher's writings. Kitcher is well known as a defender of the theory of explanation as unification. Kitcher's first article where generalization and explanation are thematized is an article on Bolzano's philosophy of mathematics from 1975. As part of his analysis of Bolzano, Kitcher argued against the thesis that 'a deductive argument is explanatory if and only if its premises are at least as general as its conclusion.' Against the thesis he raised the following objection:

> In any case, the generality criterion is ill-adapted to the case of mathematics. There is a very special difficulty with derivations in arithmetic, namely that many proofs use the method of mathematical induction. Suppose that I prove a theorem by induction, showing that all positive integers have property F. This is accomplished by showing
>
> a) 1 has F
> b) if all numbers less than n have F then n has F
>
> (Of course there are other versions of the method of induction). It would seem hard to deny that this is a genuine proof. [...] Further, this type of proof does not controvert Bolzano's claim that genuine proofs are explanatory; we feel that the structure of the positive integers is exhibited by showing how 1 has the property F and how F is inherited by successive positive integers; and, in uncovering this structure, the proof explains the theorem. But proofs by induction do violate the generality criterion[...] Whatever account Bolzano gives and whatever generality it achieves for arithmetic, Bolzano would surely be hard put to avoid the consequence that the proposition expressed by '1 has F' is less general than that expressed by 'Every number has F'. (Kitcher, 1975, p. 266)

I believe that this argument is too quick and that it suffers from an unfruitful attempt to use the complexity of the logical formulas as a measure of generality. The first problem is that it is easy to reformulate the two premises of induction as a single universal sentence, thereby eliminating '1 has F' as an independent premise. Secondly, not everybody agrees that proofs by induction are explanatory (see Mancosu (2001, note 11)). Thus, I find Kitcher's argument to be neither here nor there.

A related context in which Kitcher addresses the issue of generalization is in his book *The Nature of Mathematical Knowledge* from 1984. This is a complex book and I will not attempt to give a general overview of its contents. However, one of the main questions Kitcher raises is: How does mathematics grow? What are the patterns of change which are typical of mathematics? Is the process of growth a rational one? In Chapter 9 of his book he sets the goal as follows:

> Here I shall be concerned to isolate those constituent patterns of change and to illustrate them with brief examples. I shall attempt to explain how the activities of question-answering, question-generation, generalization, rigorization and systematization yield rational interpractice transitions. When these activities occur in a sequence, the mathematical practice may be dramatically changed through a series of rational steps. (Kitcher, 1984, p. 194)

Let us then consider generalization:

> One of the most readily discernible patterns of mathematical change, one which I have so far not explicitly discussed, is the extension of mathematical language by generalization. (Kitcher, 1984, p. 207)

As examples Kitcher mentioned Riemann's redefinition of the definite integral, Hamilton's search for hypercomplex numbers, and Cantor's generalization of finite arithmetic. Kitcher's goal is 'to try to understand the process of generalization which figures in these episodes and to see how the search for generalization may be rational' (Kitcher, 1984, p. 207). However, not all generalizations are significant. In fact it is easy to concoct trivial generalizations. What distinguishes the trivial generalizations from the significant ones? This is where explanation comes in again:

> significant generalizations are explanatory. They explain by showing us exactly how, by modifying certain rules which are constitutive of the use of some expressions of the language, we would obtain a language and a theory within which results analogous to those we have already accepted would be forthcoming. From the perspective of the new generalization, we see our old theory as a special case, one member of a family of related theories. (Kitcher, 1984, pp. 208–209)

Building upon such considerations, Kitcher tries to distinguish between the rationally acceptable generalizations and those that are not so:

> Those 'generalizing' stipulations which fail to illuminate those areas of mathematics which have already been developed are not rationally acceptable. (1984, p. 209)

In other words, to account for the rationality of processes of generalization in mathematics we need an account of mathematical explanation. Moreover, one of Kitcher's envisaged benefits of his analysis in terms of explanatory power is that through it one might be able to provide an account for certain value judgements given by mathematicians when they rave about the aesthetic quality of a piece of mathematics or about its 'interest' (see p. 232).

Thus, it should be obvious how the need for a theory of mathematical explanation emerges out of these considerations on generalization. Only significant generalizations can account for rational change in mathematics and those are the explanatory ones. It would however be a mistake to think that generalization is the only pattern of mathematical change that can give us explanations. In addition to generalization, Kitcher discusses rigorization and systematization as sources of understanding and explanation (p. 227). In his later work, such as (Kitcher, 1989), he uses unification as the overarching model for explanation both in science and mathematics:

> The fact that the unification approach provides an account of explanation, and explanatory asymmetries, in mathematics stands to its credit. (Kitcher 1989, p. 437)

Let me just refer here to the article 'Explanatory Unification and the Causal Structure of the World' for a few final quotes. At one point of the discussion, Kitcher is engaged in showing the limitations of a theory of explanation which takes causality as the central concept for the account. He objects that in formal syntax and mathematics one has explanations which are not causal. He mentions Bolzano's proof of the intermediate value theorem as an explanatory proof of a theorem whose previous proofs were not explanatory and a case dependent on alternative axiomatizations of group theory. After the discussion he concludes:

> Moreover, in this case [axiomatizations of group theory] and in that discussed in A [Bolzano's proof], it is not hard to see a reason for the distinguishing of the derivations: the preferred derivation can be generalized to achieve more wide-ranging results. (Kitcher 1989, p. 425)
>
> In both instances, the explanatory derivation is similar to derivations we could provide for a more general result; the nonexplanatory derivation cannot be generalized, it applies only to the local case. (Kitcher 1989, p. 425)

This brief mention of passages in Kitcher concerning generality and explanation is only meant to recall the importance of generality in this context. Lack of space does not allow me to discuss how the issue of generality plays a role in Steiner's model. While Steiner rejects the thesis that explanatoriness can be accounted for in terms of generality he also incorporates appeal to generality in his theory of explanation by requiring that explanatory proofs be generalizable

('It is not, then, the general proof which explains; it is the generalizable proof.' Steiner (1978a); see Hafner and Mancosu (2005) for an extended discussion)).

5.5 Conclusion

The topic of mathematical explanation offers a vast and virgin territory to exploration. Most of the work remains to be done. We need to discuss and analyze mathematical explanations in the sciences and mathematical explanations within pure mathematics. These case studies have then to be used to test a variety of theories of scientific explanation and theories of mathematical explanation. That in turn will be instrumental in the attempt of giving broader and more encompassing theories of scientific explanation (or perhaps in showing that no such thing is to be had). Finally, all this work will have a major impact on broader philosophical problems such as, among others, accounting for mathematical applications and indispensability arguments in ontology, and in providing a richer epistemology for mathematics.

Bibliography

BAKER, Alan (2005), 'Are there Genuine Mathematical Explanations of Physical Phenomena?', *Mind*, 114, 223–238.
BATTERMAN, Robert (2001), *The Devil in the Details* (Oxford: Oxford University Press).
COLYVAN, Mark (2001), *The Indispensability of Mathematics* (Oxford: Oxford University Press).
——(2002), 'Mathematics and Aesthetic Considerations in Science', *Mind*, 11, 69–78.
DUMMETT, Michael (1978), 'Platonism', in Michael Dummett (ed.), *Truth and Other Enigmas* (London: Duckworth) pp. 202–204.
FEFERMAN, Solomon (1964), 'Systems of Predicative Analysis', *The Journal of Symbolic Logic*, 29, 1–30.
——(2004), 'Comments on "Predicativity as a Philosophical Position" by G. Hellman', *Revue Internationale de Philosophie*, 229(3), 313–323.
HAFNER, Johannes and MANCOSU, Paolo (2005), 'The Varieties of Mathematical Explanation', in P. Mancosu, K. Jørgensen, and S. Pedersen (eds.), *Visualization, Explanation and Reasoning Styles in Mathematics* (Dordrecht: Springer) pp. 215–250.
HARARI, Orna (2008), 'Proclus' Account of Explanatory Demonstrations in Mathematics and its Context', forthcoming in *Archiv für Geschichte der Philosophie*.

HELLMAN, Geoffrey (2004), 'Predicativity as a Philosophical Position', *Revue Internationale de Philosophie*, 229(3), 295–312.
KITCHER, Philip (1975), 'Bolzano's Ideal of Algebraic Analysis', *Studies in History and Philosophy of Science*, 6, 229–269.
—— (1984), *The Nature of Mathematical Knowledge* (Oxford: Oxford University Press).
—— (1989), 'Explanatory Unification and the Causal Structure of the World', in P. Kitcher and W. Salmon (eds.), *Scientific Explanation* (Minneapolis: University of Minnesota Press) pp. 410–505.
LENG, Mary (2005), 'Mathematical Explanation', in C. Cellucci and D. Gillies (eds.), *Mathematical Reasoning and Heuristics* (London: King's College Publications) pp. 167–189.
LIPTON, Peter (2004), 'What Good is an Explanation?', in J. Cornwell (ed.), *Explanations. Styles of Explanation in Science* (Oxford: Oxford University Press) pp. 1–21.
MADDY, Penelope (1990), *Realism in Mathematics* (Oxford: Oxford University Press).
—— (1997), *Naturalism in Mathematics* (Oxford: Oxford University Press).
MANCOSU, Paolo (1996), *Philosophy of Mathematics and Mathematical Practice in the Seventeenth Century* (Oxford: Oxford University Press).
—— (1999), 'Bolzano and Cournot on Mathematical Explanation', *Revue d'Histoire des Sciences*, 52, 429–455.
—— (2000), 'On Mathematical Explanation', in E. Grosholz and H. Breger (eds.), *The Growth of Mathematical Knowledge* (Dordrecht: Kluwer) pp. 103–119.
—— (2001), 'Mathematical Explanation: Problems and Prospects', *Topoi*, 20, 97–117.
MANDELBROJT, Stephan (1952), 'Pourquoi Je Fais des Mathématiques', *Revue de Metaphysique et de Morale*, 422–429.
MELIA, Joseph (2000), 'Weaseling Away the Indispensability Argument', *Mind*, 109, 455–479.
—— (2002), 'Response to Colyvan', *Mind*, 111, 75–79.
MORDELL, Louis (1959), *Reflections of a Mathematician* (Montreal: Canadian Mathematical Congress).
NERLICH, Graham (1979), 'What Can Geometry Explain?', *British Journal for the Philosophy of Science*, 38, 141–158.
RESNIK, Michael and KUSHNER, David (1987), 'Explanation, Independence, and Realism in Mathematics', *British Journal for the Philosophy of Science*, 38, 141–158.
SANDBORG, David (1997), *Explanation and Mathematical Practice*, Ph.D. thesis, University of Pittsburgh.
—— (1998), 'Mathematical Explanation and the Theory of Why-Questions', *British Journal for the Philosophy of Science*, 49, 603–624.
SHAPIRO, Stewart (2000), *Thinking About Mathematics* (Oxford: Oxford University Press).
SOBER, Elliott (1993), 'Mathematics and Indispensability', *The Philosophical Review*, 102, 35–57.
STEINER, Mark (1978*a*), 'Mathematical Explanation', *Philosophical Studies*, 34, 135–151.

STEINER, Mark (1978b), 'Mathematics, Explanation, and Scientific Knowledge', *Nous*, 12, 17–28.

TAPPENDEN, Jamie (2005), 'Proof Style and Understanding in Mathematics I: Visualization, Unification and Axiom Choice', in P. Mancosu, K. Jørgensen, and S. Pedersen (eds.), *Visualization, Explanation and Reasoning Styles in Mathematics*, (Dordrecht: Springer) pp. 147–214.

WEBER, Erik and VERHOEVEN, Liza (2002), 'Explanatory Proofs in Mathematics', *Logique et Analyse*, 179–180, 299–307.

6

Beyond Unification

JOHANNES HAFNER AND PAOLO MANCOSU

As pointed out in the introduction there are at the moment two theories of mathematical explanation on offer. The first is due to Steiner (1978) and has been extensively discussed in Resnik and Kushner (1987), Weber and Verhoeven (2002), and Hafner and Mancosu (2005). The second theory is due to Philip Kitcher, who is a well-known defender of an account of scientific explanation as theoretical unification. Kitcher sees as one of the virtues of his account that it can also be applied to explanation in mathematics, unlike other theories of scientific explanation whose central concepts, say, causality or laws of nature, do not seem relevant to mathematics. In this paper we are going to discuss Kitcher's account of explanation in the context of a test case from real algebraic geometry.

6.1 Kitcher's theory of explanation

Kitcher has not devoted any single article exclusively to mathematical explanation and thus his position can only be gathered from what he says about mathematics in his major articles on scientific explanation. In his later work, such as Kitcher (1989), he uses unification as the overarching model for explanation both in science and mathematics:

> The fact that the unification approach provides an account of explanation, and explanatory asymmetries, in mathematics stands to its credit. (Kitcher, 1989, p. 437)

Unlike Steiner's model of mathematical explanation, Kitcher's account of mathematical explanation has not been extensively discussed. A general discussion is found in Tappenden (2005) but not a detailed analysis. Our aim here is threefold. We will present in outline Kitcher's theory of explanation as unification. Then we will look at a particular case of mathematical explanation coming from real algebraic geometry. Finally, we will check Kitcher's account of explanation against our case study. By doing this we will heed Kitcher's advice in his paper 'Explanatory unification':

> Quite evidently, I have only *sketched* an account of explanation. To provide precise analyses of the notions I have introduced, the basic approach to explanation offered here must be refined against concrete examples of scientific practice. What needs to be done is to look closely at the argument patterns favored by scientists and attempt to understand what characteristics they share. (Kitcher, 1981, p. 530)

6.1.1 Kitcher's account of explanation as unification

We will follow here the account of explanation given by Kitcher in his paper 'Explanatory unification and the causal structure of the world' (1989). Kitcher claims that behind the account of explanation given by Hempel's covering law model—the official model of explanation for logical positivism—there was an unofficial model which saw explanation as unification. What should one expect from an account of explanation? Kitcher in 1981 points out two things. First, a theory of explanation should account for how science advances our understanding of the world. Secondly, it should help us in evaluating or arbitrating disputes in science. He claims that the covering law model fails on both counts and he proposes that his unification account fares much better.

The basic intuition. Kitcher found inspiration in Friedman's article of 1974, 'Explanation and scientific understanding' where Friedman put forward the idea that understanding of the world is achieved by science by reducing the number of facts we take as brute:

> this is the essence of scientific explanation—science increases our understanding of the world by reducing the total number of independent phenomena that we have to accept as ultimate or given. A world with fewer independent phenomena is, other things equal, more comprehensible than one with more. (Friedman, 1974, p. 15)

Already Friedman had tried to make this intuition more precise by substituting for the notions of phenomena and laws linguistic descriptions of such. Kitcher disagrees with the specific details of Friedman's proposal but thinks that the general intuition is correct. He modifies Friedman's proposal by emphasizing that what lies behind unification is the reduction of the number of argument patterns used in providing explanations while being as comprehensive as possible in the number of phenomena explained:

> Understanding the phenomena is not simply a matter of reducing the 'fundamental incomprehensibilities' but of seeing connections, common patterns, in what

initially appeared to be different situations. Here the switch in conception from premise–conclusion pairs to derivations proves vital. Science advances our understanding of nature by showing us how to derive descriptions of many phenomena, using the same patterns of derivation again and again, and, in demonstrating this, it teaches us how to reduce the number of types of facts that we have to accept as ultimate (or brute). So the criterion of unification I shall try to articulate will be based on the idea that $E(K)$ is a set of derivations that makes the best tradeoff between minimizing the number of patterns of derivation employed and maximizing the number of conclusions generated. (Kitcher, 1989, p. 432)

We will come back to the distinction between arguments and derivations and to a clarification of what $E(K)$ is below.

Local vs. global notions of explanation. In many accounts of explanation, including the Hempelian one, explanations are arguments. Arguments are identified with pairs of premises and conclusions and can be assessed individually with respect to explanatoriness. Following Friedman, we would like to say that whether an argument is an explanation is a local property, i.e. it does not depend on more global constraints. Kitcher rejects both the identification of explanations as arguments conceived as above and the local characterization of explanation. The informal idea is that explanations qualify as such because they belong to the best systematization of our beliefs. Moreover, explanations are not pairs of premises–conclusions, as in Hempel, but rather derivations:

> On the systematization account, an argument is considered as a derivation, as a sequence of statements whose status (as a premise or as following from previous members in accordance with some specified rule) is clearly specified. An ideal explanation does not simply list the premises but shows how the premises yield the conclusion. (Kitcher, 1989, p. 431)

6.1.2 The formal details of the model

Let us make this more formal. Let us start with a set K of beliefs assumed consistent and deductively closed (informally one can think of this as a set of statements endorsed by an ideal scientific community at a specific moment in time (Kitcher, 1989, p. 431)). A *systematization* of K is any set of arguments which derive some sentences in K from other sentences in K. The *explanatory store* over K, $E(K)$, is the best systematization of K (Kitcher here makes an idealization by claiming that $E(K)$ is unique). Corresponding to different systematizations we have different degrees of unification. The highest degree of unification is that given by $E(K)$. But according to what criteria can a systematization be judged to be the best?

Kitcher's criteria for systematizations. Kitcher lists three criteria for judging the quality of a systematization Σ, although as it turns out only two of them will be relevant to our discussion. In order to introduce them we need to clarify three notions: argument pattern, generating set for Σ, and conclusion set for Σ.

Argument pattern: Kitcher derives this notion from specific examples taken from the natural sciences, e.g. Newtonian mechanics and Darwin's theory of evolution. We will illustrate it after giving the definitions with an example from mathematics. Let us begin with the notion of *schematic sentence*. This is an expression obtained by replacing some or all of the non-logical expressions in a sentence by dummy letters.

A set of *filling instructions* tells us how the dummy letters in a schematic sentence are to be replaced.

A *schematic argument* is a sequence of schematic sentences.

A *classification* for a schematic argument is a set of sentences which tells us exactly what role each sentence in a schematic argument is playing, e.g. whether it is a premise, which sentences are inferred from which and according to what rules, etc.

A *general argument pattern* $\langle s, f, c \rangle$ is a triple consisting of a schematic argument s, a set f of sets of filling instructions, and a classification c for s.

An example from mathematics

Consider the problem of determining the equation of the line tangent to the parabola $y = 2x^2 + 3x + 1$ at point (1,6). We can solve the problem using derivatives as follows.

1. $[2x^2 + 3x + 1]' = 4x + 3$
2. $[4x + 3]_{x=1} = 7$
3. Thus the tangent line to $2x^2 + 3x + 1$ at (1, 6) is $(x - 1)7 = (y - 6)$.

A schematic argument for determining the tangent line to a differentiable curve $f(x)$ at a point (x_0, y_0) can be obtained from the above as follows.

1S. $[f(x)]' = g(x)$
2S. $[g(x)]_{x=x_0} = c$
3S. Thus the tangent line to $f(x)$ at (x_0, y_0) is $(x - x_0)c = (y - y_0)$.

Filling instructions:

Replace $f(x)$ by a description of the function under consideration.
Replace $g(x)$ by a description of the derivative of $f(x)$.
Replace c by the value of $g(x)$ at x_0.

Classification:

1S and 2S are premises. 3S follows from 1S and 2S by calculus.

This example was fashioned in exact analogy to the one Kitcher gives from Newtonian mechanics. Kitcher remarks on the difference with respect to purely logical patterns:

> Whereas logicians are concerned to display all the schematic premises which are employed and to specify exactly which rules of inference are used, our example allows for the use of premises (mathematical assumptions) which do not occur as terms of the schematic argument, and it does not give a complete description of the way in which the route from [1 and 2 to 3] is to go. (Kitcher, 1981, p. 517f)

Having defined the notion of general argument pattern we now need to capture formally the notion of explanatory store over K, $E(K)$, which informally will turn out to be 'the set of derivations that makes the best tradeoff between minimizing the number of patterns of derivation employed and maximizing the number of conclusions generated' (Kitcher, 1989, p. 432).

A set of derivations is *acceptable* relative to K just in case every step in the derivations is deductively valid and each premise of each derivation belongs to K.

A *generating set* for a set of derivations Σ is a set of argument patterns Π such that each derivation in the set Σ instantiates some pattern in the generating set Π.

A generating set Π for Σ is *complete* with respect to K if and only if every derivation which is acceptable relative to K and which instantiates a pattern in Π belongs to Σ.

Informally, the determination of $E(K)$ will proceed through the following steps:

(1) Select among all possible systematizations of K only the acceptable systematizations of K, i.e. those sets of derivations that are acceptable relative to K.

(2) To each acceptable systematization selected in (1) associate the collection of the generating sets for that systematization that are complete with respect to K.

(3) For each acceptable systematization select now a basis, where a basis is an element of the collection of the generating sets for that systematization which ranks best according to the criterion of paucity of patterns.[1]

(4) Finally, rank all bases of the acceptable systematizations according to their unifying power.

---

[1] Here we leave out the criterion of *stringency* of patterns, since Kitcher does not provide a sufficiently worked out account of how the criterion is supposed to figure in determining the unifying power of a systematization. Thus it is not at all clear how to apply it in our context. We will discuss this problem in more detail below.

The quality of a generating set is inversely proportional to the number of patterns it contains.

Thus, with paucity we have the criterion needed to proceed through step 3. In order to rank the bases in step 4 we need one more thing. Define the conclusion set of a set of arguments Σ, $C(\Sigma)$, to be the set of sentences which occur as conclusions of some argument in Σ. Kitcher concludes by giving a qualitative assessment of the unifying power of a systematization Σ. Assuming that we can even give precise numerical values to the number of patterns in the basis and the size of $C(\Sigma)$ then the degree of unification of a systematization is directly proportional to the size of $C(\Sigma)$ and inversely proportional to the number of patterns.

Exemplification. In order to connect these definitions to our example, let K be the set of true sentences of calculus. Among such truths are 'the tangent line passing through the parabola $2x^2 + 3x + 1$ at point $(1, 6)$ is $(x - 1)7 = (y - 6)$' and 'the tangent line passing through the curve x^3 at point $(1, 1)$ is $(x - 1)3 = (y - 1)$'. Obviously we have an infinitude of such truths which can be unified by the argument pattern we have given. Each instance of the argument pattern gives rise to a derivation that is acceptable relative to K. Let Σ be the entire set of derivations in the calculus which have as conclusions all the truths mentioned above ($C(\Sigma)$). Then Π, i.e. the singleton set containing our general argument pattern is a generating set for Σ. Moreover, Π is complete with respect to K.

6.2 How does Kitcher's model fit concrete cases?

In order to discuss the application of Kitcher's model to concrete cases it is important to gain an insight into the sort of examples that Kitcher thinks his model can handle. The most extended discussion of such examples is given by Kitcher in his book *The Nature of Mathematical Knowledge* (1984). In Chapter 9, 'Patterns of mathematical change', one of the patterns discussed is 'systematization'. This discussion is important for our goals as it is presented by Kitcher as an exemplification of the claims contained in 'Explanatory unification' (1981). Kitcher divides 'systematizations' into two major groups: systematizations by axiomatization and systematizations by conceptualization. In the case of systematization by conceptualization he mentions the improvement obtained by Viete's algebra by treating systematically classes of equations of the same degree, say, degree 3, in contrast to the case by case analysis typical of Cardan's Ars Magna. Similarly, Lagrange—through his analysis of resolvent equations and permutation of equations—represents a huge step forward in our understanding of why the solution of certain equations can be reduced to the solution of equations of lower degree:

> In both these cases, one adopts new language which allows for the replacement of a disparate set of questions and accepted solutions with a single form of question and a single pattern of reasoning, which subsume the prior questions and solutions. Generally, systematization by conceptualization consists in modifying the language to enable statements, questions, and reasonings which were formerly treated separately to be brought together under a common formulation. The new language enables us to perceive the common thread which runs through our old problem solutions, thereby encreasing our insight into why those solutions worked. This is especially apparent in the case of Lagrange, where, antecedently, there seems to be neither rhyme nor reason to the choice of substitutions and thus a genuine explanatory problem. (Kitcher, 1984, p. 221)

In 'Explanatory unification and the causal structure of the world' (1989), we are also presented with several examples. While many of the examples there are only meant to bring home the point that there are non-causal explanations, some of them are relevant to our discussion. Let us mention in particular, Kitcher's mention of Galois theory which shows 'why equations in these classes [2, 3, and 4] permit expressions of the roots as rational function of the coefficients' (Kitcher, 1989, p. 425).

We have spent so much time reviewing these examples given by Kitcher because they allow us to say something more precise about his formal models of unification. Recall that Kitcher starts from a consistent class of statements K closed under logical consequence. Then $E(K)$ is the best systematization of K. But here immediately a problem arises. For while this approach might fit some simple cases of axiomatization it does not, as it stands, apply to all cases of axiomatization and, we claim, to most cases of systematization by conceptualization. The problem is this. Let us envisage first a case in which Kitcher's approach would apply well. Suppose we have an alternative axiomatization to Euclid, given by Euclid*, which systematizes exactly the same body K of sentences and uses different axioms (already in K). Then we could compare the two axiomatizations with the tools given by Kitcher. We skip here over a number of other problems such as the fact that it is not clear what role the argument patterns are playing here as both systematizations will require argument patterns of arbitrary complexity, unless we reduce the notion of argument pattern to that of being a logical argument that only makes use of the axioms of either Euclid's axiomatization or of Euclid*'s axiomatization.

Be that as it may, any axiomatization that needs axioms formulated in a richer language than those of the class K of sentences being systematized will end up being incomparable to an axiomatization that only uses sentences coming from K. This is because Kitcher's model only deals with systematizations of K and the arguments in K must have as premises and conclusions statements from K. The problem we would like to point out can be explained as follows. Kitcher is very explicit about the fact that the new unification provided by Lagrange (or Galois) in the theory of equations must use a richer language, new concepts, and new properties of these concepts. But where earlier on we had to account for a set K of sentences formulated in language $L(K)$ now we have a set K^* of sentences formulated in language $L(K^*)$. We would still like to say that we have a better explanation of K offered by the new systematization. But Kitcher's model seems to force us to compare only systematizations of K among each other and such a systematization has to appeal to sentences from K only. Thus, in order to make Kitcher's system more adequate to the actual situation we face when making evaluative judgements of explanatoriness we modify his model in such a way that a systematization of K can appeal to a class of sentences larger than K. Indeed, such a move also received textual support from Kitcher himself, who seems to have recognized such need in his 1989 article in Section 4.4 (p. 435).

The new definition would thus require the following modifications:

A *systematization* of K is any set of arguments which derive some sentences in K from other sentences of K^*, where K^* is a consistent superset of K (possibly identical to K) and where K^* can be rationally accepted by those who accept K.

A set of derivations is *acceptable* relative to K just in case the conclusion of each derivation belongs to K and every step in each derivation is deductively valid and each premise of each derivation belongs to K^*, where K^* is a consistent superset of K (possibly identical to K) and where K^* can be rationally accepted by those who accept K.

The advantage of this modification, to reiterate the point, is that it allows us to exploit Kitcher's machinery in a variety of situations which are very common in mathematics and in science.

6.3 A test case from real algebraic geometry

In his monograph on partially ordered rings and semi-algebraic geometry Gregory W. Brumfiel contrasts different methods for proving theorems about real closed fields. One of them relies on a decision procedure for a particular axiomatization of the theory of real closed fields. By this method one can find elementary proofs of sentences formulated in the language of that theory—at least in principle, since, as Brumfiel remarks, '[i]t certainly might be very tedious, if not physically impossible, to work out this elementary proof' (Brumfiel, 1979, p. 166).

Another proof method consists in using a so-called transfer principle which allows to infer the truth of a sentence for all real closed fields from its being true in *one* real closed field, like the real numbers. Despite the fact that the transfer principle is a very useful proof method, Brumfiel does not make any use of it, and he is very clear about this.

> In this book we absolutely and unequivocally refuse to give proofs of this second type. Every result is proved uniformly for all real closed ground fields. Our philosophical objection to transcendental proofs is that they may logically *prove* a result but they do not *explain* it, except for the special case of real numbers. (Brumfiel, 1979, p. 166)

Brumfiel prefers a third proof method which aims at giving non-transcendental proofs of purely algebraic results. This does not mean that he restricts himself to just elementary methods; he does use stronger tools, but it is crucial that they apply uniformly to all real closed fields. It is also clear from the context that Brumfiel does not consider proofs obtained from applying the decision procedure for *RCF* as explanatory.[2] And in fact they are almost never carried out in practice because their length makes them unwieldy and unilluminating.

To illustrate Brumfiel's point and subsequently confront Kitcher's theory of explanation with it, we would like to give a short exposition of a theorem, which is also mentioned by Brumfiel himself (p. 207), and various (sketches of) proof types corresponding to the classification above.

6.3.1 Some concepts from semi-algebraic geometry

Let's start with the definition of 'real closed field': a field which admits a unique ordering, such that every positive element has a square root and every

---

[2] The (elementary) proofs one can construct on the basis of the decision procedure prove their conclusion uniformly for all real closed fields because the decision procedure is carried out independently of the ground field. Despite their uniformity, however, Brumfiel does not take this kind of proof as the optimal model which should guide mathematical research—on the contrary. First, as already mentioned these proofs are not always feasible; due to their size it may in certain cases not even be physically possible to construct them. Second, Brumfiel as a rule prefers different uniform (in general non-elementary) proof techniques not only for studying the rich non-elementary theory of real closed fields but also for dealing with elementary sentences (i.e. first-order sentences in the language of ordered fields), even in cases where one could, in principle, find an elementary proof, i.e. 'even if a statement turns out to be equivalent to an elementary statement, it may be unnatural to dwell on this fact, and even worse to be forced to depend upon it' (Brumfiel, 1979, p. 166).

polynomial of odd degree has a root. More formally, a real closed field is defined by the following axioms (the complete list can be found in the Appendix to this chapter).

(i) Axioms for fields
(ii) Order axioms
(iii) $\forall x \exists y (x = y^2 \vee -x = y^2)$
(iv) For each natural number n, the axiom
$\forall x_0 \forall x_1 \ldots \forall x_{2n} \exists y (x_0 + x_1 \cdot y + x_2 \cdot y^2 + \ldots + x_{2n} \cdot y^{2n} + y^{2n+1} = 0)$

The theory of real closed fields, RCF, is the deductive closure of the above axioms. A (first-order) sentence formulated in the language of RCF is called an 'elementary sentence'. An important result about RCF is that there is a decision procedure due to Tarski and Seidenberg (Tarski, 1951; Seidenberg, 1954), i.e. given any elemantary sentence φ the algorithm outputs 1, if RCF proves φ and 0, if RCF proves $\neg \varphi$. Also, this decision procedure works uniformly for every sentence φ in the language of RCF.

Examples of real closed fields include the set of real numbers **R** and the set of real algebraic numbers $\mathbf{R}_{alg}$, i.e. the set of all roots of the nonzero polynomials with rational coefficients. The latter field is *not complete*, i.e. not every Cauchy sequence converges. The set **C** of complex numbers is not a real closed field as it does not admit of an ordering. There is an abundance of real closed fields even just within **R**. Artin and Schreier showed that there are uncountably many pairwise non-isomorphic real closed subfields of **R** whose algebraic closure is isomorphic to **C** (cf. Brumfiel 1979, p. 131).

For a real closed field R we denote the ring of polynomials in n variables with coefficients from R by $R[X_1, \ldots X_n]$.

A subset of R^n, the affine n-space over R, is called *semi-algebraic* if it belongs to the smallest family of subsets of R^n containing all sets of the form

$$\{x \in R^n \mid f(x) > 0\}, \quad f(x) \in R[X_1, \ldots X_n]$$

and which is closed under taking finite intersections, finite unions, and complements.

Semi-algebraic subsets of R (i.e. one-dimensional) are exactly the finite unions of points and open intervals (bounded or unbounded).

Let $A \subset R^m$ and $B \subset R^n$ be two semi-algebraic sets. A mapping $f : A \to B$ is semi-algebraic if its graph is semi-algebraic in R^{m+n}. It is clear that polynomials (on semi-algebraic domains) are semi-algebraic mappings.

R^n can be endowed with the *Euclidean topology* coming from the ordering on R. Let $x = (x_1, \ldots, x_n) \in R^n, r \in R, r > 0$. We set

$$\|x\| = \sqrt{x_1^2 + \cdots + x_n^2}$$
$$B_n(x, r) = \{y \in R^n \mid \|y - x\| < r\} \quad \text{(open ball)}$$

The *Euclidean topology* on R^n is the topology for which open balls form a basis of open subsets. Complements of open sets are closed. Polynomials are continuous with respect to the Euclidean topology.

6.3.2 A simple theorem and a variety of proofs (and systematizations)

Now we are in a position to state a simple theorem and compare different possible approaches to its proof or to ways of systematizing its instances.

Theorem. A polynomial $f(x_1, \ldots, x_n)$ assumes a maximum value on any bounded closed semi-algebraic set $S \subset R^n$.

(I) Within *RCF* the result cannot be proved as stated for we cannot quantify over polynomials and semi-algebraic sets. The best we could do is to prove its instances. For instance, *RCF* proves that x^3 has a maximum on $[0, 2]$. We could, in principle, arrive at this proof by running the Tarski–Seidenberg decision procedure and getting 1 as output. From there it would be a mechanical task to provide the explicit proof in *RCF*. And every single instance of the theorem could be obtained in this way. In other words, if we let $P \subset RCF$ be the set containing all the instances of the theorem in *RCF*, then we could thus arrive at a systematization of the whole set P on the basis of the Tarski–Seidenberg decision procedure.

(II) The following proof strategy draws on transcendental methods. This approach is based not just on the Tarski–Seidenberg decision procedure but rather on one of its consequences, namely that *RCF* is a *complete* theory. This gives rise to the following *transfer principle*. If a sentence φ in the language of *RCF* can be shown to hold in *one particular* real closed field, say, the real numbers **R**, then it must be true for *any* real closed field because of the completeness of *RCF*.

Now, the theorem can of course be established for **R** relying on, among other things, the Bolzano–Weierstrass theorem (every bounded sequence has a convergent subsequence) and the least upper bound principle.[3] These basic properties of **R** don't hold in general for real closed fields. For instance, they both fail for $\mathbf{R}_{alg}$. However, once the theorem has been proved for **R**, by whatever means, we can conclude by appeal to the transfer principle that all its instances, i.e. all sentences in P, hold for real closed fields in general.

---

[3] Cf. Courant, 1971, vol. I p. 60f, vol. II p. 86.

(III) Another way of establishing the theorem relies on purely algebraic means exploiting the fact that if $A \subset R^n$ is a closed and bounded semi-algebraic set and $g: A \to R^p$ a continuous semi-algebraic mapping, then $g(A)$ is a closed and bounded semi-algebraic set.[4]

Since a polynomial $f(x_1, \ldots, x_n)$ is a continuous semi-algebraic mapping and we assume a closed and bounded semi-algebraic set $S \subset R^n$ to be given, it follows that $f(S)$ is a closed and bounded semi-algebraic set. But since $f(S) \subset R$, it is a finite union of points and of closed and bounded intervals. And so $f(S)$ has a maximum.

6.4 Assessment of Kitcher's model

The previous section exemplified different approaches to systematizing knowledge about real closed fields. More precisely, in the following we will be concerned with systematizations of the theory RCF, which, to recall, is the consistent and deductively closed, in fact complete, set of elementary sentences true in any real closed field (or, equivalently, following from the axioms of RCF).

Before entering into the discussion of how Kitcher's account would rank different systematizations of RCF according to their unifying power let's pause for a moment to address a worry one might have, at the very outset, concerning the choice of RCF as the body of statements to be systematized. To be sure, mathematicians working in semi-algebraic geometry are as a rule *not only* interested in (systematizing) *elementary* sentences but also sentences that go beyond RCF by, for instance, quantifying over functions and sets, or sentences coming from a wider framework in which RCF is embedded like real analysis or category theory (we'll return to this below). The question might thus be raised how faithful to mathematical practice our focus on RCF indeed is.[5] However, this worry can easily be dispelled.

First of all, the choice of some non-elementary context for the study of real closed fields—be it a more expressive language or certain more encompassing mathematical theories, etc.—doesn't make the elementary part of semi-algebraic geometry disappear or irrelevant. Whatever the context, RCF still forms a (precisely definable) subset of K, the totality of accepted sentences in the respective context. Hence any systematization of K has to

---

[4] Cf. Brumfiel, 1979, p. 207; or Bochnak *et al.*, 1998, p. 40f.
[5] This point has been urged particularly by Jeremy Avigad.

cover also the sentences in *RCF*. Moreover, the elementary (sub)theory is by no means rendered peripheral. Indeed, the very notion of *semi-algebraic set*, which is fundamental to semi-algebraic geometry, is characterized in terms of first-order formulas. It is a salient fact, an immediate consequence of the Tarski–Seidenberg theorem, that the semi-algebraic sets coincide with the sets which are *definable* by formulas in the language of *RCF*. In other words, any construction out of semi-algebraic sets that can be expressed in terms of an elementary sentence yields again a semi-algebraic set. This fact is of major importance in the light of its 'consequences which are hard to obtain if one uses the definition of a semialgebraic set directly' (Andradas *et al.*, 1996, p. 9). Even G. W. Brumfiel, who decidedly rejects the Tarski–Seidenberg *transfer principle* as a proof method, agrees that *this* application of the Tarski–Seidenberg theorem is 'a very efficient tool' (Brumfiel, 1979, p. 165) and he uses it himself occasionally in later parts of his book.[6]

It is clear that elementary formulas demand special attention of anybody who relies on the Tarski–Seidenberg transfer principle as an essential tool in the study of semi-algebraic geometry (cf. Bochnak *et al.*, 1998; Andradas *et al.*, 1996). Yet, they are not just 'interesting' for the purely methodological reason that the transfer principle is applicable only to elementary formulas. Even Brumfiel, who is not in any way restricted by such methodological considerations, is well aware of the fact that many of the theorems he proves (by his preferred methods) can be expressed by elementary sentences.[7] In short, the theory *RCF* is anything but trifling.

Let us now return to the task of ranking different systematizations of *RCF* and the assessment of how Kitcher's model fares in this respect. Our starting point, the set to be systematized, is the theory *RCF*. To be in line with Kitcher's terminology let us call this set K. As was pointed out above, systematizations of a (consistent and deductively closed) set of sentences S may have to go beyond S and draw on premises from S^*, a consistent superset of S. That's what necessitated a slight modification of Kitcher's original model

---

[6] Concerning the use of the Tarski–Seidenberg theorem to show that a set defined in terms of semi-algebraic sets by an elementary sentence is semi-algebraic Brumfiel emphasizes the following. '[T]his type of application actually provides a *proof* that the asserted set is semi-algebraic, simultaneously for all real closed fields, in fact an elementary proof. The reason is, any single elementary sentence is just a special case of the theorem' (Brumfiel, 1979, p. 165). In contrast, proofs based on the *transfer principle* lack precisely this kind of uniformity. However, as pointed out above already Brumfiel is *not* an advocate of developing real algebraic geometry exclusively by elementary methods let alone by relying on the Tarski–Seidenberg decision procedure.

[7] 'We admit that many of our proofs are long and could be replaced by the single phrase "Tarski–Seidenberg and true for real numbers". However, we feel the effort is worthwhile' (Brumfiel, 1979, p. 166).

and the different systematizations connected with **(I)**, **(II)**, and **(III)** are cases in point.

Systematization **(I)** uses some metatheoretical machinery in addition to K ($= RCF$). The decision algorithm for sentences in K is not itself a theorem of that theory but a statement in its metatheory. Let K_I^* be K together with some metatheory of K. We leave it open to some extent how much metatheory K_I^* has to contain—certainly at least the decision algorithm but perhaps more, like induction, to actually establish the main properties of the algorithm. In **(I)** we indicated how the set P, i.e. the set of all instances in K of the theorem, could be construed as the conclusion set of a certain systematization based on the Tarski–Seidenberg decision algorithm. This approach can be generalized to apply to *all* of K, since the decision algorithm works for all elementary sentences. Somewhat streamlined we thus have the following systematization of K relative to K_I^* which is given by a simple argument pattern:

(1) $F(\varphi) = x$
(2) ψ

Here 'φ', 'x', and 'ψ' are dummy letters, whereas F denotes (some fixed specification of) a decision algorithm for K.

Filling instructions:

> Replace 'φ' by a sentence s in the language of RCF.
> Replace 'x' by the value, 0 or 1, of F at s.
> Replace 'ψ' by s, if $F(s) = 1$ and replace 'ψ' by $\neg s$, if $F(s) = 0$

Classification:

> The sentence (1) is a premise.
> The sentence (2) follows from (1) by metatheory, i.e. by using facts about the functioning of F.

The proof of the theorem along the lines of **(II)** employs transcendental methods, like the use of completeness and compactness, which hold for **R** but not for real closed fields in general. Beyond that the transfer principle is used to argue that the instances of the theorem hold in *every* real closed field. Hence in this case the superset K_{II}^* includes some part of the theory of the classical real numbers and a formulation of the Tarski–Seidenberg transfer principle. Because this requires the use of model-theoretic concepts and methods in addition to syntactical ones in order to talk about truth in certain real closed fields, notably the real numbers, the metatheoretic part of K_{II}^* is more encompassing than the one of K_I^*. Again, however, we don't need to specify the necessary metatheory precisely but can leave its concrete delimitation somewhat open.

The proof in **(III)** belongs to a systematization of K relative to a wider framework, K^*_{III}, of real algebraic geometry. Brumfiel stresses that K^*_{III} is not confined to elementary methods. 'In fact, we use Dedekind cuts, total orders, and signed places[8] repeatedly. The point is, in the form we use these concepts they apply uniformly to all real closed fields. One advantage to developing such techniques is precisely that one is not tied down to "elementary sentences"' (Brumfiel, 1979, p. 166).

Now we can ask what the best systematization of K is. If we can find a systematization that makes use of only one argument pattern to generate all of K, then any other systematization which uses a greater number of patterns is inferior. It turns out, then, that the best systematization of K is the one provided by the Tarski–Seidenberg decision procedure for RCF, i.e. the one exemplified by **(I)**. The uniformity of the procedure shows that we have in principle a single argument pattern which we can use to generate all of K. Comparing this situation with those in which we try to prove arbitrary elementary sentences of K by means of proofs such as those given in **(II)** and **(III)** makes it obvious that we can't get by with just one pattern of argument. Consider proofs of type **(III)**. The specific argument used to determine the validity of the proposition asserting that a polynomial assumes a maximum on any arbitrary closed and bounded semi-algebraic set would be useless for deriving, say, the algebraic form of Brouwer's Fixed Point Theorem.[9]

Similarly in the case of proofs of type **(II)**. For in this case we need to determine the truth of φ under consideration in **R**, or some other real closed field. And only after this task has been accomplished can we appeal to the completeness of RCF. But the first task, the determination of the truth value of φ in some real closed field, is not a uniform process; rather we need, as pointed out with respect to case **(III)**, different argument patterns for different classes of φ. In short, whereas with respect to the specific problem concerning maxima of polynomials all three systematizations exhibit a certain uniformity in the treatment of sentences that are instances of this problem, only the strategy that relies on the Tarski–Seidenberg decision procedure can be extended to

---

[8] Cf. Brumfiel, 1979, p. 144f: Let Δ be a totally ordered field. We adjoin symbols ∞ and $-\infty$ to Δ and extend the operations of addition, subtraction, multiplication, and division between ∞, $-\infty$ and elements $a \in \Delta$ (e.g. $\infty + \infty = \infty$, $\frac{a}{\infty} = 0$, etc. Some expressions, like $0 \cdot \infty$, remain undefined). If K is a field by a *signed place*, with values in Δ, we mean a function $f : K \to \Delta \cup \{\infty, -\infty\}$, such that $f(x+y) = f(x) + f(y), f(xy) = f(x)f(y)$, and $f(1) = 1$, whenever the terms are defined.

[9] Let S be a closed, bounded, convex, semi-algebraic set in R^n, for some real closed field R. Then every continuous rational function mapping S to itself has a fixed point in S.

all of K in such a way that a single argument pattern suffices for the generation of all of K. Hence Kitcher's model of explanation would declare the set of all instantiations of this single argument pattern as the *explanatory store over K*, i.e. the set of—explanatory—arguments which best unifies K. This result clearly conflicts with mathematical practice since Kitcher's model ends up positing as the best systematization one which in practice does not enjoy the properties of explanatoriness that Kitcher's model would seem to bestow upon it. Even worse, not only do arguments in this 'explanatory store' in general fail to be considered as paradigm explanations, they are hardly ever used at all by working mathematicians because of the limited feasibility of the decision algorithm.

It remains to compare the systematizations of K relative to K_{II}^* and K_{III}^*. As we have seen, Brumfiel emphatically rejects as nonexplanatory proofs obtained along the lines of **(II)** and strongly urges proofs of type **(III)** as the ones which not only *prove* a result but also *explain* it. It is interesting to note that according to Brumfiel one of the salient virtues of systematizations within K_{III}^* in contrast to those within K_{II}^* is that the argument patterns apply *uniformly* to all real closed fields. This notion of uniformity can also be seen as a type of unification. Hence Kitcher's theory appears particularly promising and suitable for adjudicating, or making pertinent contributions to the assessment of, this dispute (and related ones[10]) over the choice of proof methods. So let's see how Kitcher fares.

The first thing to observe is that both systematizations of K, the one relative to K_{II}^* as well as the one relative to K_{III}^* seem to require an infinite number of argument patterns. Because the patterns required in each theory for proving respectively, say, Brouwer's Fixed Point Theorem, the solution to Euler's problem,[11] or that x^3 has a maximum on $[0, 2]$ cannot be reduced to a single pattern, unless we take the notion of argument pattern in a theory so broadly as to identify it with any derivation from the axioms of that theory. But such an identification would not square with either one of the systematizations under

---

[10] Similar methodological issues have previously been raised in algebraic geometry. Seidenberg quotes A. Weil's critical remark concerning 'the taboo against "nonalgebraic", i.e., topological and function-theoretic methods' (Seidenberg, 1958, p. 686). On the other hand, Brumfiel points out that 'there is already a respectable tradition in this century of finding non-transcendental proofs of purely algebraic results concerning algebraically closed fields. We think real closed fields deserve (at least) equal time and effort' (Brumfiel, 1979, p. 166).

Interestingly, Seidenberg's comments on Weil and Lefschetz also involve a decision method, again due to Tarski, and a corresponding transfer principle for algebraically closed fields (of a given characteristic p).

[11] How many points at a distance at least r from one another can there be on the surface of a sphere of radius r, twelve or thirteen?

consideration. If the number of patterns in both accounts is infinite, and given that both accounts generate the set K, i.e. the respective conclusion sets are identical, we have to conclude that Kitcher's model is unable to differentiate between the two accounts. In particular, Kitcher's model cannot even start to explain or reflect in any way Brumfiel's rather pronounced methodological position *vis-à-vis* the use of transcendental methods in real algebraic geometry.

However, at this point it might be objected that we jumped to the conclusion and the negative diagnosis is incorrect. And one might point out, on behalf of Kitcher, the fact that even in the case of infinite sets of argument patterns Kitcher's model can still make a decision, provided that one set turns out to be a subset of the other—being smaller in terms of the subset relation means to be more explanatory. Kitcher explicitly states this possibility of ranking systematizations Σ, Σ' of K as one of two important corollaries to his account.[12]

> If the conclusion sets $C(\Sigma)$ and $C(\Sigma')$ coincide yet the basis of Σ' is a proper subset of the basis of Σ then $\Sigma \neq E(K)$. (cf. Kitcher, 1981, p. 522)

And the defence of Kitcher could go on: upon closer examination it might well turn out that $E_{III}(K)$, the best systematization of K relative to K^*_{III}, is (or can be construed as) a subset of $E_{II}(K)$, the best systematization of K relative to K^*_{II}, and the same holds with respect to their bases. In this case $E_{III}(K)$ would be ranked higher than $E_{II}(K)$, which would nicely account for Brumfiel's position after all. Of course, in order to substantiate this beyond mere speculation $E_{II}(K)$ and $E_{III}(K)$ and, more importantly, their respective bases B_{II} and B_{III} would have to be precisely specified. Yet, in any case, nothing can rule out *in principle* that Kitcher's model thus succeeds in ranking the systematizations under consideration after all.

How good a defence of Kitcher is this? It is certainly true that in the particular situations which are covered by the mentioned corollaries Kitcher's model can also compare infinite sets of argument patterns and/or infinite conclusion sets. However, as it will turn out this is of no help in our case. In order to rank $E_{II}(K)$ and $E_{III}(K)$ according to the corollary either $B_{II} \subset B_{III}$ or $B_{III} \subset B_{II}$ has to obtain. But neither one can hold as we shall see below. (And this can even be established *a priori*, regardless of how B_{II} and B_{III} are composed in detail.) We shall consider the two cases in turn.[13]

---

[12] Kitcher does not state these corollaries explicitly in terms of *infinite* sets but they are certainly also applicable in cases where infinite sets of argument patterns and/or infinite conclusion sets are involved.

[13] In the following we skip over some further complications. In particular, a comparison of B_{II} and B_{III} (and of the corresponding systematizations) has to take place in a broader framework K^* which encompasses both K^*_{II} and K^*_{III}. On the other hand one still needs to be able to distinguish within K^* between systematizations of K with respect to K^*_{II} and systematizations of K with respect to K^*_{III}. It shouldn't be very difficult, if tedious, to work out the details.

First consider the case $B_{II} \subset B_{III}$. Argument patterns in B_{II} incorporate, in general, transcendental methods for **R** and the Tarski–Seidenberg transfer principle. Both of them are completely absent from patterns in B_{III} since, to recall, Brumfiel 'absolutely and unequivocally' refuses to use them in his proofs.[14] Hence B_{II} cannot form a subset of B_{III} (even though B_{II} and B_{III} need not be *disjoint* as *some* argument patterns in B_{II} might be purely algebraic). In other words, we have to exclude this case of relating B_{II} and B_{III} as a possibility.

However, it is instructive to explore this a little further and consider relaxing the subset relation in favor of some kind of 'embedding', thus generalizing Kitcher's corollary. Because a defender of Kitcher's model might suggest that in cases where no subset relation between two bases B, B' for systematizations Σ, Σ' (respectively) obtains, a ranking of Σ and Σ' could still be achieved if, say, the argument patterns in B' could be brought into a 'correspondence' with the argument patterns in a proper subset $S \subset B$ such that each argument pattern in B' has a corresponding, in a certain sense equivalent, analog in S under some 'translation'. Such an embedding of B' into B would provide grounds for ranking Σ' higher than Σ.

Let's take up this suggestion and assess its contribution to the ranking problem. While it is very difficult to give general precise definitions of the notions of 'embedding', 'translation', 'equivalent', etc., which are involved here, an obvious minimal constraint on the embedding relation certainly is that conclusion sets have to be preserved. That is, if B' is embedded into B such that all the 'translations' of argument patterns from B' form a set $S \subset B$, then B' and S must give rise to the same conclusion set. Preserving conclusion sets is a necessary adequacy condition on an embedding; apart from that it can be left open which other features of (sets of) argument patterns have to be preserved by an embedding (and in what way). For our purposes we can thus leave the notion of 'embedding' for the most part at an informal, intuitive level; it will turn out that even without more detailed specification of the conditions on embeddings we'll have a workable account. Now we can generalize Kitcher's corollary.[15]

> Given two systematizations Σ, Σ' of K. If $C(\Sigma) = C(\Sigma')$ and the basis of Σ' can be embedded into a proper subset of the basis of Σ then $\Sigma \neq E(K)$.

---

[14] It is worth noting that Brumfiel categorically rejects *all* uses of transcendental methods pertaining to the real numbers not just when they are employed in connection with the transfer principle. '[O]ne of our central themes is that the real numbers are totally irrelevant in algebraic topology, so it would not do to rely on them at some point in our chain of reasoning' (Brumfiel, 1979, p. 166).

[15] The corollary in its original formulation results then from requiring that *all* features of argument patterns have to be preserved by the translation of patterns, i.e. by taking 'translation' as the identity function and 'embedding' thus as the subset relation.

Now, can there be an embedding of B_{II} into B_{III}? Without further information on the argument patterns in the two bases we are in no position to devise any specific translation or embedding. Yet, that certainly doesn't rule out the existence of such an embedding. Above, in **(II)** and **(III)** we sketched two different types of proof of the statement that some given polynomial assumes a maximum on a given closed and bounded semi-algebraic set. The respective argument patterns of which these proofs are instantiations might be seen as correlated by the kind of translation we are looking for. And if this kind of correlation could be extended to supply for each pattern in B_{II} a corresponding one in B_{III} (such that—at least—conclusion sets are preserved), then this would indeed yield an embedding of B_{II} into B_{III}. However, it is crucial to notice that while at this abstract level an embedding might in principle be possible along the indicated lines, an embedding of B_{II} into a *proper subset* of B_{III} is definitely excluded. The reason is this. Assume that B_{II} embeds into a proper subset $S \subset B_{III}$. Let Σ_S be the set of all instantiations of the argument patterns in S. (Σ_S is a subset of $E_{III}(K)$.) Since B_{II} is the basis for $E_{II}(K)$ and the embedding preserves conclusion sets, we have $C(E_{III}(K)) = K = C(\Sigma_S)$. Hence all of K is systematized by Σ_S and, moreover, Σ_S is generated by a *proper* subset, S, of B_{III}. This contradicts the fact that $E_{III}(K)$ is the *best* systematization of K (relative to K_{III}^*), i.e. the systematization whose basis ranks best with respect to the criterion of paucity of patterns.[16] There must be redundancies in B_{III} if the same conclusion set can be gotten from a proper subset of it. In other words, B_{III} must be inflated and hence could not have the greatest unifying power among all bases and, in turn, $E_{III}(K)$ would not be the set of arguments which best unifies K (relative to K_{III}^*).

Turning now to the case $B_{III} \subset B_{II}$ the reasoning is very similar. In principle, no methodological constraint precludes the possibility that B_{III} can be embedded into B_{II} (or even, in this case, that B_{III} is a subset of B_{II}). As Brumfiel himself indicates one can indeed find a certain correspondence between his own, algebraic, approach and the approach that uses transcendental methods. 'We admit that many of our proofs are long and could be replaced by by the single phrase "Tarski–Seidenberg [transfer principle] and true for the real numbers"' (Brumfiel, 1979, p. 166). This might be taken, in our context, as a hint towards the possibility of constructing a correlation between patterns

---

[16] In case S is not only a generating set for Σ_S but also the basis for it, then the assumption that S is a proper subset of B_{III} leads directly to a contradiction by applying Kitcher's corollary (in its original formulation). Because in this case we have two systematizations, $E_{III}(K)$, Σ_S, of K such that $C(E_{III}(K)) = C(\Sigma_S)$ and the basis for Σ_S is a proper subset of the basis for $E_{III}(K)$. Hence we have $E_{III}(K) \neq E_{III}(K)$.

in B_{III} and patterns in B_{II}. But even so, by the same token as before B_{III} *cannot*, on pain of contradiction, be embedded into a *proper subset* of B_{II} (nor can it actually *be* a proper subset of B_{II}). Because this would contradict the status of $E_{II}(K)$ as the *best* systematization of K (relative to K_{II}^*). The argument is exactly the same as the one in the preceding paragraph (*mutatis mutandis*).

The upshot, then, is this. The two rival systematizations $E_{II}(K)$ and $E_{III}(K)$ turn out to be *incomparable* within Kitcher's framework. Neither his corollary nor our generalization of it can discriminate between them, i.e. rank them according to their unifying or explanatory power. Faced with a genuine issue that arose within mathematical practice Kitcher's model remains silent. It fails to account for and possibly even confirm Brumfiel's position. Yet, on the other hand, it doesn't succeed either in revealing, as the case might be, Brumfiel's underlying intuitions as wrongheaded by ranking the other systematization as in fact more explanatory after all. If we had a well-established theory of mathematical explanation it would also possess critical or corrective potential *vis-à-vis* mathematical practice. Unfortunately, at this point Kitcher's model is far from well-confirmed and uncontroversial. Moreover, it doesn't even reach a decision concerning the given issue. There is a certain irony in this since Brumfiel champions a kind of unification of real algebraic geometry by insisting on proofs that exhibit a 'natural' explanatory uniformity. Yet, despite its focus on unification Kitcher's account of explanation apparently does not have the resources to provide insight into the controversy over the 'right' proof methods or at least enhance our understanding of Brumfiel's motivations. One of the reasons for Kitcher's failure may lie in the fact that his account, although much more sophisticated than Friedman's model, still shares the latter's basic intuition, namely that unifying and explanatory power can be accounted for on the basis of *quantitative* comparisons alone.[17] However, in the controversy over the use of transcendental methods in real algebraic geometry the point at issue concerns *qualitative* differences in the proof methods. The features of Brumfiel's unification which he regards as explanatory escape Kitcher's model. So we have to conclude that even under the assumption that an account of explanation as unification is, in principle, on the right track, Kitcher's model doesn't tell the whole story yet. In general there is more to explanation than

---

[17] At some point Kitcher does consider a modification of his account which goes beyond mere quantitative comparisons, 'so that, instead of merely counting the number of different patterns in a basis, we pay attention to similarities among them' (Kitcher, 1981, p. 521). And he mentions the existence of a common core pattern as another criterion that determines the best systematization. However, he makes no attempt to precisely specify this criterion nor the ways in which it is to be balanced against the other criteria. So it is not clear at all how it should be implemented.

unification in Kitcher's sense, a more fine-grained analysis of different types of unification seems to be needed.

6.5 Stringency and spurious unification

Let's now return to $E_1(K)$, the systematization of K whose basis contains only a single argument pattern and which is thus ranked best by Kitcher's model. We want to discuss (and dispel) a worry one might have concerning this ranking and the resulting criticism of Kitcher's model. In fact, this kind of worry or objection which invokes the notion of *stringency* has been voiced on several occasions when this paper was presented.

So far in our exposition and discussion of Kitcher's theory the stringency criterion has played no role. What does it consist in? Kitcher proposes 'that scientists are interested in *stringent* patterns of argument, patterns which contain some non-logical expressions and which are fairly similar in terms of logical structure' (Kitcher, 1981, p. 518). The stringency of an argument pattern is determined by two different constraints concerning two respects in which argument patterns may be similar.

> Derivations may be similar either in terms of their logical structure or in terms of the nonlogical vocabulary they employ at corresponding places. The notion of a general argument pattern allows us to express the idea that derivations similar in either of these ways have a common pattern. [...] To capture the notion that one pair of arguments is more similar than another pair, we need to recognize the fact that general argument patterns can demand more or less of their instantiations. If a pattern sets conditions on instantiations that are more difficult to satisfy than those set by another pattern, then I shall say that the former pattern is more *stringent* than the latter. The stringency of an argument pattern is determined in part by the classification, which identifies a logical structure that instantiations must exhibit, and in part by the nature of the schematic sentences and the filling instructions, which jointly demand that instantiations should have common nonlogical vocabulary at certain places. (Kitcher, 1989, p. 433)

Apart from these rather general remarks Kitcher does not give us much to go on. He just sketches the main idea 'without trying to provide an exact analysis of the notion of stringency' (Kitcher, 1981, p. 518; cf. also Kitcher, 1989, p. 433). In spite of this vagueness stringency is supposed to be one of the factors which determine the unifying power of a basis, since that unifying power is supposed to vary directly with the stringency of the patterns in the basis (cf. Kitcher, 1981, p. 520; Kitcher, 1989, p. 435). However, in the absence of an exact definition of stringency and a precise, workable account of how stringency should be weighted against the other criteria, size of conclusion set and size of the basis, it is impossible to incorporate the stringency criterion into the determination of the best systematization of a given K. At one point Kitcher even suggests that the notion of stringency could be clarified through the notions of explanation and unification (cf. Kitcher, 1981, p. 519), thus courting circularity. Due to all of these difficulties the stringency criterion could not be employed in our discussion so far.

The only context in which Kitcher actually puts to use the concept of stringency and formulates a workable requirement in terms of it is his discussion of 'spurious unifications' which threaten to trivialize his account. The problem is posed by trivial yet maximally encompassing systematizations of our beliefs on the basis of single patterns like 'from $\alpha \& B$ infer α' or, even more simple, 'from α infer α', where 'α' is to be replaced by any sentence we accept. In order to exclude such patterns Kitcher invokes the criterion of stringency. As he points out, the mentioned systematizations are indeed highly successful according to the criteria of generating a large number of beliefs on the basis of a small number of patterns but they fail badly in terms of stringency. That is, 'both of the above argument patterns are very lax in allowing any vocabulary whatever to appear in the place of α' (Kitcher, 1981, p. 527). Hence they are (intuitively) non-stringent and should be excluded.

Now, on the face of it, it might seem that a very similar case could be made out against our systematization $E_I(K)$. To recall, $E_I(K)$ unifies the set K by using the following single argument pattern, '*TSP*' for short, which relies on the Tarski–Seidenberg decision algorithm.

(1) $F(\varphi) = x$
(2) ψ

Here 'φ', 'x', and 'ψ' are dummy letters, whereas F denotes (some fixed specification of) a decision algorithm for K.

Filling instructions:

> Replace 'φ' by a sentence s in the language of *RCF*.
> Replace 'x' by the value, 0 or 1, of F at s.
> Replace 'ψ' by s, if $F(s) = 1$ and replace 'ψ' by $\neg s$, if $F(s) = 0$

Classification:

> The sentence (1) is a premise.
> The sentence (2) follows from (1) by metatheory, i.e. by using facts about the functioning of F.

$E_I(K)$ comes out as the best systematization of K on Kitcher's model, as pointed out above, contrary to how it is considered from the perspective of actual mathematical practice. This reveals a serious defect of Kitcher's model.

Against this result one may object as follows. The account of Kitcher's model on which our ranking of $E_I(K)$ is based leaves out completely the stringency criterion. Yet, the argument pattern *TSP* has exactly the same flaw as the trivial argument patterns considered—and rejected—above. It allows any vocabulary whatever (from the language of ordered fields) to appear in the place of 'φ' and 'ψ'. Hence it, too, should be rejected as non-stringent. In other words, taking into account the stringency criterion not only blocks $E_I(K)$ from being ranked higher than $E_{II}(K)$ and $E_{III}(K)$ but it even excludes $E_I(K)$ altogether from the 'competition' as an inadmissible, *spurious unification*. Hence it poses no threat to Kitcher's model after all.

To see how much weight this objection indeed carries we have to evaluate it not at the level of our intuitive concept of stringency but in light of the explicit requirement Kitcher formulates to screen out spurious unifications.

> If the filling instructions associated with a pattern *P* could be replaced by different filling instructions, allowing for the substitution of a class of expressions of the same syntactic category, to yield a pattern *P′* and if *P′* would allow the derivation of any sentence, then the unification achieved by *P* is spurious. (Kitcher, 1981, p. 527f)

This requirement in fact identifies as spurious the previously considered trivial argument patterns since the filling instruction ''α' is to be replaced by any sentence we accept' can be generalized to ''α' is to be replaced by any sentence' thus allowing the derivation of *any* sentence. Kitcher further illustrates and motivates his new requirement with respect to the following argument pattern which might be used by 'a group of religious fanatics' to explain and unify their beliefs about the world (cf. Kitcher, 1981, p. 528).

(1) God wants it to be the case that α.
(2) What God wants to be the case is the case.
(3) α

> *Filling instruction:* 'α' is to be replaced by any sentence describing the physical world.

This pattern is also identified by the new requirement as spurious since it can be trivialized by changing the filling instruction to ''α' is to be replaced by any sentence'. And Kitcher continues:

> Why should patterns whose filling instructions can be modified to accommodate any sentence be suspect? The answer is that, in such patterns, the nonlogical

vocabulary which remains is idling. The presence of that nonlogical vocabulary imposes no constraints on the expressions we can substitute for the dummy symbols. (Kitcher, 1981, p. 528)

Now, is the pattern *TSP* in the same ball park as the other spurious unifications we discussed? Does it, for instance, parallel in all the relevant respects the pattern of the religious fanatics? Apparently *TSP* can in fact be ruled out on the basis of the new requirement by a move which is analogous to the ones previously considered. If we change the filling instruction

Replace 'ψ' by s, if $F(s) = 1$ and replace 'ψ' by $\neg s$, if $F(s) = 0$.

to

Replace 'ψ' by any sentence (in the language of ordered fields).

we obtain a pattern which, apparently, allows the derivation of any old sentence (in the given language). The problem with such a quick dismissal of *TSP* as achieving merely a spurious unification is that this argument pattern imposes more constraints on 'ψ' over and above what is expressed by the filling instruction concerning this dummy letter. Hence the simple modification of the filling instruction above is *not* sufficient to license the derivation of any old sentence. Even if the filling instruction concerning 'ψ' is relaxed completely, the part of the *classification* of the argument pattern is still in place which specifies that *sentence (2), i.e. ψ, follows from (1) by metatheory*. And this requirement is certainly not fulfilled just by any arbitrary sentence ψ. By definition, the classification identifies the logical structure that instantiations of a pattern must exhibit, which, to recall, is one of the two constraints that determine the *stringency* of a pattern. Hence, in short, *TSP* is *too stringent* to be ruled out by the new requirement.[18] A valid complaint one might have about *TSP*, however, is that it is too compressed, its classification somewhat opaque, leaving in a perhaps misleading way certain elucidations to the filling instructions. This can be remedied to some extent by bringing out more clearly the structure and functioning of the argument pattern in the following reformulation *TSP'*.

(1) F is some fixed specification of a decision algorithm for *RCF*.
(2) $F(\varphi) = x$

---

[18] If—on a mistaken interpretation of the workings of filling instructions—one allowed the possibility that the modified filling instruction *overrides* the classification in any given conflict between them, such that indeed *any* sentence ψ could be derived, then this would yield in effect an *incoherent* argument pattern or, rather, one wouldn't be left with an argument pattern any more (in the original sense of the word). If, on the other hand, the (modified) filling instruction does not interfere with the classification—as it should be—then the modification has no effect on the functioning of the pattern. In particular, this modification cannot render the pattern spurious.

(3) $F(\varphi) = 0 \to \neg\varphi$
(4) $F(\varphi) = 1 \to \varphi$
(5) ψ

Here 'φ', 'ψ', and 'x' are dummy letters.

Filling instructions:

> Replace 'φ' by a sentence in the language of *RCF*.
> Replace 'ψ' by a sentence in the language of *RCF*.
> Replace 'x' by '0' or '1'.

Classification:

> Sentence (1) is a premise.
> Sentence (2) follows from (1) by metatheory, i.e. by using facts about the functioning of *F*.
> Sentences (3) and (4) also follow from (1) using facts about the functioning of *F*.
> Sentence (5) follows from (2) together with either (3) or (4) by Modus Ponens.

In this reformulation the filling instructions only specify the syntactic categories of the expressions to be substituted for the dummy letters without making any reference to semantic, logical, or epistemic features of these expressions. There has to be a rigorous separation between, on the one hand, filling instructions *qua* purely syntactical constraints on substitutions for dummy letters and the classification, on the other hand, which specifies inferential characteristics of the schematic argument, i.e. puts in effect further constraints on those substitutions based on (deductive or inductive) logical relationships between sentences in the argument and/or theories in the background. Maintaining this separation sets *TSP'* apart from argument patterns which achieve only spurious unifications. The filling instructions of, e.g. the pattern of self-derivation ('from α infer α') or the theological pattern involve the notion of the *acceptance* of sentences. Thus all (or almost all) of the work is done by these kind of filling instructions, which ensure that a particular type of conclusion (for instance, accepted sentences describing the physical world) is generated by such patterns. Neither the structure of the schematic argument nor the classification really contribute anything to this 'unification'. As Kitcher put it, the remaining nonlogical vocabulary in such patterns is *idling*. The real work of unifying beliefs is only *mimicked* by the filling instructions, i.e. by appropriate restrictions on the sentences to be substituted for the dummy letters (cf. Kitcher, 1981, p. 527). And that's precisely why such patterns can be trivialized by varying (i.e. generalizing) their filling instructions. In sharp contrast, the nonlogical vocabulary in *TSP'*, the name '*F*' of the decision algorithm for *RCF*, is *not* idling, the presence of it does pose constraints on the (combination of) the expressions we can substitute for 'φ', 'ψ', and 'x'. Hence *TSP'* cannot be turned into a pattern that yields any sentence whatsoever by modifying the filling instructions. Its classification does not allow that. In fact, the filling instructions, due to their purely syntactic character, are already stated in the utmost generality and cannot be generalized any further (as could easily be done in case of the spurious unifications).

To sum up, in Kitcher's framework $E_1(K)$ cannot be dismissed as spurious because of non-stringency; it clearly passes the test based on the new requirement. The problem posed by this systematization for Kitcher's model of mathematical explanation won't thus go away.

Acknowledgments. We would like to thank Jeremy Avigad, Christian Fermüller, Philip Kitcher, Dana Scott, Jamie Tappenden, and Mark Wilson for many useful comments which improved the final version of this chapter.

Appendix

Axioms for real closed fields

(i) Axioms for fields
$\forall x \forall y \forall z (x + (y + z) = (x + y) + z)$
$\forall x \forall y (x + y = y + x)$
$\forall x (x + 0 = x)$
$\forall x \exists y (x + y = 0)$
$\forall x \forall y \forall z (x \cdot (y \cdot z) = (x \cdot y) \cdot z)$
$\forall x \forall y (x \cdot y = y \cdot x)$
$\forall x (x \cdot 1 = x)$
$\forall x \forall y \forall z (x \cdot (y + z) = (x \cdot y) + (x \cdot z))$
$\forall x (x \neq 0 \to \exists y (x \cdot y = 1))$
$0 \neq 1$

(ii) Order axioms
$\forall x \forall y \forall z (x \leq y \& y \leq z \to x \leq z)$
$\forall x \forall y (x \leq y \& y \leq x \to x = y)$
$\forall x (x \leq x)$
$\forall x \forall y (x \leq y \vee y \leq x)$
$\forall x \forall y \forall z (x \leq y \to x + z \leq y + z)$
$\forall x \forall y \forall z (x \leq y \& 0 \leq z \to x \cdot z \leq y \cdot z)$

(iii) $\forall x \exists y (x = y^2 \vee -x = y^2)$

(iv) For each natural number n, the axiom
$\forall x_0 \forall x_1 \ldots \forall x_{2n} \exists y (x_0 + x_1 \cdot y + x_2 \cdot y^2 + \ldots + x_{2n} \cdot y^{2n} + y^{2n+1} = 0)$

Bibliography

ANDRADAS, C., BRÖCKER, L., and RUIZ, J. M. (1996), *Constructible Sets in Real Geometry* (Berlin: Springer).

BOCHNAK, J., COSTE, M., and ROY, M.-F. (1998), *Real Algebraic Geometry* (Berlin: Springer).

BRUMFIEL, Gregory W. (1979), *Partially Ordered Rings and Semi-Algebraic Geometry* (Cambridge: CUP).

COURANT, Richard (1971), *Vorlesungen über Differential- und Integralrechnung*, 2 vols. 4th edn (Berlin: Springer).

FRIEDMAN, Michael (1974), 'Explanation and scientific understanding', *Journal of Philosophy*, 71, 5–19.

HAFNER, Johannes and MANCOSU, Paolo (2005), 'The varieties of mathematical explanation', in P. Mancosu, K. Jørgensen, and S. Pedersen (eds.), *Visualization, Explanation and Reasoning Styles in Mathematics* (Dordrecht: Springer), 215–250.

KITCHER, Philip (1981), 'Explanatory unification', *Philosophy of Science*, 48, 507–531.

——(1984), *The Nature of Mathematical Knowledge* (Oxford: OUP).

——(1989), 'Explanatory unification and the causal structure of the world', in P. Kitcher and W. Salmon (eds.), *Scientific Explanation, vol. XIII of Minnesota Studies in the Philosophy of Science* (Minneapolis: University of Minnesota Press), 410–505.

RESNIK, Michael D. and KUSHNER, David (1987), 'Explanation, independence, and realism in mathematics', *British Journal for the Philosophy of Science*, 38, 141–158.

SEIDENBERG, A. (1954), 'A new decision method for elementary algebra', *Annals of Mathematics*, 60, 365–374.

——(1958), 'Comments on Lefschetz's principle', *The American Mathematical Monthly*, 65, 685–690.

STEINER, Mark (1978), 'Mathematical explanation', *Philosophical Studies*, 34, 135–151.

TAPPENDEN, Jamie (2005), 'Proof-style and understanding in mathematics I: visualization, unification and axiom choice', in P. Mancosu, K. Jørgensen, and S. Pedersen (eds.), *Visualization, Explanation and Reasoning Styles in Mathematics* (Dordrecht: Springer), 147–214.

TARSKI, Alfred (1951), *A Decision Method for Elementary Algebra and Geometry*, 2nd edn (Berkeley: University of California Press).

WEBER, Erik and VERHOEVEN, Liza (2002), 'Explanatory proofs in mathematics', *Logique et Analyse*, 45(179–180), 299–307 [appeared 2004].

7

Purity as an Ideal of Proof

MICHAEL DETLEFSEN

7.1 The Aristotelian ideal of purity

The topic of this chapter—'purity' of proof—received its classical treatment in the writings of Aristotle. It was part of his theory of demonstration, which was set out in the *Posterior Analytics*.

Aristotle presented purity as an ideal. Proofs lacking it were not necessarily worthless, but they did not provide highest or best knowledge of their conclusions. It was thus a quality of highest or best proof.

Aristotle presented the ideal in the form of a prohibition against what he termed *metabasis ex allo genos*, or crossing from one genus to another in the course of a proof.

> To sum up, then: demonstrative knowledge must be knowledge of a necessary nexus, and therefore must clearly be obtained through a necessary middle term; otherwise its possessor will know neither the cause nor the fact that his conclusion is a necessary connexion. ...
>
> It follows that we cannot in demonstrating pass from one genus to another. We cannot, for instance, prove geometrical truths by arithmetic.[1]
>
> Aristotle (anc1), 75a29–75b12

The prohibition against crossing from one genus to another in the course of an argument had various motives. Among them were Aristotle's opposition to (i) the Pythagoreans' reduction of all things to number (cf. Aristotle (anc3), 1036b8–21), (ii) the related view that all things ultimately have but a single basic Form (*loc. cit.*), and (iii) Plato's conception of dialectic as a kind of 'master science' (cf. Plato (anc2), 55d–59d; Plato (anc3), 533b, c; Aristotle (anc1),

---

[1] In a later passage (cf. Aristotle (anc1), 76a22–25), Aristotle sanctioned the use of metabasis between a science and its *subordinates*. Thus, geometrical argument could be used in mechanics and optics, and arithmetical arguments in harmonics.

76a37–40, 77a26–35, 78a10–13). Perhaps most fundamental, though, was Aristotle's view of the preconditions required for the extension of knowledge through argumentation.

> All instruction given or received by way of argument proceeds from pre-existent knowledge. ... The mathematical sciences and all other speculative disciplines are acquired in this way ...
>
> The pre-existent knowledge required is of two kinds. In some cases admission of the fact must be assumed, in others comprehension of the meaning of the term used, and sometimes both assumptions are essential.
>
> Aristotle (anc1), 71a1–4; 12–14

For Aristotle, then, the development of mathematical knowledge was in large part development by reasoning and such development depended on prior knowledge of its topic or subject. Specifically, it demanded knowledge of the 'what' of its topic.

Conservation of topic was thus built into Aristotle's conception of how knowledge develops through inference and was itself an important aspect of purity. There was in addition, however, another basis for purity, namely, the need for a necessary connection between the subject and predicate of a theorem.

> We ... possess unqualified scientific knowledge of a thing ... when we ... know the cause (*aitia*) on which the fact depends as the cause of the fact and of no other, and, further, that the fact could not be other than it is.
>
> Aristotle (anc1), 71b9–12

To secure such a connection, mathematical demonstrations had ultimately to be based on knowledge of their subjects' essences.

> The 'why' is referred ultimately ... in mathematics ... to the 'what', to the definition [*horos*] of straight line or commensurable or the like ...
>
> Aristotle (anc4), 198a16–18[2]

In Aristotle's view, then, purity increased epistemic quality. A pure proof provided knowledge that the predicate of its conclusion (the minor term of the proof) held of its subject (the major term) solely because of *what* the subject in itself was. It showed the very *whatness* (i.e. the essence) of the subject of a theorem to be the 'cause' of its having the property expressed by its predicate.

---

[2] See also Aristotle (anc2), 90a31–33, where it is stated that:

> ... to know what a thing is [*ti estin*] is the same as to know why it is [*dia ti estin*] ... and this is equally true of things in so far as they are said without qualification to be as opposed to being possessed of some attribute, and in so far as they are said to be possessed of some attribute such as equal to right angles, or greater or less.

There are wider indications of concern to avoid metabasis in the writings of ancient mathematicians. Perhaps most notable in this connection are the cautions sounded repeatedly by Archimedes in his *Method*.[3] There he states that though his mechanical arguments are useful for discovering truths and their proofs, they themselves are not proofs.

> I thought fit...to explain...a certain method, by which it will be possible...to investigate...problems in mathematics by means of mechanics. This procedure is...no less useful even for the proof of the theorems themselves; for certain things became clear to me by a mechanical method, although they had to be demonstrated by geometry afterwards because their investigation by the said method did not furnish an actual demonstration.
>
> Archimedes (anc), 13

Archimedes thus appears to have accepted mechanical methods as useful instruments of discovery in geometry—discovery both of their truth and of their proper geometric proofs. He was therefore in substantial agreement with Aristotle as regards the place and significance of purity as a condition on demonstration.[4]

7.2 Neo-Aristotelian ideals of purity

Aristotle's 'causal' conception of proof was influential for a long time, and among both mathematicians and philosophers. A prime example is Leibniz, who echoed Aristotle in distinguishing two types of reasons or grounds, one a ground of belief, the other a ground of truth.

> A reason is a known truth whose connection with some less well-known truth leads us to give our assent to the latter. But it is called a 'reason', especially and *par excellence*, if it is the cause not only of our judgment but also of the truth itself—which makes it what is known as an '*a priori reason*'. A *cause* in the realm of things corresponds to a *reason* in the realm of truths, which is why causes themselves—and especially final ones—are often called 'reasons'.
>
> Leibniz (1764), 476 (Bk. IV, ch. xvii, §3)

---

[3] Pappus' classification of geometrical problems into linear, planar, and solid should also be mentioned here since it reflected a similar (and earlier) concern for purity among mathematicians.

> It appears to be no small error for geometers when a plane problem is solved by conics or other curved lines, and in general when any problem is solved by an inappropriate kind, as in the problem concerning the parabola in the fifth book of the *Conics* of Apollonius.
>
> Pappus (anc), 351

[4] For further discussion see Livesey (1982, Ch. 2).

The objective grounding relationship induced an objective (partial) ordering of truths, and it was this ordering that Leibniz was interested in.

> ... we are not concerned here with the sequence of our discoveries, which differs from one man to another, but with the connection and natural order of truths, which is always the same.
>
> Leibniz (1764), 412 (Bk. IV, ch. vii, §9)

Such an ordering of truths suggests a parallel conception of purity: pure proof is proof which recapitulates a segment of the natural, objective ordering of truths concerning a given subject.

In the early years of the 19th century, Bolzano also articulated such an idea and applied it to the reformation of mathematics generally, and particularly to analysis.[5] It comprised, indeed, a prime motive of his early attempts to 'arithmetize' analysis. This arithmetization, or, perhaps better, this *de-geometrization*, was wanted in order to combat what Bolzano saw as a pervasive type of circularity in proofs in analysis—a circularity borne of impurity.[6]

The impurity represented by the importation of geometrical considerations into the proofs of genuinely algebraic or analytic theorems had serious consequences. In particular, it inverted the objective ordering of truths. In so doing, it introduced circularities of reasoning into analysis.[7] More particularly, a geometrical proof presented a theorem of analysis θ_1 as depending on a geometrical theorem θ_2 when, in fact, the opposite is true.

---

[5] Bolzano also vigorously pursued reform in geometry, including, perhaps especially, elementary geometry. In his view, not even a proper theory of triangles and parallels had been given. This was because a proper theory—that is, a properly *pure* theory—would be based solely on a theory of the straight line. Yet all past attempts had presupposed *axioms of the plane*, axioms the proper founding of which would itself require a theory of triangles. Bolzano was thus convinced that

> ... the first theorems of geometry have been proved only *per petitionem principii*; and even if this were not so, the *probatio per aliena et remota* [proof by alien and remote elements, MD]... must not be allowed.
>
> Bolzano (1804, 174)

Bolzano's thinking here is reminiscent of the ancient division of geometrical problems into linear, planar, and solid that Pappus emphasized. Bolzano also opposed the long standing common practice of appealing to motion in the proof of 'purely geometrical truths' (*op. cit.*, 173). See Mancosu (1996, Ch. 1) for a useful discussion of the early modern controversy concerning such appeals.

[6] At approximately the same time, Lagrange pursued a different purification program, one which urged the liberation of analysis from the kind of dependency on motion encouraged by Newton's conception of fluxions. He claimed that to introduce motion into a calculus that had only algebraic quantities as its objects was to introduce an alien or extraneous idea (*une idée étrangère*). Cf. Lagrange (1797, 4).

[7] Not every 'departure from' an ordering need be an 'inversion of' certain of its elements. Bolzano's argument seems, however, to require this latter, narrower conception of 'departing' from an ordering.

Analysis, or the general theory of quantity, was more basic than geometry, the science of spatial quantities. In many instances, then, a proper (i.e. a properly pure) proof of a geometrical theorem θ_2 would derive from a more basic, general theorem of quantity θ_1. In such cases, an impure proof of the more general θ_1 from the more particular θ_2 would, in effect, amount to a circular grounding of θ_2 on itself.

Bolzano was thus convinced that, in analysis, 'geometrical proof is, ... in most cases, really circular' (Bolzano, 1817a, 228). He therefore set out to purge analysis of its geometrical contaminations. His best known work in this connection is his purely analytic proof (*rein analytischer Beweis*) of the *intermediate value theorem*—the theorem that for a real function f continuous on a closed bounded interval [a, b], for every $\mu, f(a) < \mu < f(b)$, there is a ν, $a < \nu < b$, such that $f(\nu) = \mu$. There are other important examples as well.[8]

Bolzano termed 'purely analytic' those methods by which one function is derived from others by means of rules that are completely independent of the *particular* natures of the quantities involved (cf. Bolzano (1817b), English translation in Ewald (1996), vol. I, 225, fn. a). Pure proof, he maintained, is proof in which the generality of the methods used fits the generality of the theorem proved. The intermediate value theorem expresses a general truth concerning quantities, not one that holds merely for spatial or geometrical quantities. A proper proof of it ought not, therefore, appeal to truths peculiar to the latter.

The usual proofs of the theorem did just that, however. In particular, they made use of the following 'truth borrowed from geometry' (Bolzano, 1817a, 228): *every continuous line of simple curvature of which some ordinate values are positive and some negative must intersect the x-axis at a point between the positive and negative ordinate values.*

Bolzano acknowledged that this was an evident truth of geometry. Indeed, he described it as 'extremely evident' (cf. Bolzano, 1817a, 228). This notwithstanding, he believed that it had no place in a proof of the intermediate value theorem. It was, he maintained,

> ...an intolerable offense against correct method to derive truths of pure (or general) mathematics (i.e. arithmetic, algebra, analysis) from considerations which belong to a merely applied (or special) part, namely, geometry. Indeed, have we not felt and recognized for a long time the incongruity of such *metabasis eis allo genos*? Have we not already avoided this whenever possible in hundreds of other

---

[8] Cf. *op. cit.*, §13, where Bolzano proves a version of the Bolzano–Weierstrass theorem which, in its contemporary analytic formulation, says that every bounded infinite sequence has a convergent subsequence.

cases, and regarded this avoidance a merit? ... if one considers that the proofs of the science should not merely be *confirmations* (*Gewissmachungen*), but rather *justifications* (*Begründungen*), i.e. presentation of the objective reason for the truth concerned, then it is self-evident that the strictly scientific proof, or the objective reason, of a truth which holds equally for *all* quantities, whether in space or not, cannot possibly lie in a truth which holds merely for quantities which are in *space*.

Bolzano (1817a, 228)

By eliminating such appeals to geometric truths, Bolzano's proof was, as he saw it, more than 'a mere confirmation (*blosse Gewissmachung*)' (Bolzano, 1810, 233) of the intermediate value theorem. It not only made that theorem certain, it gave its 'objective justification' (*objektive Begründung*) (*loc. cit.*). In so doing, it achieved the ideal of 'genuinely scientific' (*echt wissenschaftlich*) (*loc. cit.*) proof.

Bolzano's determination to free analysis of impure geometrical influences recalls similar earlier themes of Descartes and Wallis. Wallis, for example, defended his use of algebraic methods in geometry on grounds of superior objectivity. This objectivity was, he believed, borne of the fact that some of the properties of geometrical figures (specifically, in Wallis' view, properties concerning their rectification, quadrature, and cubature) reflect properties that the figures have *in themselves*, so to say, and independently of how they might be constructed (cf. Wallis, 1685, 291, 292–293, 298–299). In Wallis' view, then, objectivity in geometry actually required the use of algebraic methods.

Bolzano held an in some ways similar view, arguing that the persistent misuse of geometrical reasoning in analysis was due to a lingering confusion between *theoretical* and *practical* mathematics and the associated mistake of taking a mathematical object to be *real* (*wirklich*) only to the extent that it was constructible by certain means (cf. Bolzano, 1810, 218). In this respect, then, he was more neo-Platonist than neo-Aristotelian.

Bolzano also pointed out (cf. Bolzano, 1817a, 227, 232) similar misgivings in Gauss, misgivings which Gauss directed at the use of geometrical reasoning in his first (1799) proof of the fundamental theorem of algebra.[9] Gauss later expressed similar concerns regarding his geometrical interpretation of the complex numbers.

> The representation (*Darstellung*) of the imaginary quantities as relations of points in the plane is not so much their essence (*Wesen*) itself, which must be grasped in a higher and more general way, as it is for us humans the purest or perhaps a uniquely and completely pure example of their application.
>
> Gauss (1870–1927, vol. X, 106)[10]

---

[9] Gauss (1799).   [10] Letter to Moritz Drobisch, 14 August 1834.

Later still, and in terms strongly reminiscent of Bolzano's earlier assessment of the usual proofs of the intermediate value theorem, Gauss further observed that

> ... the real content (*eigentliche Inhalt*) of the entire argumentation belongs to a higher realm of the general, abstract theory of quantity (*abstracten Grössenlehre*), independent of the spatial. The object of this domain, which tracks the continuity of related combinations of quantities, is a domain about which little, to date, has been established and in which one cannot maneuver much without leaning on a language of spatial pictures (*räumlichen Bildern*).
>
> Gauss (1870–1927, vol. III, 79)[11]

Like Bolzano, then, Gauss too was concerned for purity in analysis. Less clear is whether he shared Bolzano's views concerning the historical reasons for impurity and the circularities of reasoning he took it to represent.[12]

In truth, though, avoidance of circularity was not Bolzano's only reason for advocating purity. Indeed, it was not his principal reason. More fundamental was his acceptance of the traditional distinction between two types of reasoning in mathematics and in science generally. These were (i) confirmatory reasoning, or reasoning which convinces *that* (what Bolzano termed *Gewissmachung*), and (ii) reasoning which reveals the objective reasons for truth (*objektive Begründung*).[13] The former, he observed, falls short of the latter: 'the *obviousness* (*Evidenz*) of a proposition does not absolve me of the obligation to look for a proof of it' (Bolzano, 1804, 172).

Bolzano pointed to the earliest instances of proof as his inspiration. Thales, he said, did not settle for knowledge *that* the angles at the base of an isosceles triangle are equal, though this was doubtlessly evident to him. Rather, he pressed on to understand *why*. In so doing, he was rewarded by an extension of his knowledge. Specifically, he obtained knowledge of those truths that implicitly underlay common-sense belief in the theorem. Since these were 'new truths which were not clear to common sense' (*op. cit.*, 173), his knowledge was thereby extended.

In addition, it promoted the further extension of his knowledge, both by increasing its reach and by improving its efficiency. It increased its reach because 'if... first ideas are clearly and correctly grasped then much more can

---

[11] This is from the *Jubiläumschrift* of 1849, where Gauss was commenting on the use of geometrical methods to prove the existence of roots of equations.

[12] Also problematic is how to harmonize Gauss' view of the impurity caused by the use of geometrical reasoning in analysis with the contrast he drew between the *created* character of number and the *objective* character of space (cf. 1817 letter to Olbers, Gauss (1976), 651–652; 1829 letter to Bessel, Gauss (1870–1927), vol. VIII, 200; Gauss (1832), 313, fn.1).

[13] See Cicero (anc., 459–460); Ramus (1574, 17, 71, 93–94); Viéte (1983, 28–29); Arnauld (1662, 302); Wallis (1685, 3, 290, 305–306), among others, for earlier statements of this distinction.

be deduced from them than if they remain confused' (Bolzano (1804), 172). It improved efficiency because learning is easier when concepts are 'clear, correct, and connected in the most perfect order' (*loc. cit.*). We must therefore

> ...regard the endeavour of unfolding all truths of mathematics down to their ultimate grounds, and thereby providing all concepts of this science with the greatest possible clarity, correctness, and order, as an endeavour which will not only promote the *thoroughness* of education but also make it *easier*.
>
> Bolzano (1804, 172)

The discipline of purity thus promised to increase the extent of our knowledge while at the same time increasing our capacity to extend it further. We are thus to imitate Thales. That is, we are to look 'inside' theses, to analyze their concepts into constituent concepts and to excavate the basic truths that concern these sub-concepts. We are not to make use of conceptual resources outside those of the given thesis. Bolzano's basic recommendation was thus that we should not rest content with a proof

> ...*if it is not... derived from concepts* which the thesis to be proved contains, but rather makes use of some fortuitous alien, instrumental concept (*Mittelbegriff*)[14], which is always an erroneous *metabasis eis allo genos*.
>
> Bolzano (1804, 173)

All in all, then, Bolzano saw purity as serving a variety of epistemic ends. First, by forcing us to look behind the 'obvious', it revealed hitherto unnoticed truths, the revelation of which increased the extent of our knowledge. Secondly, it promised to improve efficiency through the clarification of the concepts that appear in a thesis. Concepts are easier to grasp (and therefore easier to develop) the clearer they and their connections with other concepts are. Thirdly, purity protected against the circularities, and the attendant epistemic futility, that marked so many of the proofs of 18th and 19th century analysis and traditional elementary geometry.

Later 19th century mathematicians also acknowledged the need, or at least the virtue of purity. They acknowledged, in particular, the importance of purging geometrical reasoning from analysis. Moreover, their reasons were, if anything, even more pressing than Bolzano's. For while Bolzano accepted the *reliability* of ordinary geometrical reasoning (at least in his early writings), later thinkers did not, or at least not generally. Weierstrass' example of an everywhere continuous but nowhere differentiable function cast doubt on

---

[14] The translation in the Ewald volume by Stephen Russ renders '*Mittelbegriff*' as 'intermediate concept'. I think it is more in keeping with Bolzano's ideas to take '*Mittel*' in the sense of a 'means' or 'instrument'.

even the reliability of geometrical thinking as regards continuity phenomena.[15] Analysts of the latter half of the 19th century thus had, if anything, even stronger reasons to try to rid analysis of geometrical thinking.

An important example is Dedekind, who expressed his commitment to purity this way:

> In discussing the notion of the approach of a variable magnitude to a fixed limiting value ... I had recourse to geometric evidences. Even now such resort to geometric intuition ... I regard as exceedingly useful, from the didactic standpoint.... But that this form of introduction into the differential calculus can make no claim to being scientific, no one will deny. For myself this feeling of dissatisfaction was so overpowering that I made the fixed resolve to ... find a purely arithmetical and perfectly rigorous foundation for the principles of infinitesimal analysis.
>
> Dedekind (1872, 1–2)

Elsewhere Dedekind stated as the first procedural demand on his theory of algebraic integers that 'arithmetic remain free from intermingling with foreign elements (*mélange d'éléments étrangers*)' (Dedekind, 1877, 269). This meant for him that 'the definition ... of irrational number ought to be based on phenomena one can already define clearly in the domain R of rational numbers' (*loc. cit.*).

Dedekind's fellow logicist, Frege, also accepted purity as an ideal. Indeed, he described it as a principal motive for his logicist program, one which set it apart from the misbegotten attempt by Gauss to found complex arithmetic on geometry.[16]

> The overcoming of this reluctance [to accept complex numbers, MD] was facilitated by geometrical interpretations; but with these, something foreign was introduced into arithmetic. Inevitably there arose the desire of ... extruding these geometrical aspects. It appeared contrary to all reason that purely arithmetical theorems should rest on geometrical axioms; and it was inevitable that proofs which apparently established such a dependence should seem to obscure the true state of affairs. The task of deriving what was arithmetical by purely arithmetical means, i.e. purely logically, could not be put off.
>
> Frege (1885, 116–117)[17]

---

[15] Bolzano himself discovered a function of this type in 1834. The work in which this was done (namely, Bolzano, 1834, §75), however, was not published until 1930.

[16] As noted above, Gauss himself conceded the impurity of his geometrical interpretation in his later writings. In his 1831 essay, however, he credited it with providing a sensible representation of the complexes 'that leaves nothing to be desired' (Gauss, 1832, 311). '[O]ne needs', he said, 'nothing further to bring this quantity [i.e. $\sqrt{-1}$, MD] into the domain of the objects of arithmetic' (Gauss, 1832, 313). It should perhaps also be noted here that, towards the end of his life, Frege returned to the idea of a geometrical foundation for arithmetic.

[17] See Frege (1884, §§103–104) for similar statements.

Purity was thus widely accepted as an ideal of proof in 19th century mathematics. This notwithstanding, it was not without its detractors. These included such eminences as Hilbert and Klein. Hilbert viewed purity as a 'subjective' ideal (cf. Hilbert, 1899, 106–107).[18] The important thing, he said, is that we learn as much as we can about the different means by which a given theorem or body of theorems can be proved. Pure proofs—that is, proofs that concern themselves only with the concepts contained in the theorems proved—are therefore part of what we want to know. They are, however, only part and are not inherently more interesting or valuable than proofs that draw upon other conceptual resources.

Klein believed that much of mathematics, including much of analysis, ought in many instances to be confirmed by geometrical intuition. In addition, he believed that without 'constant use' of geometrical intuition the proof of some of the most interesting and important analytic results concerning continuity would be impossible, at least practically speaking (cf. Klein, 1894, 45).

He also challenged purity in the other direction. That is, he proposed a system of axioms for geometry that made use of analytic ideas and methods, and, generally speaking, he recommended this system for its efficiency. At the same time, he decried the resistance of geometers to this simple approach, a resistance he also attributed to traditional ideas of purity, in this case those embodied in the retrograde refusal to use numbers in geometry (cf. Klein, 1925, 160).

Promoting efficiency as a more compelling ideal than purity was hardly a new attitude, however. It was, indeed, at the core of 17th, 18th, and 19th century debates concerning the use of algebra in geometry (cf. Wallis, 1685, 117–118, 305–306; MacLaurin, 1742, 37–50) and, relatedly, the division of mathematical reasoning into methods of discovery and methods of demonstration.

7.3 Purity as a contemporary ideal

Purity has thus come into conflict with other ideals of mathematical reasoning, conflicts which, at best, have been only partially resolved. This unresolved

---

[18] The subdivision of axioms in Hilbert's *Grundlagen* bespeaks a moderate concern for purity, and one that's reflected in the larger organization of the book. Thus, for example, Ch. I, §4 is entitled 'Consequences of the axioms of incidence and order' and Ch. I, §6 'Consequences of the axioms of congruence'. His intention, he said, was to deduce the most important theorems in such a way as to bring to light 'the meaning of the various groups of axioms, as well as the significance of the conclusions that can be drawn from the individual axioms' (Hilbert, 1899, 2).

conflict has not, however, driven purity into exile. Indeed, it remains a common ideal of proof today. Somewhat more accurately, there are different ideals of (and, not coincidentally, motives for) purity today, and these are reflected in actual mathematical practice. I'll now briefly survey the more important of these.

7.3.1 Topical purity

A natural point of departure is the prime number theorem[19] and the well-known search for its 'elementary' proof, a search which culminated in the more or less independently developed proofs of Selberg and Erdös in 1948 (see Selberg (1949) and Erdös (1949)). The theorem was, of course, proved a half century earlier (1896) by non-elementary means, again independently, by Hadamard and de la Vallée Poussin. Both their proofs relied heavily on methods from complex analysis.

The proofs of Selberg and Erdös avoided such methods, and it was this avoidance which, at least in Selberg's view (Selberg, 1949, 305), qualified them as *elementary*.

Gian-Carlo Rota gave a useful digest of the developments leading up to Selberg's and Erdös' proofs in Rota (1997b). Among other things, he pointed out that it should be seen as a natural continuation of earlier work of Norbert Wiener's, work which suggested that there might be a '*conceptual underpinning* to the distribution of the primes' (Rota, 1997b, 115, emphasis added). This, he said, was the primary motivation for looking for an elementary proof since it encouraged the idea that the prime number theorem might have a proof that proceeds from the analysis of the concept of prime number itself.

Rota thus proposed the following understanding of the notion of *elementary* proof in the setting of the prime number theorem.

> What does it mean to say that a proof is 'elementary?' In the case of the prime number theorem, it means that an argument is given that shows the 'analytic inevitability' (in the Kantian sense of the expression) of the prime number

---

[19] The prime number theorem is the theorem that for any real number x, the number of primes not exceeding x (commonly expressed by the so-called 'prime counting' function and written '$\pi(x)$') is asymptotic to $\frac{x}{\ln x}$.

The reader should recall that for all x, y, $\ln x = y$ iff $e^y = x$, where $e = \lim_{n\to\infty} \left(1 + \frac{1}{n}\right)^n$. To say that $\pi(x)$ is asymptotic to $\frac{x}{\ln x}$ is to say that the ratio of $\pi(x)$ to $\frac{x}{\ln x}$ approaches 1 as x approaches infinity. As Selberg pointed out in his paper, it is possible to eliminate all references to limits in the proof he gave. That done, he also noted, the appeal to limits in the statement of the theorem would also have to change.

theorem on the basis of an analysis of the concept of prime without appealing to extraneous techniques.

Rota (1997b, 115)

Parallels to the Aristotelian and neo-Aristotelian conceptions of purity could hardly be more striking, and they suggest that purity of a broadly Aristotelian type continues to function as an ideal among contemporary mathematicians. This is also suggested by the notice that was taken of Selberg's and Erdös' proofs, notice which resulted in Selberg's being awarded the Fields Medal in 1950. The theorem itself was old.[20] What was new was the elementary proof, where 'elementary', in this case, meant something like what we mean by 'topically pure'.

The proof took what had been a mystery (the relationship of the density of the primes to the Riemann zeta function, as presented in the proofs of Hadamard and de la Vallée Poussin) into something rooted in the concept of a prime number itself.[21] The result was thought to be remarkable.

Such views of the significance of the elementary proof of the prime number theorem have been voiced by mathematicians other than Rota. Thus, in a review of Selberg's and Erdös' work, A. E. Ingham described their proofs as 'not depending on ... ideas remote from the problem itself' (Ingham, 1949, 595). Indeed, well before their work he challenged the use of analytic methods in proving such theorems on the grounds that they 'introduce ideas very remote from the original problem' (Ingham, 1932, 5). In a similar spirit, H. G. Diamond praised the use of elementary methods in solving problems concerning the distribution of the primes more generally because 'they do not require the introduction of ideas so remote from the arithmetic questions under consideration' (Diamond, 1982, 556).

The elementary proof of the prime number theorem is thus a potent illustration of the continuing concern for broadly Aristotelian purity among mathematicians. It is, however, by no means unique. In fact, purity seems to be a common concern among combinatorists generally. A good example is a recent paper by Stanton and Zeilberger in which they motivate their project by observing that '[t]o a true combinatorialist, a combinatorial result is not properly proved until it receives a *direct combinatorial proof*' (Stanton and Zeilberger, 1989, 39). They also offer an interesting suggestion concerning the benefit of such proof—namely, the added 'insight' it gives to the combinatorist.

The idea seems to be that (i) a specialist is most likely to do something significant with any solution to a problem in her specialty, that (ii) this

---

[20] As Rota put it, it 'had been cooked in several sauces' (Rota, 1997b, 114).

[21] Or this concept plus, perhaps, certain related concepts necessary in order to see it as a genuinely *combinatory* problem.

advantage is enhanced if a solution is given in terms of concepts most familiar to her, and that (iii) the purer the proof the more it conforms to this type of solution. Viewed this way, purity is a *pragmatic* virtue, albeit one which serves epistemic ends (namely, the effective utilization of knowledge to produce more knowledge). It does not itself constitute an epistemic virtue; a pure proof is not, sheerly by dint of its purity, a better justification of the theorem it proves. It is, however, a more effective instrument for gaining further knowledge. We see here, then, a suggestion of the idea that purity generally increases the effectiveness of divisions of epistemic labor based on specialization.[22]

Purity is thus a common concern among combinatorialists. It is not, however, unique to them. Indeed, it extends across a wide range of mathematical fields, and has been of concern both to researchers and educators throughout the 20th century.

A simple illustration is the search for a pure proof of the Erdös–Mordell theorem. (Let ABC be a triangle and P be a point in its interior. Construct perpendiculars from P to each of the sides, letting the points of intersection be A', B', C', respectively. The Erdös–Mordell theorem says that $PA + PB + PC \geq 2(PA' + PB' + PC')$.) Finding an elementary proof of this theorem was a going concern in the middle decades of the 20th century, one that ended in success in the late 1950s (cf. Kazarinoff, 1957; Bankoff, 1958).

Still other examples include Formanek (1973), Edmonds (1986), Gilmer and Mott (1971), and Woo (1971). The first gives a proof of a theorem of algebra (the Eakin–Nagata theorem) the virtue of which is said to be that it uses only *definitions* of terms contained in the theorem.[23] The second presents a purely topological proof of a topological theorem and, like the case of Stanton and Zeilberger mentioned above, it expresses the hope that such a proof will make the theorem more 'accessible' to specialists.

The third concerns a theorem first proved by Abraham Robinson using model-theoretic methods. This theorem, though algebraic in character, was said by Robinson not to have an accessible purely mathematical proof independent of his model-theoretic approach. The authors challenge this by providing a purely algebraic proof.[24]

I mention this because it points to a larger concern regarding purity, namely, whether pure proofs always exist. This question, in turn, points to

---

[22] Recall here the remark of Bolzano's (from Bolzano, 1804, 172) quoted in Section 2 concerning the advantages of purity not only for purposes of making education more thorough but of making it easier and more efficient.

[23] The theorem states that if T is a commutative Noetherian ring finitely generated as a module over a subring R, then R is also Noetherian.

[24] See Friberg (1973, 421) for a different example concerning algebraic purity.

another basic question—namely, how the notion of *topic* ought rightly to be conceived.

One facet of this question concerns how to judge what it is that a problem is *about*. Suppose, for instance, that a problem is formulated as a problem concerning the real roots of a certain type of polynomial equation. Suppose, in addition, that these roots can only be found by methods that make ineliminable use of complex numbers.[25] What should we then say our problem is *about*? Since all the roots we're interested in are reals, this suggests that the problem is *about* real numbers. That is, it is about those things to which the terms appearing in it refer.

On the other hand, we might rather take our problem to be about those things and concepts we seemingly must bring into our reasoning in order to solve it.[26]

The fourth paper mentioned above (i.e. Woo 1971) gives a proof of a theorem from measure theory (the Lebesgue Decomposition Theorem) and claims that this proof is purer than previous proofs in that it makes no appeal to measure theory beyond the definitions needed to *state* the theorem. Here again, though, it's not entirely clear what the motive is. Specifically, it's not clear whether purity is being assumed to provide an epistemically superior justification for the theorem proved, or a more effective instrument for furthering certain larger epistemic (including, perhaps, pedagogic) projects.[27]

In all these cases we see a concern for purity of a broadly topical type. The motives are not always the same, though they seem generally to be directed towards epistemic concerns—concerns, particularly, for the quality of solutions offered to problems and the efficacy of such solutions as a foundation for the further development of our mathematical knowledge.

7.4 Conclusion

The ideal of purity that seems to figure most significantly in modern mathematical practice is what I've referred to here as *topical* purity. This

---

[25] I'm obviously thinking of the *casus irreducibilis* or *irreducible case* of the general cubic here.

[26] There are, of course, questions to be asked concerning the meaning of 'must' and 'solve'. That this is so, though, doesn't make the question any less relevant, only more intricate.

[27] As regards pedagogic projects, the idea seems roughly to be that proofs ought to be conducted with as little overhead as possible. Aristotelian purity would seem to constitute a minimum here. Surely it would make no sense to try to prove a theorem if not enough defining had been done to make the content of the theorem intelligible. On the other hand, once such a point has been reached, a proof built upon exactly that learning would represent a remarkable pedagogical efficiency. For examples linking purity with such motives see, Mullin (1964), Luh (1965), Spitznagel (1970), and Jenkins (1982).

is purity which, practically speaking, enforces a certain symmetry between the conceptual resources used to prove a theorem and those needed for the clarification of its *content*. The basic idea is that the resources of proof ought ideally to be restricted to those which determine its content.

There are two main views concerning the significance of such symmetry. For Aristotelians, neo-Aristotelians (e.g. Leibniz and, to some extent, Bolzano) and some neo-Platonists (e.g. Wallis and, to some extent, Bolzano), including many working mathematicians (those mentioned plus, for example, Rota and Ingham), it represents (or has represented) an epistemic ideal, a view concerning the *qualitative type* of knowledge that proof ought, or at least might ideally provide. According to this view, proof is at its best when it provides knowledge of the most basic reasons why the proposition proved is true.

For others, including many working mathematicians, it has been more a strategic or pragmatic ideal, albeit one serving epistemic ends. By this I mean that it has not so much been prized for the knowledge it itself *constitutes*, as for the knowledge it in some broadly pragmatic sense provides for.

In this latter connection, it has been characterized as 'good discipline for the mind' (cf. Dieudonné, 1969, 12), a mental training that increases the prover's potential for future epistemic development. It has also been said to improve efficiency in classroom learning by decreasing the time and the conceptual distance that separates the definition of terms needed to understand a proposition from the demonstration of that proposition (cf. fns. 22, 27, and the quote from Bolzano (1804, 172) on p. 186). Finally, it has been thought to improve the epistemic efficiency of a community by making better use of the way(s) in which it divides labor. Pure proofs put theorems at the disposal of those who, in terms of their training and expertise, are in the best position to use them to develop further knowledge of the concepts involved.

Even for those who take a basically pragmatic view of the value of purity, it is therefore generally true that they see it as serving epistemic ends, whether of increased extent or of improved efficiency.

Bibliography

ARCHIMEDES (2002), *The Method of Archimedes*, in *The Works of Archimedes*, ed. Sir Thomas Heath (Mineola, NY: Dover).
ARISTOTLE (anc1), *Posterior Analytics*, Oxford Translations, ed. W. D. Ross and J. A. Smith, trans. G. R. G. Mure (Oxford: OUP) 1908–1954.
_____ (anc2), *Posterior Analytics*, *The Complete Works of Aristotle*, vol. I. Revised Oxford Translation, ed. J. Barnes (Oxford: OUP) 1984.

ARISTOTLE (anc3), *Metaphysics*, Oxford Translations, ed. W. D. Ross and J. A. Smith, trans. W. D. Ross (Oxford: OUP) 1908–1954.

——— (anc4), *Physics*, Oxford Translations, ed. W. D. Ross and J. A. Smith, trans. R.P. Hardie and R. K. Gaye (Oxford: OUP) 1908–1954.

ARNAULD, A. *The Art of Thinking*, English translation of the 1662 French original by J. Dickoff & P. James (Indianapolis: Bubbs-Merrill, 1964).

BANKOFF, L. (1958), 'An elementary proof of the Erdös-Mordell theorem', *American Mathematical Monthly*, 65, 521.

BOLZANO, Bernhard (1804), *Betrachtungen über einige Gegenstände der Elementargeometrie* (Prague: Karl Barth). English translation of the preface in Ewald (1996), vol. I. Page references are to this translation.

——— (1810), *Beiträge zu einer begründeteren Darstellung der Mathematik*,(Prague: C. Widtmann). English translation in Ewald (1996), vol. I. Page references are to this translation.

——— (1817a), 'Purely analytic proof of the theorem that between any two values which give results of opposite sign there lies at least one real root of the equation'. English translation of 1817 German original in Ewald (1996).

——— (1817b), 'Die drey Probleme der Rectification, der Complanation und der Cubierung, ohne Betrachtung des unendlich Kleinen, ohne die Annahme des Archimedes und ohne irgend eine nicht streng erweisliche Voraussetzung gelöst; zugleich als Probe einer gänzlichen Umgestaltung der Raumwissenschaft allen Mathematikern zur Prüfung vorgelegt' (Leipzig: Gotthelf Kummer).

——— (1930), *Funktionlehre*, K. Rychlik (Prague: Königliche Bömische Gesellschaft der Wissenschaften). Written in 1834.

——— (2004), *The Mathematical Works of Bernard Bolzano*, trans. S. Russ (Oxford: OUP).

CICERO, *Treatise on Topics* in *The Orations of Marcus Tullius Cicero*, trans. C. D. Yonge, 4 vols., vol. IV, 458–486 (London: G. Bell and Sons) 1894–1903.

DEDEKIND, Richard (1872), 'Continuity and irrational numbers', in *Essays on the Theory of Number* (Eng. trans. by W. W. Beman of the 1872 German original) (Dover: New York, 1963).

——— (1877), 'Sur la théorie des nombres entiers algébriques', *Bulletin des Sciences mathématiques et astronomiques*, 1re série, tome XI, 2e série, tome II, 1876, 1877. Reprinted in *Gesammelte mathematische Werke*, vol. 3, ed. Richard Dedekind, Robert Fricke, Emmy Noether, Øystein Ore (Braunschweig: Vieweg, 1930).

DIAMOND, Harold (1982), 'Elementary methods in the study of the distribution of prime numbers', *Bulletin of the American Mathematical Society*, 7, 553–589.

DIEUDONNÉ, Jean (1969), *Linear Algebra and Geometry* (New York: Houghton Mifflin) Engl. trans. of *Algèbre Linéaire et Géométrie Élémentaire* (Paris: Hermann, 1964).

EDMONDS, Allan (1986),'A topological proof of the equivalent Dehn lemma', *Transactions of the American Mathematical Society*, 297, 605–615.

ERDÖS, Paul (1949),'On a new method in elementary number theory which leads to an elementary proof of the prime number theorem', *Proceedings of the National Academy of Sciences of the United States of America*, 35, 374–384.

EWALD, William (1996), *From Kant to Hilbert: A Source Book in the Foundations of Mathematics*, 2 Vols. (Oxford: OUP).

FORMANEK, Edward (1973), 'Faithful Noetherian Modules', *Proceedings of the American Mathematical Society*, 41, 381–383.

FREGE, Gottlob (1884), *Die Grundlagen der Arithmetik, eine logisch-mathematische Untersuchung über den Begriff der Zahl* (Breslau: W. Koebner). Page references are to the English trans. by J. L. Austin, *The Foundations of Arithmetic, a Logico-Mathematical Enquiry into the Concept of Number*, 2nd rev. edn (Evanston, IL: Northwestern University Press).

—— (1885), 'Formal theories of arithmetic'. English trans. of 1885 German original in *Collected Papers on Mathematics, Logic and Philosophy*, ed. B. McGuinness (Oxford: Blackwell).

FRIBERG, Bjorn (1973), 'A topological proof of a theorem of Kneser', *Proceedings of the American Mathematical Society*, 39, 421–426.

GAUSS, Carl (1799), *Demonstratio Nova Theorematis, omnem functionem algebraicam rationalem integram unius variabilis in factores reales primi vel secundi gradus resolvi posse* (Helmstadii: C. G. Fleckeisen). <http://dochost.rz.hu-berlin.de/dissertationen/historisch/gauss-carolo/HTML/gauss2.gif>.

—— (1870–1927), *Werke*, 12 vols (Göttingen: Königlichen Gesellschaft der Wissenschaften).

—— (1976), *Briefwechsel mit H. W. M. Olbers* (Hildesheim: Georg Olms).

—— (1832), *Theoria residuorum biquadraticorum, commentatio secunda (Anzeige)*. Reprinted in Gauss (1870–1927), vol. II. English translation in Ewald (1996), vol. I. Page references are to this translation.

GILMER, Robert and MOTT, Joe (1971), 'An algebraic proof of a theorem of A. Robinson', *Proceedings of the American Mathematical Society*, 29, 461–466.

HILBERT, David (1899), *The Foundations of Geometry*. English trans. by L. Unger of 10th edn of *Die Grundlagen der Geometrie* (Stuttgart: Teubner, 1968). First edition published by Teubner, Leipzig, 1899.

HOBSON, Ernest W. (1927), *The Theory of Functions of a Real Variable and the Theory of Fourier's Series*, vol. I, 2nd edn (Cambridge: CUP).

INGHAM, Albert (1932), *The Distribution of Prime Numbers* (Cambridge: CUP).

—— (1949), Review of Selberg (1949) and Erdös (1949), in *Mathematical Reviews*, 10, 595–596.

JENKINS, James (1982), 'A uniqueness result in conformal mapping. II', *Proceedings of the American Mathematical Society*, 85, 231–232.

KAZARINOFF, D. K. (1957), 'A simple proof of the Erdös–Mordell inequality for triangles', *Michigan Mathematics Journal*, 4, 97–98.

KLEIN, Felix (1894), *The Evanston Colloquium: Lectures on Mathematics* (New York: Macmillan & Co.).

—— (1925), *Elementary Mathematics from an Advanced Standpoint: Geometry* (Mineola: Dover). English trans. by E. Hedrick and C. Noble of the 1925 3rd edn of the 1908 German original *Elementarmathematik vom höheren Standpunkte aus*, vol. 2 (Berlin: J. Springer, 1924–1928).

LAGRANGE, Joseph-Louis (1797), *Théorie des fonctions analytiques: contenant les principes du calcul différentiel, dégagés de toute considération d'infiniment petits ou d'évanouissans, de limites ou de fluxions, et réduits à l'analyse algébrique des quantités finies*, De l'imprimerie de la République, Paris, 1797. Available online at the Gallica website: <http://visualiseur.bnf.fr/Visualiseur?Destination=Gallica&O=NUMM-88736>.

LEIBNIZ, Gottfried W. F. (1764) *New Essays on Human Understanding*, trans. P. Remnant and J. Bennett (Cambridge: CUP, 1981).

LIVESEY, Steven (1982), *Metabasis: The Interrelationship of the Sciences in Antiquity and the Middle Ages*. Doctoral dissertation, Department of History, UCLA.

LUH, Jiang (1965), 'An elementary proof of a theorem of Herstein', *Mathematics Magazine*, 38, 105–106.

MACLAURIN, Colin (1742), *A Treatise of Fluxions*, 2 vols. (Edinburgh: Ruddimans).

MANCOSU, Paolo (1996), *Philosophy of Mathematics and Mathematical Practice in the Seventeenth Century* (Oxford: OUP).

MULLIN, R. C. (1964), 'A combinatorial proof of the existence of Galois fields', *American Mathematical Monthly*, 71, 901–902.

PAPPUS, *Collection*, vol. iv, 36, 57–59. Reprinted and translated in Thomas (1939), 347–353. Page references are to this reprinting.

PLATO *Parmenides* (anc1), in *Plato: The Collected Dialogues*, ed. Hamilton and Cairns, Bollingen Series LXXI (Princeton University Press, 1961).

——— *Philebus* (anc2), in *Plato: The Collected Dialogues*, ed. Hamilton and Cairns, Bollingen Series LXXI (Princeton University Press, 1961).

——— *Republic* (anc3), in *Plato: The Collected Dialogues*, ed. Hamilton and Cairns, Bollingen Series LXXI (Princeton University Press, 1961).

RAMUS, Petrus (1574), *The Logike of the Moste Excellent Philosopher P. Ramus Martyr*. English trans. by M. Roll. Imprinted at London by Thomas Vantroullier dwelling in the Blackefrieres, 1574.

ROTA, Gian-Carlo (1997a), *Indiscrete Thoughts* (Boston: Birkhäuser).

——— (1997b), 'The phenomenology of mathematical truth', in Rota (1997a), ch. IX.

SELBERG, Atle (1949), 'An elementary proof of the prime number theorem', *Annals of Mathematics*, 50, 305–313.

SPITZNAGEL, Edward (1970), 'An elementary proof that primes are scarce', *American Mathematical Monthly*, 77, 396–397.

STANTON, Dennis and ZEILBERGER, Doron (1989), 'The Odlyzko conjecture and O'Hara's unimodality proof', *Proceedings of the American Mathematical Society*, 107, 39–42.

THOMAS, Ivor (1939), *Greek Mathematical Works: Thales to Euclid*, Loeb Classical Library (Cambridge, MA: Harvard University Press). Reprint of 1980 revision of 1939 original.

——— (1941), *Greek Mathematical Works: Aristarchus to Pappus*, Loeb Classical Library (Cambridge, MA: Harvard University Press). Reprint of 1993 revision of 1941 original.

VIÈTE, Francois (1983), *Introduction to the Analytic Art*, ed. and trans. T. R. Witmer (Ohio: Kent State University Press).

WALLIS, John (1685), *A Treatise of Algebra, both historical and practical: shewing the original, progress, and advancement thereof, from time to time, and by what steps it hath attained to the height at which it now is*. Printed for Richard Davis by John Playford, London, 1685.

WOO, Joseph (1971), 'An elementary proof of the Lebesgue decomposition theorem', *American Mathematical Monthly*, 78, 783.

8

Reflections on the Purity of Method in Hilbert's *Grundlagen der Geometrie*

MICHAEL HALLETT

8.1 Introduction: The 'Purity of Method' in the Grundlagen

The publication of Hilbert's monograph *Grundlagen der Geometrie* in 1899 marks the beginning of the modern study of the foundations of mathematics. In the *Schlusswort* to the book, Hilbert explicitly mentions the 'purity of method' question. He states first that his book was guided by 'the basic principle' of elucidating given problems in such a way as to decide whether they can be answered in a '*prescribed way with certain restricted methods*', or not so answered, and then designates this as a general principle governing the search for mathematical knowledge:

> This basic principle seems to me to contain a general and natural prescription. In fact, whenever in our mathematical work we encounter a problem or conjecture a theorem, our drive for knowledge [*Erkenntnistrieb*] is only then satisfied when we have succeeded in giving the complete solution of the problem and the rigorous proof of the theorem, or when we recognise clearly the grounds for the

---

This paper is a much expanded version of talks given to the Logic Group of the Department of Philosophy at Stanford in April 2004, and to the Annual Meeting of the Association of Symbolic Logic in Pittsburgh in May 2004. An abbreviated talk based on it was given at the HOPOS Biennial Conference in Paris in June 2006. I am grateful to the various participants for discussion and comments, especially Yvon Gauthier, and to Bill Demopoulos and Dirk Schlimm, who suggested helpful modifications to an earlier written version. I am also grateful to my fellow General Editors of Hilbert's unpublished lectures on foundational subjects, William Ewald, Ulrich Majer, and Wilfried Sieg, for discussions on these and other issues over many years. The generous financial support of the Social Sciences and Humanities Research Council of Canada is gratefully acknowledged.

impossibility of success and thereby the necessity of the failure. (Hilbert (1899, p. 89), p. 525 of Hallett and Majer (2004))

Hilbert goes on to stress the importance of impossibility proofs:

> Hence, in recent mathematics the question as to the impossibility of certain solutions or problems plays a very prominent role, and striving to answer a question of this kind was often the stimulus for the discovery of new and fruitful domains of research. By way of example, we recall *Abel's* proof for the impossibility of solving the equation of the fifth degree by taking roots, further the establishment of the unprovability of the Parallel Axiom, and *Hermite's* and *Lindemann's* proofs of the impossibility of constructing the numbers e and π in an algebraic way. (*Loc. cit.*, p. 89, p. 526 of Hallett and Majer (2004))

Hilbert now associates the basic principle explicitly with the idea of the 'purity of method':

> This basic principle, according to which one ought to elucidate the possibility of proofs, is very closely connected with the demand for the 'purity of method' of proof methods stressed by many modern mathematicians. At root, this demand is nothing other than a subjective interpretation of the basic principle followed here. (*Loc. cit.*, pp. 89–90; p. 526 of Hallett and Majer (2004))

Hilbert does not say what he means by the search for 'purity of method', nor who the 'many modern mathematicians' are, nor why he thinks the search for 'purity' is subjectively coloured. However, part of what he means by questions of purity is this: one can enquire of a given proof or of a given mathematical development whether or not the means it uses are 'appropriate' to the subject matter, whether one way of doing things is 'right', whereas another, equivalent way is 'improper'. Is it 'appropriate' to use complex function theory in proofs in number theory, as had become common following the work of Riemann and Dirichlet, or transfinite numbers in proofs in analysis or point-set theory?[1] Is it 'right' or 'better' to pursue geometry synthetically rather than analytically? Hilbert's reaction to this question would be to say that neither way is 'right', that each is the right way to do things if certain purposes are kept in mind, and the 'right' approach is to embrace both developments. What is genuinely explanatory for one mathematician might be simply opaque for another, and to insist on one method would be regarded as unjustifiably idiosyncratic, or (as he says here) 'subjective'. No one kind of mathematical knowledge is, in general, superior to another. Again, the 'appropriateness' of proof methods might easily be taken as a matter of personal taste. More importantly, to put

---

[1] For analysis of the use of transfinite numbers in the original proof of the Heine–Borel Theorem, see Hallett (1979*a, b*).

all these questions on an objective footing requires gathering certain kinds of information. For example, we require answers to questions of the following sort: Can two different ways of developing a theory match each other in the kinds of theorem they can prove? Can a certain theorem *only* be proved by using the extended theoretical means? Are the connections so established accidental, or can one find a deeper theoretical reason for them?[2] To answer such questions as these, as Hilbert suggests, is just to pursue the 'basic principle' set out in the *Grundlagen der Geometrie*. Certainly this is what Hilbert does *say* with respect to the principle's application in the case of his work on geometry:

> In fact, the geometric investigation carried out here seeks in general to cast light on the question of which axioms, assumptions or auxilliary means are necessary in the proof of a given elementary geometrical truth, and it is left up to discretionary judgement [*Ermessen*] in each individual case which method of proof is to be preferred, depending on the standpoint adopted. (*Loc. cit.*, pp. 89–90, p. 526 in Hallett and Majer (2004))

In other words, in geometry, no general decision is to be made as to what is to be preferred and what not. The purpose of the foundational investigation, neutrally stated as it is in the *Grundlagen der Geometrie*, is therefore to assess what we might call the 'logical weight' of the axioms and central theorems. This, in any case, is what there is in abundance in Hilbert's work on geometry. We will come to some examples in due course.

The *Grundlagen* itself was immediately preceded by a long series of lectures held in 1898/1899, which Hilbert entitled 'Elemente der Euklidischen Geometrie'. These lectures contain most of what is novel in the *Grundlagen*, but they contain also many more philosophical and informal remarks, and are very differently arranged from the presentation in the *Grundlagen der Geometrie*. The notes for the lectures exist in two different forms: 110 pages of Hilbert's own notes, and then a beautifully executed protocol of the lectures following the notes very carefully (in German, an '*Ausarbeitung*'), which Hilbert had commissioned from his first doctoral student at Göttingen, Hans von Schaper, whose own field of research was analytic number theory, specifically the Prime Number Theorem.[3] Towards the end of these notes, there are several passages which are clearly the origin of the citations from the *Grundlagen* given above.

---

[2] Recall the emphasis in Hilbert's paper of 1918 on the '*Tieferlegung der Fundamente*'. See Hilbert (1918).

[3] For a more detailed description of the relationship between the 1898/1899 lectures and the *Grundlagen* of 1899, as well as that between the notes and the *Ausarbeitung*, see my Introduction to Chapter 4 in Hallett and Majer (2004).

At the end of the main part of the *Ausarbeitung*, there is the following statement about unprovability results:

> An essential part of our investigation consisted in *proofs of the unprovability* of certain propositions; in conclusion, we recall that proofs of this kind play a large role in modern mathematics, and have shown themselves to be fruitful. One only has to think of the squaring of the circle, of the solution of equations of the fifth degree by extracting roots, Poincaré's theorem that there are no unique integrals except for the known ones, etc. (Hilbert (*1899, p. 169), p. 392 of Hallett and Majer (2004))[4]

In his own notes for these lectures, Hilbert writes:

> The subjects we deal with here are old, originating with Euclid: the principle of the proof of unprovability is modern and arises first with two problems, the squaring of the circle and the Parallel Axiom. <[*Interlineated addition:*] Thus, solution of a problem impossible or impossible with certain means. With this is connected the demand for the purity of method.> However, we wish to set this as a modern principle: One should not stand aside when something in mathematics does not succeed; one should only be satisfied when we have gained insight into its unprovability. Most fruitful and deepest principle in mathematics. (Hilbert (*1898/1899, p. 106), p. 284 in Hallett and Majer (2004))

Moreover, earlier in his notes Hilbert had written the following (the particular example concerned will be discussed in Section 8.4.1; for the moment the details do not matter):

> Thus here for the first time we subject *the means of carrying out a proof* to a critical *analysis*. It is modern everywhere to guarantee the *purity* of method. Indeed, this is quite in order. In many cases our understanding is not satisfied when, in a proof of a proposition of arithmetic, we appeal to geometry, or in proving a *geometrical truth* we draw on *function theory*.[5]

But Hilbert immediately goes on to say this:

> Nevertheless, drawing on differently constituted means has frequently a *deeper and justified* ground, and this has uncovered beautiful and *fruitful relations*; e.g. the prime number problem and the $\zeta(x)$ function, potential theory and analytic functions, etc. In any case, one should never leave such an occurrence of the mutual interaction of different domains unattended. (Hilbert (*1898/1899, p. 30), p. 237 of Hallett and Majer (2004))

Thus, even while he acknowledges the epistemological disquiet behind many purity questions, Hilbert admits that 'impure' mixtures might point

---

[4] Bibliographical items with the date preceded by an asterisk were unpublished by Hilbert.
[5] This remark has a special interest in view of one of the examples I will present later, namely that concerning the analysis of the proof of the elementary Isoceles Triangle Theorem.

to something important and deep. Thus, the first lesson which might be drawn is that a standard 'purity' question ('Is this means necessary to this end?') is often an occasion for a *foundational* investigation, and this is not carried out to show that a certain kind of preferred knowledge is or is not sufficient, but rather for reasons of *mathematical* productivity and logical clarity, and that answering such possibility and impossibility questions was frequently the occasion for opening up 'new and fruitful domains of research', as happened, say, with Abel's investigation.

The heuristic value of unsolved problems was stressed in a powerful way by Hilbert in his famous lecture on mathematical problems (Hilbert, 1900c) held a little over a year later. Hilbert stresses that the lack of a solution to a problem might well raise the suspicion that it is indeed insoluble as stated, or with the expected means, and that one might then seek a demonstration of the relevant *unprovability*. Such a demonstration for Hilbert counts as 'a fully satisfactory and rigorous solution' of the problem at hand (for example, the problems concerning whether we can prove the Parallel Axiom or square the circle), 'although in a different sense from that originally intended'; see Hilbert (1900c, p. 261), English translation, p. 1102. Settling such problems in this way is, as Hilbert says, in no large part responsible for the legendary mathematical optimism he displays, i.e. for his 'conviction' of the solvability of every mathematical problem. Hilbert says of the achievement of impossibility proofs:

> It is probably this remarkable fact alongside other, philosophical, reasons which gives rise in us to the conviction (shared by every mathematician, but which, at least hitherto, no one has supported by a proof) that every definite mathematical problem must necessarily be susceptible of exact settlement, be it in the form of an answer to the question as first posed, or be it in the form of a proof of the impossibility of its solution, whereby it will be shown that all attempts must necessarily fail. ...
>
> This conviction of the solvability of every mathematical problem is a powerful spur in our work. We hear within us the perpetual call: *There is the problem; seek its solution. You can find it by pure thought, for in mathematics there is no ignorabimus!* (Hilbert, 1900c, pp. 261−262)

This is not merely a peculiarity characteristic of mathematical thought alone, but rather what he calls a 'general law' (or an 'axiom') inherent in the nature of the mind, that all questions 'which it [the mind] asks must be answerable'. This is an important remark, and I will return to it. For the moment, suffice it to say that this is why Hilbert seeks 'projection onto the conceptual level', a term we will elucidate in Section 8.3.1 (see especially p. 217).

These remarks suggest why Hilbert says that seeking insight into apparent cases of unprovability is the 'most fruitful and deepest principle in mathematics', and partly explain therefore his interest in purity questions. But it would be misleading to suggest that Hilbert regards 'purity' questions as *merely* heuristic in this way, at least judging by his foundational work on geometry, which Hilbert took to be the paradigm of foundational analysis. Purity of method investigations, in so far as they concern not just open problems but also the analysis of the sources of mathematical knowledge, *was* fundamentally important to Hilbert. But to understand exactly how, it is necessary to recognize the novelty of Hilbert's approach to foundational issues, and to see the considerable effect this had on the type of mathematical knowledge obtained through a purity investigation. There is a sense in which mixture of mathematical domains is intrinsic to Hilbert's investigations; on the one hand, higher-level mathematics is essential to foundational investigation of theories, even relatively low-level ones, and this higher level also instructs the fundamental source of geometrical knowledge. On the other hand, the examples given in Sections 8.4.1–8.4.3 make it clear that, while 'purity' results *do* throw light on on the question of the appropriate sources of knowledge for geometry, parallel to this are abstract mathematical/logical results at what Hilbert calls the 'conceptual level'. Not only does this level largely emancipate mathematics from the epistemological constraints of the 'appropriate', but it is an essential part of what is attained by a 'purity' result. But there is a twist, which we will consider at the end in Section 8.5.

8.2 The foundational project

Before we go further, it is worth pointing out a number of things about Hilbert's approach to geometry.

In the first place, Hilbert's axiomatic presentation as it appeared in the *Grundlagen* of 1899 (and, to some extent, the preceding lectures) builds on the tradition of synthetic geometry, a tradition which saw a strong revival through the work of Monge and von Staudt in the middle of the 19th century. In particular, Hilbert attempted to avoid where possible the direct intrusion of numerical elements. Underlying this, at least in part, was a view that geometry is of empirical and intuitive origin, and concerns 'the properties of things in space'.[6] This is concisely summarized in Hilbert's introduction to his 1891 lectures on projective geometry:

---

[6] This view is what underlies the *Vorlesungen* of Pasch from 1882. It is, of course, of much older provenance.

> I require *intuition and experimentation*, just as with the founding of physical laws, where also the *subject matter* [Materie] *is given through the senses.*
>
> In fact, therefore, the *oldest geometry* arises from *contemplation of things* in space, as they are given *in daily life*, and like all science at the beginning had posed *problems of practical importance*. It also rests on the *simplest kind of experimentation* that one can perform, namely on *drawing*. (Hilbert (*1891, Introduction, p. 7), p. 23 in Hallett and Majer (2004))

But he goes on to say that Euclidean geometry had '... an *essential defect*: it had no *general method*, without which a *fruitful further development* of the science is impossible' (*ibid.*, p. 8). This defect was rectified through the invention of analytic (Cartesian) geometry, which indeed provided a powerful, unified method. Nevertheless, this brought its own disadvantages:

> As important as this *step forward* was, and as *wonderful as the successes were*, nevertheless *geometry* as such *in the end suffered* under the *one-sided development of this* method. One *calculated* exclusively, *without having any intuition of what was calculated*. One lost the *sense for the geometrical figure*, and for *the geometrical construction*. (Hibert (*1891, p. 10), p. 24 in Hallett and Majer (2004))

In what follows, Hilbert makes it clear that he sees the movement in the 19th century to promote synthetic geometry as at least in part a reaction to this. This movement concentrated on projective geometry (what was often called '*Geometrie der Lage*'), but Hilbert's aim was to reformulate and restructure full Euclidean geometry itself as far as possible in an essentially synthetic way. In doing so, he develops geometry in a modern axiom system building up from the simplest possible projective framework (an incidence and order geometry), and arriving at full Euclidean geometry with, for example, the standard results from a Euclidean theory of congruence, of proportions, area (or surface measure), and parallels, all of which is sometimes called by Hilbert 'school geometry', and developed (where possible) before continuity is broached. This synthetic restructuring is much clearer in the 1898/1899 lectures which preceded the *Grundlagen* than it is in the *Grundlagen* itself.

The tone of the remarks from 1891 leaves the impression that, according to Hilbert, Euclidean geometry represents knowledge of a certain kind, an impression which is strengthened by Hilbert's repeated declaration that an axiomatisation in any area of science always begins with a certain domain of 'facts [*Tatsachen*]'. By this, Hilbert does not mean just facts in the sense of empirical facts or even established truths, though such things might be included, but simply what over time has come to be accepted, for example, from an accumulation of proofs or observations. Geometry, of course, is the central example: there are empirical investigations, over 2,000 years of mathematical development, including 300 years' experience with analytic geometry, and the 19th-century growth of this into function-theoretic investigations, the discovery of various 'impossibilities', and above all the independence of Euclid's Axiom of Parallels. Within this domain, certain 'basic facts' are isolated (different sets of 'basic facts' could be chosen, depending on the purpose), and an axiomatization based on these is designed, central assumptions being identified and grouped.

This grouping, fundamental to Hilbert's mature presentation of geometry, is based on a natural conceptual division (e.g. all the congruence assumptions are grouped together), but this conceptual division also in its turn corresponds to different levels of empirical/intuitive justification. For instance, in his short holiday lecture course from 1898, Hilbert says, when presenting Euclidean geometry:

> 4.) 5.) do not have the same *empirical*, constructible character as 1.) 2.) 3.), these being established through a finite number of experiments [*Versuchen*]. (Hilbert (*1898, p. 18), p. 171 in Hallet and Majer (2004))

Axiom groups 1.)–3.) here are the incidence, order, and congruence axioms, while 4.) and 5.) are the Parallel and Archimedean axioms respectively. In the 1898/1899 lecture notes, he writes:

> A *general remark on the character of our axioms I–V* might be pertinent here. The axioms I–III [incidence, order, congruence] state very simple, one could even say, original facts; their validity in nature can easily be demonstrated through experiment. Against this, however, the validity of IV and V [parallels and continuity in the form of the Archimedean Axiom] is not so immediately clear. The experimental confirmation of these demands a greater number of experiments. (Hilbert (*1899, pp. 145–146), p. 380 in Hallett and Majer (2004))

For Hilbert, central independence results concerning the axioms underline the conceptual division, for they show that different levels of empirical/intuitive support are essential. For example, Hilbert regarded the Archimedean Axiom, interpreted as about actual space, as something like an empirical principle (or 'highly intuitive'), since it encapsulates the assumption that all distances, the cosmic and the sub-atomic, can be measured along the same scale. But its independence shows that it has to be supported on grounds separate from the arguments for the more elementary axioms, e.g. congruence, since it is not a logical consequence of them, and 'logic does not demand it'.[7]

---

[7] See Hilbert (*1917/1918, 18; 1918, 408–409; 1919, 14–15, book p. 11) and also Hilbert (*1905, 97–98).

Similar points hold about the Parallel Axiom. In his 1891 lectures on projective geometry, Hilbert remarks in a note that:

> This *Parallel Axiom* is furnished by intuition. *Whether this latter is innate or nurtured,* whether *the axiom corresponds to the truth,* whether it *must be confirmed by experience,* or *whether this is not necessary,* none of this concerns us here. We treat of intuition, and this demands the axiom. (Hilbert (*1891, p. 18), p. 27 in Hallett and Majer (2004). See also the 1894 lectures, pp. 88–89, pp. 120–121 of (Hallett and Majer, 2004).

In the *Ausarbeitung* of the 1898/1899 lecture notes, Hilbert is somewhat more circumspect:

> ...the question as to whether our intuition of space has an *a priori* or empirical origin remains elucidated. (Hilbert, *1899, p. 2) p. 303 in Hallett and Majer (2004).

But the proof of the independence of the Axiom shows again that a distinct justification has to be given:

> Even the philosophical value of the investigation should not be underestimated. If we wish to apply geometry to reality [*Wirklichkeit*], then intuition and observation must first be called on. It emerges as advantageous to take certain small bodies as points, very long things with a small cross-section, like for instance taught threads etc., as straight lines, and so on. Then one makes the observation that a straight line is determined by two points, and in this way one observes [as correct] the other facts expressed in the Axioms I–III of the schema of concepts. Non-Euclidean geometry, i.e. the axiomatic investigation of the Parallel Proposition, states then that to know that the angle sum in a triangle is 2 right angles a new observation is necessary, that this is no way follows from the earlier observations (respectively from their idealised and more precise contents). (Hilbert (*1905, pp. 97–98))

For Hilbert, this shows that Euclid's instincts had been quite right, that the axiom is *required* as a new assumption to prove certain central, intuitively correct 'facts' such as the angle-sum theorem or the existence of a rectangle:

> Even if Euclid did not state these axioms [incidence, order, and congruence] completely explicitly, nevertheless they correspond to what was intended by him and his successors down to recent times. However, when Euclid wanted to prove further propositions immediately furnished by intuition, propositions such as the presence of a quadrilateral with four right-angles, so he recognised that these axioms do not suffice, and therefore erected his famous Parallel Axiom. ... The brilliance it demanded to adopt this proposition as an axiom can best be seen in the short historical sketch: Stäckel–Engel Parallellinien, Teubner, 1895 (Stäckel and Engel, 1895. (Hilbert's lecture notes for 1898/1899, p. 70; p. 261 in Hallett and Majer (2004))

Two pages later, Hilbert remarks:

> One can indeed say that Gauss was the *first one* in about *2100 years* who first *understood and completely grasped* why Euclid adopted the Axiom of Parallels as one of the axioms. (*Ibid.*, p. 74; p. 263 in Hallett and Majer (2004))[8]

The empirical/intuitive admixture and the conceptual division and its corresponding demands for different levels of empirical justification is a large part of what lies behind the statement in the published *Grundlagen* (p. 4) that each of the Axiom Groups represents 'certain connected facts of our intuition' (Hallett and Majer, 2004, p. 437).

Just as important as the acknowledgement of the empirical/intuitive roots of geometry is a precise assessment of the relation of Hilbert's work to Euclid's. An important component of Euclid's work in the opening books of the *Elements* was the desire to show that certain central propositions can be established using only 'restricted methods', and thus embodies a certain 'purity of method' concern. By this is not merely meant that Euclid's system was an axiomatization; it is illustrated rather by the way in which Euclid withholds deployment of the Parallel Postulate until rather late on in Book I, and by the way he restricts use of congruence arguments using the translation and flipping of figures in the plane, even when proofs would be easier when these techniques are employed. Some of Hilbert's work in geometry is a direct continuation of this kind of investigation. For one thing, Hilbert's lectures of 1898/1899 which immediately precede the *Grundlagen*, are on Euclidean geometry, concentrating on, and analyzing, the theoretical results in the early part of the *Elements*, in particular on congruence, proportion and area, the involvement (or exclusion) of the Archimedean Axiom, and the Parallel Axiom itself. Hilbert shows, for example, that we can dispense with what is initially adopted in his lectures as Axiom 9, added to the Euclidean system by many of Hilbert's predecessors in the 19th century, stating the assumption that all 'straight angles' (angles on a straight-line) are congruent. More significantly, Hilbert also shows that a Euclidean theory of linear proportion and of surface content (roughly, polygonal area) can be developed without the implicit assumption that the content measures or the 'lengths' are themselves magnitudes, an assumption he directly criticizes Euclid for.[9] Connected to this is Hilbert's proof of what

---

[8] The examination of the Parallel Axiom is an excellent example of what Hilbert calls 'analyzing intuition': independence confirms Euclid's 'intuition' to use the axiom. We will comment on 'analyzing intuition' below; see the remarks on p. 292.

[9] Hilbert's student Dehn showed that the same does *not* hold for tetrahedral/polyhedral volume; the Archimedean Axiom is required. Hilbert posed the problem in the late 1890s; see the 1898 *Ferienkurs* (p. 26), and the notes for the 1898/1899 lectures (pp. 106 and 169 respectively), all in Hallett and Majer

he calls the Pascal Theorem (usually called Pappus's Theorem) using just the plane part of the elementary axioms together with congruence, but without any involvement of continuity, a result new to Hilbert. More generally, the whole of 'school geometry' can be developed without continuity assumptions, showing that Euclid was right not to include explicit continuity principles among his axioms.

Of course, there are ways in which Hilbert's treatment is not 'purely' or historically Euclidean, for example in developing the core of projective geometry on the basis of the order and incidence axioms alone (i.e. before the Congruence and Parallel Axioms have been introduced), and in its further concentration on the Desargues and Pascal theorems. An underlying concern here is the relationship between Euclidean geometry and the coordinate structure of analytic geometry, for part of what Hilbert investigates are the hidden field properties in segment structure showing that the basic magnitude principles (represented by the core of the ordered field axioms) are true of linear segments once addition and multiplication operations have been defined for them in a reasonable way. Part of the point is surely to defend Euclid, in that Hilbert shows that the 'theory of magnitudes' arises intrinsically, and does not have to be imposed from without by some extra assumptions, but another motive is clearly to show that the central guiding assumption of analytic geometry, coordinatization by real numbers, is not *ad hoc*, a central concern of Hilbert's since at least the 1893/1894 lectures. The guiding insight is clearly that, since analytic geometry was, at the time Hilbert was writing, the pre-eminent way to pursue Euclidean geometry, careful analysis of any suitable synthetic replacement should reveal some central conceptual parallels with analytic structures.[10] Indeed, despite the desire to keep synthetic Euclidean geometry as far as possible independent of analytic geometry, Hilbert did impose a strong adequacy condition, namely 'completeness' with respect to analytic geometry, i.e. the demand that a satisfactory synthetic axiomatization should be able to prove all the geometrical results that analytic geometry could.[11]

Nevertheless, despite these modern flourishes, it is clear (above all from the 1898/1899 lectures) that Hilbert's own investigations were profoundly influenced by Euclid's. In the Introduction to his 1891 lectures on projective geometry, Hilbert gives a short but highly illustrative survey of geometry. He

---

(2004, pp. 177, 284, 392 respectively). The problem reappears as Problem 3 in Hilbert's famous list of mathematical problems set out in 1900; see Hilbert (1900c, 266–267).

[10] Desargues's Theorem is essential to this: see Section 8.4.1.

[11] For a discussion of what Hilbert meant by completeness in 1899 when calling for a 'complete' axiomatization of geometry, see Section 5 of my Introduction to Chapter 5 in Hallett and Majer (2004, 426–435).

divides geometry into three domains: (1) intuitive geometry, which includes 'school geometry', projective geometry, and what he calls 'analysis situs'; (2) axiomatic geometry; and (3) analytic geometry. Axiomatic geometry, according to Hilbert, 'investigates which axioms are used in the garnering of the facts in intuitive geometry, and sets up systematically for comparison those geometries in which various of these axioms are omitted', and its main importance is 'epistemological'.[12] The description of axiomatic geometry is a reasonably good, rough description of much of what Hilbert actually carries out systematically in the period from 1898 to 1903. He does indeed investigate the 'facts' obtained from intuition; in the *Grundlagen* (p. 3) he actually describes his project as being fundamentally a 'logical analysis of our spatial intuition' (Hallett and Majer, 2004, p. 436), a description which also appears in the *Ausarbeitung* of the 1898/1899 lecture notes (p. 2), where he says 'we can outline our task as constituting a *logical analysis of our faculty of intuition [Anschauungsvermögens]*', and where 'the question of whether spatial intuition has an apriori or empirical character is not hereby elucidated' (see Hallett and Majer 2004, p. 303). Hilbert *does* consider the geometries one obtains when various axioms are 'set aside', non-Euclidean geometry, non-Archimedean geometry, non-Pythagorean geometry, etc., and we will see some examples later on. Such investigations necessarily embrace questions of what can be, and cannot be, proved on the basis of certain central propositions, either axioms or central theorems. That the main benefit of doing this is said by Hilbert to be 'epistemological' is also understandable. If empirical investigation and geometrical intuition are the first sources of geometrical knowledge, then Hilbert's dissection of Euclidean geometry is indeed an 'analysis' of this source, revealing the propositions responsible for various central parts of our intuitive geometrical knowledge.

In short, surely some of Hilbert's work can be seen as stemming from a rather straightforward epistemological concern with the purity of method, namely showing that P (or some theoretical development T) can be deduced solely using some specified axioms Σ (or more generally $\Sigma - \Gamma$), and this is designed at least to explore the epistemological underpinnings of the axioms.[13] This kind of investigation fits with the *traditional* conception of axiomatic

---

[12] See Hilbert (*1891, 3–5), pp. 21–22 in Hallett and Majer (2004).

[13] Another geometrical example outside the framework of Euclidean geometry might be Hilbert's analysis of Lie's work, where Lie is criticized for the way he uses the full theory of differentiable functions to analyze the concept of motion. Hilbert showed that much weaker assumptions than Lie's (assumptions approximating more to the 'ancient Euclid', as he puts it) will suffice, Lie's assumptions being 'foreign to the subject matter, and because of this superfluous'. For further details, see Hallett and Majer (2004, 9).

investigation, and might be used to describe, say, Pasch's work on projective geometry in his (Pasch, 1882), an important influence on Hilbert's work, and also Frege's work on arithmetic, for Frege attempted to show that natural number arithmetic can be derived from purely logical principles alone, with the help only of appropriate definitions. Both Frege's and Pasch's projects attempt to show that, when properly reduced or reconstructed, the respective mathematical theories represent knowledge of a certain, definite kind, logical and empirical respectively. The projects thus have conscious epistemological aims, and are very broadly 'Euclidean' in the general sense that they attempt to demonstrate that certain bodies of knowledge can be *deduced* from a stock of principles circumscribed in advance, a demonstration which can only be effected by the actual construction of the deductions.

While these *are* very important elements in Hilbert's treatment of Euclidean geometry, they are by no means exhaustive.

For one thing, the description of Hilbert's geometrical work as an extension of the Euclidean project does nothing to explain its entirely novel contribution to foundational study. The novelty comes in treating the *converse* of the Euclidean foundational investigation, asking in addition the question: Can we show that a given proposition P (or some theoretical development T) *cannot* be deduced (carried out) solely using the precisely specified 'restricted methods' Σ?[14] In other words, as Hilbert puts it in the *Grundlagen* (pp. 89–90), we seek to 'cast light on the question of which axioms, assumptions or auxilliary means are necessary in the proof of certain elementary geometrical truths'. It is important to see that Hilbert's investigations are not just *complementary* to the Euclidean ones; indeed, to describe them so would be to underemphasize their novelty. The very pursuit of this new line of investigation involved Hilbert in a radical transformation of the axiomatic pursuit of mathematics. As Hilbert expressed it in some lectures from 1921/1922 (Hilbert, *1921/1922, pp. 1–3):

> The further development of the exact sciences brought with it an essential transformation of the axiomatic method. On the one hand, one found that the propositions laid down as axioms in no sense could be held sublimely free from doubt, where no difference of opinion is possible. In particular, in geometry the

---

[14] There are two important variations to this:
1. Show that P (or some theoretical development T) *can* be deduced (carried out) using Σ, but *not* using Σ^- (slight weakening).
2. Show that P (or some theoretical development T) *cannot* be deduced (carried out) using Σ, but *can* with Σ^+ (slight strengthening).

evidence in favour of the Parallel Axiom was called into question.... In this way, there developed the view that the essential thing in the axiomatic method does not consist in the securing of absolute certainty, which is transmitted to the theorems by means of logic, but in this, that the investigation of the logical interconnections is separated off from the question of the actual truth of the axioms.[15]

This Hilbert emphasized as being the 'main service' of the axiomatic method. The interest is therefore much more logical than epistemological.

The remark of Hilbert's just quoted leads to a second sense in which Hilbert's work is new, and which has not yet become clear: part of the point of Hilbert's investigation is to effect an *emancipation* from the sources of knowledge provided by the 'facts' or by intuition. As I hope will be shown, this plays a significant part in understanding the role of purity of method investigations in Hilbert's investigation of geometry.

The radicalness of Hilbert's approach has three important features, to be dealt with briefly in the next section.

8.3 Independence and metamathematical investigation

8.3.1 Interpretation

It is clear that in his study of geometry, Hilbert's focus is less on questions of provability from a circumscribed stock of principles than on *unprovability*, in other words, on independence. The basic technique which Hilbert adopted for this investigation is that of modelling, more strictly, of translating the theory to be investigated into another mathematical theory. For this, it is essential (and Hilbert is very clear about this) that the primitive concepts employed are not tied to their usual fixed meanings; they must be free for *reinterpretation*. Indeed, in his 1898/1899 lectures Hilbert stresses that the most difficult part in carrying out the investigation will be separating the basic terms from their usual, intuitive associations.[16] It follows that the axioms cease to have fixed meaning, and thus cease to be, for someone like Frege, genuine axioms at all.

This was not just a matter of expediency for Hilbert, done for the sake of the independence proofs; along with it goes a new picture of mature mathematics,

---

[15] See also Bernays (1922, 95). Bernays's remarks are very reminiscent of the long passage from Hilbert (*1921/1922) just quoted. Bernays was the *Ausarbeiter* for the lecture notes.
[16] See p. 7 of the 1898/1899 lectures (Hallett and Majer, 2004, 223).

intrinsic to which is the view that in general no *one* interpretation of an axiom system is privileged above others, despite what might seem like the overwhelming weight of the interpretation underlying the 'facts' as originally given: thus in the case of geometry the weight of the 'intuitive' or 'empirical' origins. Thus, we see a radical departure from the kind of enterprise Frege and Pasch (and even Euclid) were engaged in, enterprises part of whose very point was to *explain* (and thereby *delimit*) the primitives. Hilbert's axiomatic method abandons the direct concern with the kind of knowledge the individual propositions represent because they are about the primitives they are, and concentrates instead on what he calls 'the logical relationships' between the propositions in a theory.

Hilbert's fundamental supposition of foundational investigation, going back to 1894 and stated repeatedly thereafter, is that a theory is only 'a schema of concepts' which can be variously 'filled with material'. He says:

> In general one must say: Our theory furnishes only the schema of concepts, which are connected to one another through the unalterable laws of logic. It is left to the human understanding how it applies this to appearances, how it fills it with material [*Stoff*]. This can happen in a great many ways. (Hilbert (*1893/1894, p. 60), or Hallett and Majer (2004, p. 104))

In the 1921/1922 lectures already referred to, Hilbert calls axiomatization of this kind a 'projection into the conceptual sphere':

> According to this point of view, the method of the axiomatic construction of a theory presents itself as the procedure of the mapping [*Abbildung*] of a domain of knowledge onto a framework of concepts, which is carried out in such a way that to the objects of the domain of knowledge there now correspond the concepts, and to statements about the objects there correspond the logical relations between the concepts.
>
> Through this mapping, the investigation is completely severed from concrete reality [*Wirklichkeit*]. The theory has now absolutely nothing more to do with the real subject matter or with the intuitive content of knowledge; it is a pure *Gedankengebilde* [construct of thought] about which one cannot say that it is true or it is false. (Hilbert (*1921/1922, p. 3), forthcoming in Ewald and Sieg (2008))

It is not that a mathematical theory in this sense has nothing to do with reality; indeed it may have *more* to do with it, for the connections might be established 'in a great many ways', to use Hilbert's phrase from 1894.

It is important to see how this view fits with the insistence in the 1890s, outlined above, that the root of geometry is empirical, and that geometry is, as Hilbert frequently put it, the 'most perfect natural science'. For Hilbert, geometry is a natural science primarily because it can be *applied* to nature to furnish a more or less accurate *description* of it. The term 'description' is important. In the *Vorrede* to his lectures on mechanics published in 1877 (Kirchhoff, 1877), Kirchhoff insists that physics should only set itself the restricted aim of describing the phenomena, and not that of trying to get at the underlying 'causes', since these frequently suffer from deep-seated 'conceptual unclarity'. In his 1894 lectures on the foundations of geometry, Hilbert refers directly to Kirchhoff's aim of 'describing' only, and states that correspondingly the aim of geometry is to 'describe' the 'geometrical facts'. See Hilbert (*1893/1894, p. 7) or Hallett and Majer (2004, p. 72). And the basis of application in this sense is *interpretation* (or, as Hilbert says, '*Deutung*'), and this is essentially inexact. In these 1894 lectures, Hilbert says:

> With the axioms given hitherto, the existence [incidence] and position [order] axioms, we can already describe a large collection of geometrical facts and phenomena. We require only to take bodies for points, straight lines and planes, for the relation of passing through, touching, for being definite, immovable or fixed (perhaps, in the unrefined sense, when nudged by the hand). The bodies we should think of as finite in number, and such that the axioms are satisfied under this interpretation [*Deutung*] (for which, as one recognises, it will be necessary to have for the bodies taken in place of points, straight lines, planes, something like grains, rods or stretched threads or wire, cardboard) and indeed precisely. Then we know that all the propositions set up so far are also satisfied, and indeed precisely satisfied.

The direct continuation of this passage also makes it clear that applications only hold approximately, and this is stated in one breath with the claim that theories are only schemata of concepts, represented here by the ellipsis:

> If one finds that, with an application, the propositions are not satisfied (or not precisely satisfied), this arises because an inappropriate application has been taken, i.e. the bodies, movement, touching do not agree with our scheme of axioms. In this case it will be necessary to replace the things: bodies, movable, touching, by others, perhaps by smaller grains, blots [*Klexe* (sic)], tips, thinner wires, thinner cardboard, touching with firmer contact, movability [of the bodies] even when we blow on them [*Anpusten*], in such a way that the axioms are satisfied. Then we know that the propositions also hold (precisely). ... But always when the axioms are satisfied, the propositions also hold. The easier and more far reaching the application, then so much better*) the theory.

---

*) All systems of units and axioms which describe the phenomena are equally justified. Show nevertheless that the system given here is in a certain respect the uniquely possible one. (Hilbert (*1893/1894, pp. 60–60A), pp. 104–105 in Hallett and Majer (2004))

Hilbert repeatedly stresses the inexactness of application, for example on p. 92 of the 1893/1894 lectures (Hallett and Majer, 2004, p. 122), or p. 106 of the

1898/1899 lectures (Hallett and Majer, 2004, p. 283). And he even states the approximate nature of the physical interpretations of geometry as a *reason* why we require the logical development of geometry separately:

> In physics and nature generally, and even in practical geometry, the axioms all hold only approximately (perhaps even the Archimedean Axiom). One must however take the axioms precisely, and then draw the precise logical consequences, because otherwise one would obtain absolutely no logical overview. Necessarily finite number of axioms, because of the finitude of our thought. (Note on the front cover of Hilbert's copy of the *Ausarbeitung*. Date unknown, but after 1899, and very probably before 1902; see Appendix to Chapter 4 in Hallett and Majer (2004, p. 401), remark [15].)

Two things are immediately clear from these passages. (1) Many interpretations are possible, and even desirable, even where we are concerned apparently with only one general area of application (the application of the *same* theory to the *same* spatial world). (2) There is an implicit assumption that the (unspecified) logical apparatus is *sound*, i.e. if the axioms hold under an interpretation, then so do the theorems. (See also Hilbert's letter to Frege of 29.xii.1899.) Most importantly, this latter means that the internal (logical) workings of the theory, in short ('pure' or 'free') mathematics, can proceed independently of any particular application, and this is the case even when one might firmly believe that application is the primary purpose. This is stressed by Hilbert in the continuation of the passage from the 1921/1922 lectures quoted on page 296:

> Nevertheless this framework of concepts has a meaning for knowledge of the actual world, because it represents a 'possible form in which things are actually connected'. It is the task of mathematics to develop such conceptual frameworks in a logical way, be it that one is led to them by experience or by systematic speculation. (Hilbert (*1921/1922, p. 3) in Ewald and Sieg (2008))

In short, interpretation (quite possibly manifold interpretations) can establish *various* connections to the world, and the more the better.[17] Thus, the mathematical theory is not *determined* by *Wirklichkeit*, it does not necessarily extend our knowledge of it (it might or might not), and is in the end not responsible to it. Mathematics can learn from intuition, observation, and empirical investigation more generally, but is not to be their slave, even when they have played a major part in the establishment of the domain of 'facts', and therefore in the axiomatization itself. A prime example is the

---

[17] See again Hilbert's letter to Frege of 29.xii.1899. In his own letter of 27.xii.1899, to which Hilbert's is a reply, Frege had objected to the very idea of considering different interpretations for geometry.

formulation of the congruence axioms. Linear congruence in geometry, both the idea of congruence, and the central propositions governing it, was originally motivated by simple observations about movement of rigid bodies in space, but Hilbert's axioms are no longer to do with movement itself, though the connection between them and the movement of rigid bodies is not hard to discern. Rather, Hilbert's view (see Hilbert, *1899, pp. 59–60) is that a proper mathematical analysis of spatial movement *requires* an independently established and neutral notion of congruence. Thus the abstract notions are to be applied in the analysis of movement, but geometry with Hilbert's congruence axioms is not dependent upon the purely empirical matter of whether or not there are in fact rigid bodies, and thus whether bodies can indeed ever be congruent in the intuitive sense. In pursuing mathematical investigations of geometry, one is investigating 'possible forms of connection' and not necessarily 'actual connections', an emancipation which reflects the ascent to what Hilbert calls the conceptual.

It follows from this account that even *if* the underlying notion of intuition were strong enough to guarantee the 'apodeictic certainty' of the axioms, it would still be possible to drop axioms, or modify them, or replace them, and it would still be part of the task of mathematics to investigate the consequences of so doing. In other words, a strong notion of intuition would not restrain Hilbert's axiomatic programme.

There is another important consequence of this, namely that theories cannot be straightforwardly true or false through correctly representing some fixed subject matter, or failing to represent it. This is stated clearly in the 1922/1923 lectures, but it is already clear in the 1898/1899 lectures and the 1899 correspondence with Frege. Theories, variously interpretable, are either 'possible' or not, and what shows their 'possibility' is a demonstration of consistency, in which case the mathematical theory 'exists'. Thus, for Hilbert, the correct account of truth and falsity with respect to mathematical theories is that of consistency/inconsistency. Derivatively a mathematical *object* exists (relative to the theory) if an appropriate existence statement can be derived within the consistent theory. As Hilbert puts it in his 1919 lectures:

> What however is meant here by existence? If one looks more closely, one finds that when one speaks of existence, it is always meant with respect to a *definite system taken as given*, and indeed this system is different, according to the theory which we are dealing with. (Hilbert, 1919, p. 147, book p. 90).

Hilbert sums this up in his 1902 lectures on the foundations of geometry:

> We must now show the *freedom from contradiction of these axioms taken together;* ... .
> In order to *facilitate* the *understanding* of this, we begin with a remark:

> The things with which mathematics is concerned are defined through axioms, *brought into life*.
>
> The axioms can be taken quite arbitrarily. However, if these axioms contradict each other, then no logical consequences can be drawn from them; the system defined then does not exist for the mathematician. (Hilbert (*1902, p. 47) or Hallett and Majer (2004, p. 563))

This complex of views is a very important foundation for Hilbert's theory of what he calls 'ideal objects', where the concepts, freed from the constraints of the actual, are 'completed'. In his 1898/1899 lectures, Hilbert notes that while the Archimedean Axiom as sole continuity axiom shows that to every point, there corresponds a real number, the converse will not generally hold:

> That to every real number there corresponds a point of the straight line does not follow from our axioms. We can achieve this, however, by the *introduction of ideal (irrational) points (Cantor's Axiom)*. It can be shown that these ideal points satisfy all the axioms I–V; it is therefore a matter of indifference whether we introduce them first here or at an earlier place. The question whether these ideal points actually 'exist' is for the reason specified completely idle [*völlig müssig*]. As far as our knowledge of the spatial properties of things based on experience is concerned the irrational points are not necessary. Their use is purely a matter of method: *first with their help is it possible to develop analytic geometry to its fullest extent.* (Hilbert (*1899, pp. 166–167); Hallett and Majer (2004, p. 391))

Thus, while one might start from the idea that the points in geometry correspond to points whose existence in actual space can be shown, perhaps through construction, this idea is left behind; as Hilbert says, it is 'idle' to consider the question of whether the new points 'actually exist'.[18] It is precisely in this context that Hilbert first introduces his new notion of mathematical existence; thus, full Euclidean geometry 'exists' because we can furnish a model for the theory by using analysis. Put more abstractly, the procedure is this. We begin with a Euclidean geometrical system without continuity axioms, and where 'real' points are instantiated by geometrical constructions. These constructions correspond to various algebraic fields over the rationals (depending on what is permissible in construction). These number fields are seen to have a maximal extension in the reals; therefore, we postulate that there are points corresponding to *all* real numbers. These new objects are then 'ideal' with respect to the original, real points. The geometrical system corresponding

---

[18] This view, that we can decide to extend the field of objects without being constrained by what 'really exists', is very much in evidence in Dedekind's 1872 memoir on continuity and irrational numbers; see Dedekind (1872, 11). The most notable difference compared with Hilbert is in the idea that objects are 'created' to fill the 'gaps'; Hilbert's reliance on consistency avoids this notion of 'creation'.

to the completion is axiomatized; this new system is consistent, since it has a model in the reals. As Hilbert says:

> *Euclidean geometry exists, so long as we take over from arithmetic the proposition that the laws of the ordinary real numbers lead to no contradiction.* With this we have shown the existence of all those other geometries which we have considered in the course of this investigation. (Hilbert (*1899, p. 167); Hallett and Majer (2004, p. 391). See also Hilbert's own lecture notes, p. 104; in Hallett and Majer (2004, p. 280).)

From the point of view of this theory, any dispute then over the 'real' existence of some points as opposed to others is 'idle'; the *theory* exists, and it has the existence of all these points as consequences; the real/ideal distinction was always only ever relative to the original domain. As Hilbert puts it later:

> The terminology of ideal elements thus properly speaking only has its justification from the point of view of the system we start out from. In the new system we do not at all distinguish between actual and ideal elements. (Hilbert, 1919, p. 149, book p. 91)[19]

Furthermore, the core system of elementary geometry possesses many different and incompatible 'ideal extensions', some of them giving rise to the 'other geometries' Hilbert mentions. All 'exist' in so far as they are consistent.

In short, advanced 'pure' mathematical knowledge is now knowledge about various conceptual schemes, and a significant part of this is logical knowledge, in the first instance, knowledge of the existence or *non*-existence of derivations from certain starting points. It is the abandonment of the idea of fixed content and the shift to the 'conceptual sphere' which allows Hilbert to introduce a sophisticated, and *relative*, conception of the 'real/ideal' distinction, one which is of great importance for his later foundational work.

The arrival at this 'pure thought form' is, I think, the key to the correct explanation for the progression from intuition to concept to ideal stated at the beginning of Hilbert's *Grundlagen* in an epigram from Kant's *Kritik der reinen Verknunft*.[20] And it is as a 'pure thought form', thus through the projection into the 'conceptual sphere', that mathematics falls under the 'general law' on the nature of the mind which Hilbert stated in his lecture on mathematical problems, namely that all questions 'which it [the mind] asks must be answerable'.

Important in all this is the relationship *between* mathematical theories. For one thing, the approximate (not to say coarse) nature of interpretation in

---

[19] See also p. 153 of the 1919 lecture notes (Hilbert, 1919, book p. 94).
[20] The epigram is the following passage: 'So fängt denn alle menschliche Erkenntnis mit Anschauungen an, geht von da zu Begriffen, und endigt mit Ideen.' (See *Kritik der reinen Vernunft*, A702/B730. A very similar remark is also to be found at A298/B355.)

the observable world means that these interpretations are less than ideal as ways of demonstrating 'possibility', and the same holds for investigations of independence. Secondly, the process of ideal extension as described above appears to rely heavily on the notion of embedding one mathematical theory in another, and indeed Hilbert himself describes the procedure in just this way in his 1919 lectures:

> Precisely stated, the method [of ideal elements] consists in this. One takes a system which behaves in a complicated way with respect to certain questions which we want to answer, and one transforms to a new system in which these problems take on a simpler form, and which in addition has the property that it contains a sub-system isomorphic to the original system. ... The theorems concerning the objects of the original system are now special cases of theorems of the new system, in so far as one takes account of the conditions which characterise the sub-system in question. And the advantage of this is that the execution of the proofs takes place in the new system where everything becomes much clearer and easier to take in [*übersichtlich*]. (Hilbert, 1919, pp. 148–149, pp. 90–91 of the book)

This procedure can be brought out, too, by looking more closely at the general form of argument in the independence and relative consistency proofs, a form first articulated by Poincaré in one of his explanations of the proof of the independence of the Parallel Postulate.[21] The general procedure is this. Proving the independence of Q from $P_1, P_2, \ldots, P_n$ with respect to a given conceptual scheme T_1, that is, showing that there cannot be a derivation of Q from $P_1, P_2, \ldots, P_n$, involves showing that, if there were, there would then also be a derivation of $\tau(Q)$ from $\tau(P_1), \tau(P_2), \ldots, \tau(P_n)$ with respect to another conceptual scheme T_2, where: (*a*) we know that each of the $\tau(P_i)$ are *theorems* of T_2; (*b*) we know that the conceptual scheme T_2 can *already* derive $\neg\tau(Q)$; (*c*) we assume (or know) that the scheme T_2 is not in fact inconsistent. Here, crucially, $\tau(P)$ is some map from formulas to formulas which preserves logical structure. (This point was first articulated in Frege (1906).) An obvious variant of this procedure can be used to prove the relative consistency of theory T_1 with respect to T_2. Hilbert puts it this way: After specifying a minimal Pythagorean (countable) field Ω of real numbers which determines an analytic model of the axioms for Euclidean geometry, he continues:

> We conclude from this that any contradiction in the consequences drawn from our axioms would also have to be recognizable in the domain Ω. (Hilbert (1899, p. 21); p. 455 in Hallett and Majer (2004))

Thus, crucial to the investigation are what we might call *base theories*, those through which the propositions of the 'home theory' are interpreted.

[21] See Poincaré (1891).

8.3.2 Non-elementary metamathematics

The second very important thing to recognize about Hilbert's foundational enterprise as developed in his study of geometry follows on from this. When looking for base theories, the natural thing is to choose those over which one has a fine degree of control, perhaps because they have only countably many elements which can be individually described.

I mentioned above that Hilbert tried to separate synthetic geometry from analytic geometry, but the separation is only at the mathematical level; at the metamathematical level, the link is retained. For one thing, as we have remarked, analytic geometry is taken as the measure of synthetic geometry (for example, via the demand for completeness). But more importantly here, real and complex analysis in its broadest sense is taken as the fundamental tool in the logical analysis of synthetic geometry. Since synthetic geometry has as one of its goals the aim of matching analytic geometry, then it should be clear that ordinary analytic geometry based on the complete ordered field of the real numbers provides a model for (disinterpreted) synthetic geometry. It is not a large leap from this to expect that one can find (or fabricate) substructures of the reals (or wider analysis) which will correspond to the variations of axioms and central propositions of synthetic geometry, not least because analytic geometry gives point-by-point and line-by-line control over the geometrical structure. For instance, in the standard arrangement, lines are given by simple linear functions of the number pairs giving the coordinates of points. But in principle, a vast range of other functions could be chosen, showing one kind of behaviour within a certain region, and a quite different behaviours outside that region. Indeed, this is the lesson taught by the various models of non-Euclidean geometry. A fundamental presupposition of Hilbert's investigation, therefore, is the presence of the full panoply of analytic techniques. Thus, while Hilbert's axiomatization of geometry distances itself from the analytic developments of the 19th century, the full range of analytic geometry is made available, not to prove results in the theory itself, but to prove results *about* the theory, and in particular to throw light upon the underlying source of knowledge. I will attempt to draw out the points made here in the examples given in Section 8.4.

8.3.3 Foundations

There is another sense in which Hilbert's project is radically different from other foundational projects at the time, projects with a Euclidean flavour, and this is that Hilbert does not automatically seek a more primitive conceptual level. Of course, this might be done for specific reasons in certain circumstances, and of course important mathematical information might be gained from doing so, whether one succeeds or fails. The point is that it is not the central goal of foundational investigation. This means that, to a large extent, Frege-style definitions can be dispensed with. They appear now rather as assignments for the purpose of modelling the principles being investigated. The prime example is perhaps Hilbert's use of the theory of real numbers (or rather a minimal Pythagorean field drawn from them) to model the axioms of elementary geometry by giving 'temporary' definitions of point, straight line, congruence, and so on; this is in place of the standard procedure behind analytic geometry, which effects a reduction of geometry to analysis by making these assignments as Frege-style definitions. Similarly, one could use the Dedekind-cut construction or the Cauchy-sequence construction to model the axioms for the theory of real numbers, rather than to define them, and likewise with the 'definitions' of the integers as equivalence classes of ordered pairs of natural numbers, complex numbers as ordered pairs of reals, and so on.[22]

There is a further important point to be made. Even if we seek a conceptual reduction, it is important to have available 'local' axioms, for instance for the natural numbers or for the reals. To take an example, in seeking to show in his 1898/1899 lectures (and later the *Grundlagen*) that line segments themselves exhibit a field structure like that of the real numbers, it was necessary for Hilbert to have a detailed axiomatization of field structure, in order to say that the fields are alike in certain respects but differ in others. This is the origin of Hilbert's axiomatisation of the real numbers in the lecture notes, published as axioms for what Hilbert calls 'complex number systems' in the *Grundlagen* of 1899, and then completed in (Hilbert, 1900a). The point is simple: to show that a conceptual 'reduction' (like Frege's) has worked, one has to be able to derive theorems which say that the defined objects (e.g. the numbers) have the right properties.[23] Thus, one has to have in effect an axiom system, and this is *overriding*.

Frege sought a reduction to fewer and fewer principles, as in a sense did Euclid; Hilbert's work on the other hand shows that, in some key cases if

---

[22] Hilbert's later paper 'Axiomatisches Denken' makes the parallel point that the uncovering of ever 'deeper' primitives is a constant theme in the development of mathematics and physics. The paper even seems to suggest that the use of the Axiomatic Method incorporates such a search; he uses the phrase '*Tieferlegung der Fundamente*'. The point to stress, though, is that there is never thought to be an ultimate conceptual layer, and the 'foundations' for a theory might be given, and productively so, in different, even incompatible, ways. As Boolos observes with respect to the Frege analysis of number and the attempt to reduce arithmetic to logic:

> Neither Frege nor Dedekind showed arithmetic to be part of logic. Nor did Russell. Nor did Zermelo or von Neumann. Nor did the author of *Tractatus* 6.02 or his follower Church. They merely shed light on it. (Boolos, 1990, 216–218).

[23] Frege seeks to establish just this in his (1884, §§78–83).

we have more axioms, then our logical analysis can be more refined. A nice example is provided by the maintenance of the Archimedean Axiom as an independent axiom while seeking an additional axiom which leads to point-completeness. A natural axiom would have been some version of the limit axiom which Hilbert had used in earlier axiomatizations (e.g. in his 1893/1894 lectures, in Hilbert (1895), and in his 1898 *Ferienkurs*), and which is even presented as a possibility in the 1898/1899 lectures. But this axiom would 'hide' the Archimedean Axiom because it would imply it. Instead, Hilbert retains the Archimedean Axiom and seeks an axiom which will complement it and boost its power to full completeness. The axiom he chooses is his own '*Vollständigkeitsaxiom*'.[24] Many of Hilbert's metamathematical investigations concern the precise role of the Archimedean Axiom in the proofs of central theorems, typically whether, when used, its use is necessary or not. One example is given by Dehn's work on the Legendre Theorems and the Angle-Sum Theorem (Dehn, 1900), another concerns the involvement of the axiom in the theories of plane area and three-dimensional volume. The axiom and the requisite analyses are also clearly important in the development of non-Archimedean geometries; and its presence *without* the Completeness Axiom is very important in the search for countable models. Yet another example is provided by Hilbert's own work on the Isoceles Triangle Theorem to be considered below in Section 8.4.2.

The point about the definitions, and this point about not reducing the conceptual level, are, of course, closely related. The creative purpose behind defining, as Frege recognized, is that with properly chosen definitions, unprovable propositions become provable, and therefore there is no need to take them (and attendant propositions) as axioms. Examples abound, but a striking one is furnished, of course, by Frege's own work. With the Frege definition of number, we can *prove* what is now known as *Hume's Principle* (*HP*),[25] the principle in effect from which Frege arithmetic is derived. Without such a proof, we would have to take the principle as a primitive axiom, and the terms '*NxFx*' (i.e. the numbers) as primitive terms, contrary to Frege's intention. However, it might be argued that the really revealing thing about Frege's work is not that, with the Frege definition, *HP* is provable, but rather that the central arithmetic principles can be proved from *HP* alone, given Frege's other definitions.

---

[24] See my *Introduction* to Chapter 5 of Hallett and Majer (2004, §5, 426–435).
[25] That is, the second-order proposition: '$\forall F, G[NxF = NxG \leftrightarrow F \approx G]$', where '*F*', '*G*' are variables for concepts, '*NxF*' is a singular term standing for 'the number of things falling under the concept *F*', and where '$\approx$' stands for the (definable) relation of equinumerosity between concepts. See Frege (1884, §63).

So, to sum up the points in these last two sub-sections: for Hilbert, we do *not* generally seek 'more primitive' conceptual levels from which a theory can be finally deduced; moreover, often conceptually 'higher' mathematics is intrinsically necessary for the investigation of conceptual schemes, even elementary ones.

8.4 The 'purity of method' reconsidered

Let us now return to the way that 'purity of method' is dealt with in Hilbert's geometrical work. I will consider three examples, concerning respectively Desargues's Theorem, the Isoceles Triangle Theorem and the Three Chord Theorem, all of them extremely elementary geometrical results, and all three of which touch centrally on the intuitive 'facts' behind geometry. This is no accident. We have seen that for Hilbert the main source of knowledge behind traditional geometry is a mixture of intuition and empirical investigation (experiment), a mixture ultimately behind the successful axiomatization. But, starting with informal 'purity' questions, Hilbert's metamathematical analysis of the 'facts' uses higher mathematics, which in turn informs elementary geometrical knowledge. None of the examples treated is fully represented in the original 1899 version of the *Grundlagen*. The central result on Desargues's Theorem (Section 8.4.1) *is* in the *Grundlagen*, but what leads up to this result, namely, the philosophical reflection and analysis undertaken in the 1898/1899 notes, is suppressed; the analysis of the Three Chord Theorem (Section 8.4.3) is an important part of the 1898/1899 lectures, but only the abstract mathematical result, and not the analysis itself, appears in the *Grundlagen*; and the analysis of the Isoceles Triangle Theorem (Section 8.4.2) makes no appearance, being first dealt with in the 1902 lectures.

8.4.1 *Desargues's Theorem*

The first example I want to consider where purity of method and the analysis of intuition play a significant role is in Hilbert's treatment of Desargues's Theorem in elementary projective geometry. Suppose given two triangles ΔABC and $\Delta A'B'C'$, not in the same plane, which are so arranged that the lines AA', BB', CC' meet at a point. Desargues's Theorem then says that the three points of intersection generated by the three pairs of straight lines AB and $A'B'$, BC and $B'C'$, AC and $A'C'$ themselves lie on a straight line. Intuition might be said to play a role from the beginning, since it is very easy to 'see' the correctness of the theorem; the intersection points must all lie in the planes of both triangles, and these planes intersect in a straight line. Cast in Hilbert's system, it is easily proved using just the full (i.e. planar and spatial) incidence and order axioms (Groups I and II). However, the theorem has a restricted version, where the triangles in question both lie in the *same* plane.[26] This version is not so easily visualizable, neither is it especially easy to prove. More importantly, the standard proof goes via the *unrestricted* version, using a point outside the plane of the triangles to reconstruct a three-dimensional Desargues's arrangement whereby the three-dimensional version of the theorem can be applied. In short, this proof, too, calls on all the axioms of I and II, even though the theorem itself appears to involve only *planar* concepts, i.e. is concerned only with the intersection of lines in the same plane. An indication of the standard way of proving the planar version of Desargues's Theorem is given by Fig. 8.1. Intuition might seem to play a role here, too, since it relies on projection of the original triangles up from the plane they lie in; in short, one can 'see' that planar Desarguesian configurations are reflected in spatial Desarguesian configurations and conversely.

This situation is remarked upon by Hilbert in his 1898/1899 lectures as raising a purity problem. Hilbert writes:

> I have said that the *content* of Desargues's Theorem is important. For now however what's important is its *proof*, since we want to connect to this a very important *consideration*, or rather *line of enquiry*. The theorem is one of plane geometry; the proof nevertheless makes use of space. The question arises whether there is a proof which uses just the *linear and planar axioms*, thus I 1–2, II 1–5. Thus here

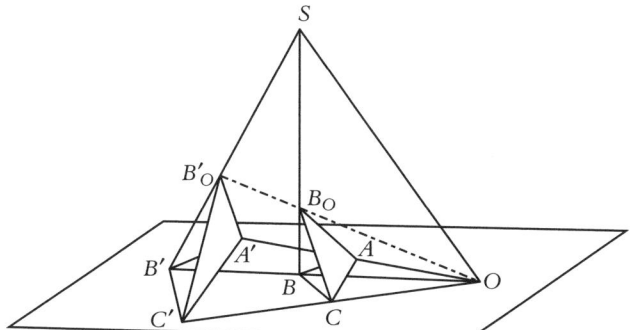

Fig. 8.1. Diagram for the usual proof of the Planar Desargues's Theorem, taken from: Hilbert and Cohn-Vossen (1932), p.108, Fig. 134. $\triangle ABC$ and $\triangle A'B'C'$ are the two triangles in the same plane, and S is an auxilliary point chosen outside this plane.

---

[26] Both versions have fully equivalent converses, where it is *assumed* that the three points of intersection lie on a line, and it is then *shown* that the lines AA', BB', CC' meet in a point.

for the first time we subject *the means of carrying out a proof* to a critical *analysis*. (Hilbert (*1898/1899, p. 30), p. 236 in Hallett and Majer (2004))

Then follow the general remarks about 'purity of method' which were quoted earlier (see p. 282). What this passage raises is the question of the *appropriateness* of a given proof, a 'purity' question *par excellence*, for here we have to do with a theorem, the Planar Desargues's Theorem, which is proved by means which at first sight might be thought inappropriate, owing to the proof's appeal to *spatial* axioms.

What we are faced with first is an independence question, and Hilbert sets out to show that the proof of the *Planar* Desargues's Theorem is *not* possible without appeal to spatial incidence axioms. As he puts it:

> We will rather show that Desargues's Theorem is *unprovable* by means of I 1–2, II 1–5. One will thus be *spared the trouble* of looking for a proof in the plane. For us, this is the first, simplest example for the proof of *unprovability*. Indeed: To satisfy us, it is necessary either *to find* a proof which operates just in the plane, or *to show that there is no such proof*. Prove so, that we specify a system of things = points and things = planes for which axioms I 1–2, II 1–5 hold, but the Desargues Theorem does not, i.e., that a plane geometry with the axioms I, II is possible without the Desargues Theorem. (Hilbert (*1898/1899, p. 30), pp. 236–237 in Hallett and Majer (2004))

To show this, Hilbert constructs an analytic plane of real numbers, with the closed interval $[0, +\infty]$ removed. The key thing is that some of the new straight lines are gerrymandered compositions: *below* the x-axis, the line is an ordinary straight (half-)line; *above* the x-axis, the line is in fact an arc of a circle uniquely determined by the conditions (*a*) that it goes 'through' the origin, O (it is open-ended here, since O is not itself in the model), and (*b*) that the given straight line below the x-axis is a tangent to it. Hilbert's figure (Fig. 8.2) illustrates the essentials of the model.[27]

The first thing to note about this is that we are here operating at one remove from the intuition ordinarily thought to underlie projective geometry. For one thing, the model produces highly unintuitive 'straight lines', even though they are pieced together using *intuitable* objects. The gerrymandered 'straight lines' are again 'intuitable' in the sense that one can easily visualize them (witness Fig. 8.2). Moreover, there is a sense in which lines made up of two distinct pieces are extremely familiar to us; think of a straight stick (a line segment) half immersed in still, clear water and then viewed from above looking down into the water. (Indeed, if it is viewed when the water is not still, then at any instant

---

[27] For full details, see Hilbert (*1898/1899, 31), or the *Ausarbeitung*, p. 28; these are pp. 237 and 316 respectively in Hallett and Majer (2004).

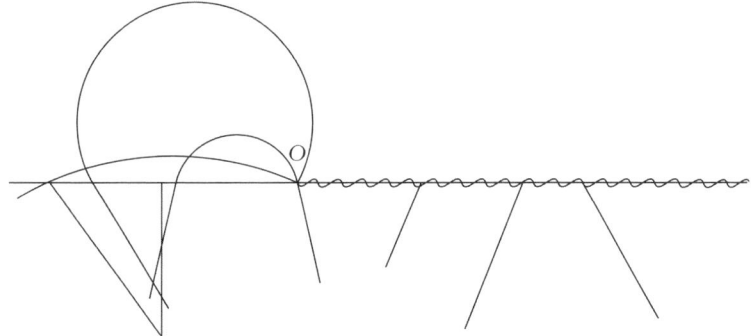

Fig. 8.2. Model for the failure of Desargues's Theorem; diagram taken from p.31 of Hilbert's 1898/1899 lecture notes on the foundations of Euclidean geometry.

the upper part will appear as an ordinary straight line, while the lower part will be a non-straight curved line. It might be said that what Hilbert has done here is to take a simple example like this and to choose a particularly simple curve.) However, while the direct perception ('intuition') of such objects is perhaps familiar, the analytic treatment is essential, and not merely a convenient way of proceeding. For one thing, the use of the algebraic manipulation is indispensable; it has to be shown that the model satisfies the plane axioms of I and II, and this is by no means trivial. For instance, in considering Axiom I$_1$, it has to be shown that any two points determine a straight line in the new sense, including the case where one point lies above the x-axis (has positive y-coordinate), and one point lies below the x-axis (has negative y-coordinate); in other words, it has to be shown that there is always a circle passing through 0 and the upper point and which cuts the interval $[-\infty, 0)$ in a point below 0 such that the tangent to the circle at that point is a straight line which passes through the given point in the lower half-plane. This, however, is correct, as careful calculation shows.[28] It is hard to see how this could be accomplished without the use of calculation. The point is that the use of the analytic plane and the accompanying algebra gives Hilbert extremely fine control over the pieces and how to glue them together in the right way, even though the result (when transferred back to the intuitive level) is fairly easily understood visually.

Hilbert's model specifies the first example of a non-Desarguesian geometry. Hilbert's treatment of the Desargues's Theorem in the 1898/1899 notes is

---

[28] Hilbert does not explicitly adddress this case, and I am grateful to Helmut Karzel for bringing it to my attention. It is more fully explained in my n. 46 to the text of Hilbert (*1898/1899) in Hallett and Majer (2004, 237).

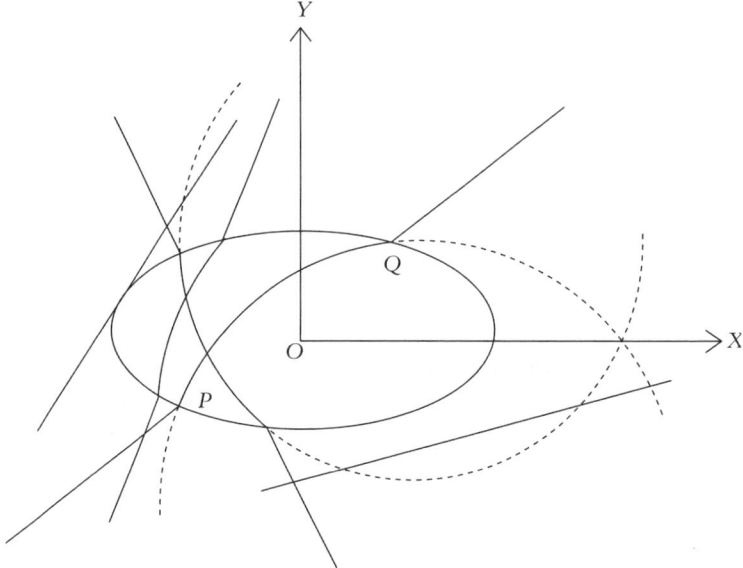

Fig. 8.3. Model for the failure of Desargues's Theorem used in the *Grundlagen der Geometrie*; diagram adapted from p. 53.

fully projective, and takes place before the congruence axioms are even stated. In the *Grundlagen*, all the axioms are set out before there is any significant development. The statement of Desargues's Theorem given there involves parallels, and the proof from the full spatial axioms (and the Parallel Axiom) is called upon, though not given (p. 49). Hilbert then remarks that one can give a simple *planar* proof provided a central result from the theory of proportions is used. Since congruence is involved, this means in effect that congruence can replace the spatial assumptions involved in the usual proof. Hilbert then gives another planar model (see Fig. 8.3), thus creating a non-Desarguean affine geometry, in which Desargues's Theorem and the appropriate congruence assumption (the Triangle Congruence Axiom, IV 6 of the *Grundlagen*, III 6 of the lectures) both fail. The investigation is thus different, and certainly adds important information to that of the lectures.[29] But the same points as were made above hold. Hilbert again takes an ordinary analytic plane, but this time one with a certain 'distorting' ellipse around the origin. Here, straight lines which would normally pass through the origin are 'distorted' by the

---

[29] As against this, though, note that the whole fascinating discussion of the import of Desargues's Theorem, of '*Reinheit der Methode*', etc. is quite lacking in the first and subsequent editions of the *Grundlagen*.

elliptical 'lens' such that in the interior of the ellipse they describe arcs of circles, and the new 'straight lines' are these straight line/circle arc/straight line composites. (See Fig. 8.3 and then Hilbert's *Grundlagen*, pp. 52–54, in Hallett and Majer (2004, pp. 488–491).) Once again the idea has connections with perceptual experience; once more, we see a case where the result issues from interplay between geometrical intuition and abstract, non-intuitive calculation. But again note that it is this calculation which shows that the model works, for once again, precision is the key. We also have a case which is similar to the Isoceles Triangle Theorem result in Section 8.4.2, for what is shown here is that spatial assumptions can be avoided if (here) we adopt the full set of axioms governing plane congruence. It is also worth bearing in mind that highly refined mathematical/numerical models like these are what show for Hilbert that certain 'intuitive' geometries 'exist'.

One of the results of Hilbert's investigation is that one need not look any further for a purely plane proof (from I 1–2, II) of the Planar Desargues's Theorem; one can, as Hilbert puts it, 'sich die Mühe sparen'. But this is by no means the end of the matter. As Hilbert states, our 'drive' for mathematical knowledge is only satisfied when one can establish *why* matters cannot go in the initially expected way. (See the passage from the *Grundlagen* cited on p. 201.) Since Desargues's Theorem *can* be proved from the axiom groups I and II, which among other things put conditions on the 'orderly' incidence of lines and planes, one might say that the theorem is a *necesssary* condition for such incidence. Hilbert now conjectures the following:

> Is the Desargues Theorem also a *sufficient* condition for this? i.e. can a system of things (planes) be added in such a way that all Axioms I, II are satisfied, and the system before can be interpreted as a sub-system of the whole system? Then the Desargues Theorem would be the very condition which guarantees that the plane itself is distinguished in space, and we could say that everything which is provable in space is already provable in the plane from Desargues. (Hilbert (*1898/1899, p. 33), p. 240 in Hallett and Majer (2004))

Hilbert shows that this conjecture is indeed corrrect, and the result is achieved by profound investigations of the relationship between the geometrical situation and the analytic one, the overall result being a re-education of our geometrical intuition, for what it reveals is that the Planar Desargues's Theorem in effect actually has spatial content. This provides an explanation of why it is (in the absence of congruence and the Parallel Axiom) that the Planar Desargues's Theorem cannot be *proved* without the use of spatial assumptions, and it provides a beautiful example of Hilbert establishing in the fullest way possible *why* 'impure' elements are required in the proof of Desargues's Theorem, the grounds for the 'impossibility of success' in trying to prove Desargues's

Theorem in the plane. When Hilbert first articulated the conjecture in the passage just quoted, he clearly had no proof of it, the same being true at the time the corresponding place in the *Ausarbeitung* was composed.[30] However, towards the very end of the *Ausarbeitung* (pp. 159–160; Hallett and Majer (2004, p. 387)), Hilbert presents a proof showing the conjecture to be correct.[31] The situation is a little complicated, but worth sketching.

(*i*) Hilbert's first step is to set up a *segment calculus* using the full axioms I–II together with IV (i.e. the axioms of incidence, order, and parallels, following the enumeration of the lectures; in the *Grundlagen*, these axioms are I–III). Segments are in effect to be treated as magnitudes, though the relevant properties are not assumed, but have to be demonstrated: addition and multiplication are defined for segments, the basis being an arbitrarily chosen system of axes; the zero and unit segments are defined; negative segments and fractional segments are then also defined, as well as an ordering relation. The Planar Desargues's Theorem is then called on in a fairly natural way to show that these operations satisfy all the axioms for a non-commutative ordered field. Hilbert also shows that the usual equation for a straight line holds, where segment 'magnitudes' are taken as what the variables vary over.

(*ii*) The Planar Desargues's Theorem is available because of the presence of the spatial incidence axioms in I. Not only are these the *only* spatial assumptions, their sole purpose here is to guarantee the presence of the Planar Desargues's Theorem. If we drop these axioms, but adopt the Planar Desargues's Theorem instead, in effect as a new axiom, then we have apparently *just* planar axioms, but importantly the results about the segment field still go through.

(*iii*) Hilbert's third step is to point out that the segment magnitudes can be used in pairs to coordinatize the plane; this is done in a very natural way, given the axes which lie at the base of the definitions of addition and multiplication. Hilbert argues now that we can use *triples* of these magnitudes as the basis of a coordinatization of *space*. The space thus determined will satisfy all the original axioms I–II, IV (I–III in the enumeration of the *Grundlagen*), and will be such that the original plane (now characterized by coordinate pairs) will be a plane *in the new space*; moreover, the fact that the right incidence

---

[30] In the *Ausarbeitung* of the lectures notes, Hilbert introduces the conjecture with the words 'The question is ...'. He then writes:

> This question is probably to be answered in the affirmative; one could then say in this case: The Desargues Theorem is the resultant of the elimination of the spatial axioms from I and II. (p. 32; Hallett and Majer, 2004, 318)

[31] For the evidence that Hilbert did not have the proof until late on in the course, see Hallett and Majer (2004, 189–190).

axioms hold for planes means that this plane will be 'properly' embedded in the space.

What this shows is that that the Planar Desargues's Theorem is a *sufficient* condition for the orderly incidence of lines and planes, in the sense that it can be used to *generate* a space. We thus have an explanation for why the Planar Desargues's Theorem cannot be proved from planar axioms alone: the Planar Desargues's Theorem appears to have spatial content. Moreover, as is clear in Hilbert's statement of his conjecture and from his construction, Desargues's Theorem could now be taken as an axiom to act *in place* of the spatial incidence axioms. The use of the theorem as an alternative axiom is quite standard in modern treatments of projective geometry, especially in treatments of affine planes; see, for example, Coxeter (1974) or Samuel (1988).

In this example, the investigation begins with a straightforward 'purity' question, which involves the analysis of an easy result underwritten, if not generated by, intuition; the analysis itself consists of an interplay between intuitive geometrical configurations and analytic geometry, though the precise results are obtained via the analytic models. The final analysis produces results which inform or educate (perhaps even challenge) our intuition.

8.4.2 The Isoceles Triangle Theorem

The Isoceles Triangle Theorem (ITT) of elementary geometry says that the base angles in an isoceles triangle are equal or, equivalently, that in a triangle whose base angles are equal, the sides opposite the equal angles are equal. The theorem figures early in Book I of Euclid's *Elements* as Proposition 5. Euclid's proof relies on the congruence of certain triangles using the side-angle-side criterion, the justification for this criterion being given in I, 4. The proof of I, 4 relies on a superposition argument, something which after this point Euclid seems keen to avoid. In Hilbert's axiomatization, the theorem is proved from incidence, order, and congruence axioms, the latter being designed to avoid physico-spatial assumptions, such as those involving rigid body movement or superposition. Hilbert's Triangle Congruence Axiom, bringing together assumptions about linear congruence and angle congruence, itself directly legitimizes the side-angle-side criterion for triangle congruence; the other usual criteria for triangle congruence are then given in the subsequent triangle congruence theorems, all of which rest ultimately on the Triangle Congruence Axiom. Hilbert's proof of the ITT, as given in his 1898/1899 lectures (the theorem is not mentioned in the *Grundlagen*) rests on a simple observation, namely that in the isoceles triangle $\triangle ABC$ (where AB and BC are taken to be the sides which are equal) we can describe the triangle in two distinct ways, namely as $\triangle ABC$ and $\triangle ACB$, and the triangles so described must be congruent by the side-angle-side criterion; consequently, the base angles must be the same. The proof appears to be, thus, little more than trivial.

Hilbert's proof can be traced to Pappus; Euclid's is different and a bit more involved. In his edition of the *Elements*, Heath notes of Pappus's proof:

> This will no doubt be recognised as the foundation of the alternative proof frequently given by modern editors [of the *Elements*], though they do not refer to Pappus. But they state the proof in a different form, the common method being to suppose the triangle to be taken up, turned over, and placed again upon *itself*, after which the same considerations of congruence as those used by Euclid in I, 4 are used over again. (Heath (1925, Volume 1, p. 254))

But Heath points out that Pappus himself avoids the assumptions about 'lifting' and 'turning', simply describing the same triangle in two ways. Hilbert thus follows Pappus, not those 'modern editors' Heath mentions.

Heath notes, though, that even Pappus's proof depends indirectly on a superposition argument, since it rests on Euclid's I, 4. Hilbert's proof does not, simply because the side-angle-side congruence criterion depends on the Triangle Congruence Axiom, not on any special proof method. It seems that matters have been shifted to the 'conceptual sphere', and no longer directly concern intuitions of space and movement in space.

Nevertheless, a 'purity of method' question can still be posed at the intuitive level: is 'flipping' and turning of the triangles, and thus a spatial dimension, *essential* to the argument for *ITT*? In some work first presented in his 1902 lecture course on the foundations of geometry, and then more fully in a paper published later in that year, Hilbert transposed this question at the intuitive level to a correlative question at the level of his axiomatization.[32] At the centre of concern is the Triangle Congruence Axiom. In its usual form, this involves no reference to what intuitively might be called the 'orientation' of the triangles involved; two triangles with matching side-angle-side combinations will be determined as congruent regardless of whether their orientation is the same or not. To mix idioms, accepting the quasi-physical interpretation of the concepts and axioms, and with the notion of congruence formulated

---

[32] The lecture notes are Hilbert (*1902), in Chapter 6 of Hallett and Majer (2004); Hilbert's paper is Hilbert (1902/1903). This paper was reprinted as Anhang II to the editions of the *Grundlagen der Geometrie* from the Second on. An Appendix (mentioned below) was made to the first reprinting in Hilbert (1903), and then a note (also mentioned below) was added to the Sixth Edition published in 1923, i.e. Hilbert (1923). The Appendix was radically revised by Arnold Schmidt for the Seventh Edition (Hilbert 1930) of the *Grundlagen*, as Hilbert makes clear in the Preface, this being the last edition of the monograph to be published in Hilbert's lifetime. We will actually concentrate here on Hilbert's presentation in his 1905 lectures on the 'Logische Principien der Mathematik', since of all the versions, this is the one where Hilbert is most philosophically expansive about the results and what they show. These lectures will appear in Ewald, Hallett and Sieg (Forthcoming).

with physical manipulation and superposition in view, triangles of different orientation can only be shown congruent by superposition after *lifting* and *turning* them in space. Thus, under such a physical interpretation, Hilbert's axiom licenses 'flipping' as a kind of displacement, and not just 'sliding' within the same plane. Seen in this way, the proofs of the *ITT* all exploit such flipping, either directly or indirectly, even, it might be said, Hilbert's in his 'neutral' system. The original 'purity of method' question, namely, 'Is flipping essential to the proof?', now has a direct correlate with respect to Hilbert's system: Is the Triangle Congruence Axiom (which can be seen to license 'flipping') essential to the proof of *ITT* in the axiomatic system, divorced as it is from the intuitive/empirical perspective? In order to pose the question more precisely, Hilbert considers a *weakened* version of the Triangle Congruence Axiom, one which insists in effect that, before the side-angle-side criterion can be used to license triangle congruence, the triangles be of the same orientation in the plane. This weakened axiom no longer underwrites the usual proofs of the *ITT*, as a quick look at the Pappus/Hilbert argument indicates; can the *ITT* nevertheless be proved? As Hilbert puts it in his presentation of the work in his 1905 lectures, the original 'purity of method' question is now transformed into a question primarily about logical relations:

> There now arises the question of whether or not the [original] broader version of the [Triangle Congruence] Axiom contains a superfluous part, whether or not it can be replaced by the restricted version, i.e. whether it is a logical consequence of the restricted version. This investigation comes to the same thing as showing whether or not the *equality of the base angles in an isoceles triangle* is provable on the basis of the restricted version of the congruence axiom. The question has a close connection with that of the validity of the theorem that the sum of two sides of a triangle is always greater than the third. (Hilbert, *1905, pp. 86–87)

This question is the beginning of Hilbert's investigations, and the consequences are fascinating, both for the abstract mathematical structure, for what the investigations tell us about 'geometrical intuition', and because of the connection Hilbert draws with the property of triangles he states, which we will refer to here as the Euclidean Triangle Property.

Hilbert first shows that the restricted congruence axiom together with the *ITT* itself implies the normal Triangle Congruence Axiom (see Hilbert, *1902, p. 32).[33] The key question, in effect first raised in Hilbert (*1902), is then

---

[33] In a note added to Appendix in the Sixth Edition of the *Grundlagen*, Hilbert points out that this is in fact only the case if one adds a further congruence axiom guaranteeing the commutativity of angle addition. See Hilbert (1923, 259–262); a suitable further axiom due to Zabel is stated on p. 259. This note does not appear in subsequent editions, but Schmidt's revised version of Appendix II adopts a similar additional congruence axiom, a weaker one attributed to Bernays. See Hilbert (1930, 134).

this: What has to be added to the usual system, with both the Parallel Axiom and the Archimedean Axiom, but only the restricted Triangle Congruence Axiom, to enable the Isoceles Triangle Theorem (or indeed the usual Triangle Congruence Axiom) to be proved?

Hilbert's first answer to this question is given in the 1902 lectures, and then in the 1902/1903 paper, namely: one *can* prove the *ITT* if one adopts alongside the Archimedean Axiom a second continuity axiom which Hilbert calls the *Axiom der Nachbarschaft*. This latter axiom (see Hilbert, *1902, p. 84) states that, given any segment AB, there exists a triangle ('oder Quadrat etc.') in whose interior there is no segment congruent to AB. Hilbert later gave other answers, focusing, in place of *Nachbarschaft*, on an axiom he calls the *Axiom der Einlagerung* (Axiom of Embedding), which says that if one polygon is embedded in another (i.e. its boundary contains interior but no exterior points of the first), then it is not possible to split the two polygons into the same number of pairwise congruent triangles. (Bernays later gave an essential simplification.)[34]

The bulk of Hilbert's 'purity' investigation is now devoted to showing that this axiom (or respectively *Einlagerung*) *and* the Archimedean Axiom are both essential if the *ITT* (respectively the broader version of the Triangle Congruence Axiom) is to be proved in this way. What Hilbert shows is that geometries can be constructed in which all the plane axioms hold (with the weaker version of triangle congruence), and where *Nachbarschaft* (respectively *Einlagerung*) holds, but where the Archimedean Axiom and the *ITT* both fail, or similarly where the Archimedean Axiom holds, but where *Nachbarschaft* (respectively *Einlagerung*) and the *ITT* fail.

In his 1902 lectures, Hilbert is very careful to set out some of the important things which can be reconstructed on the basis of the weaker congruence axiom[35], and among them is Hilbert's equivalent of Euclid's theory of linear proportion. On the other hand, Hilbert's full theory of triangular area ('surface content'), a conscious reconstruction of Euclid's, does not apparently go through. Euclid's fundamental theorem concerning this (*Elements*, I, 39) is that triangles on the same base and with the same area must have the same height. In establishing his version of this theorem (that two triangles which are *inhaltsgleich* and on the same base have the same height), Hilbert defines the

---

[34] This new 'very intuitive' requirement was first introduced by Hilbert in a section added to the first reprinting of his paper in the Second Edition of the *Grundlagen*, i.e. Hilbert (1903, 88–107). Schmidt's revision of Appendix II for the Seventh Edition (1930) omitted the consideration of the *Einlagerungsaxiom*. It was revived by Bernays in Supplements to later editions, where he points out the simplification.

[35] See §A of the 1902 lectures, which occupies pp. 26–32 of Hilbert (*1902), to be found in Hallett and Majer (2004, 553–556).

notion of the surface measure, or *Flächenmass*, of a triangle as the product of half the base and the height.[36] But in order for this definition to be a good one, it has to be shown that this quantity is independent of the choice of which side is to be the base. However, this apparently simple fact depends on the recognition that two right-angled triangles constructed within the given triangle are similar, and thus it depends on the theory of triangle congruence. But the triangles in question are not in the same orientation, and consequently it seems that the restricted version of the Triangle Congruence Axiom is not sufficiently strong for the purpose at hand, and that the demonstration required must ultimately depend on the unrestricted Triangle Congruence Axiom.

The construction of the models showing the various independence results is not so clear in the 1902 notes, but is given much more explicitly in the 1902/1903 paper, which was already clearly in preparation while the 1902 lectures were being held, and then in the 1905 lectures. We will also concentrate on the first part of Hilbert's independence investigation, where the Archimedean Axiom fails, but *Nachbarschaft* holds, for this is very revealing of Hilbert's approach to foundational questions.

To show the necessity of the Archimedean Axiom, Hilbert constructs a model in which all the axioms up to the Congruence Axioms (with, of course, the restricted Triangle Congruence Axiom), hold, *Nachbarshaft* holds, but the Archimedean Axom fails. In this model: (*i*) the Isoceles Triangle Theorem fails; (*ii*) the Euclidean triangle property fails; and (*iii*) the theory of triangle area fails to go through. The basis of this model is a non-Archimedean field, which he calls T, using as elements the power series expansions

$$a_0 t^n + a_1 t^{n+1} + \cdots$$

around a certain fixed parameter t, with real number coefficients, a_0 non-zero, and n (which can be zero) a rational number; arithmetic calculation using these expansions takes place in the standard way. The ordering on the elements of T is defined as follows: for any elements α, β in the field, $\alpha > 0$ if $a_0 > 0$, a_0 being the first coefficient of the expansion of α, and < 0 if $a_0 < 0$; $\alpha > \beta$ if $\alpha - \beta > 0$. These stipulations are enough to show that t is infinitesimal, since $1 - mt > 0$ whatever natural number m is; any α whose expansion begins with t or indeed any t^n is infinitesimal. If $\alpha = a_0 t^n + a_1 t^{n+1} + \cdots$ is such an infinitesimal element, then e^α itself will be in the field, and will be representable by a power series expansion whose first element is $1 \cdot t^0$.

---

[36] See the 1899 *Grundlagen*, p. 43, or the *Ausarbeitung* of the 1898/1899 lectures, p. 131, or pp. 372–373 and 478–479 respectively in Hallett and Majer (2004).

Suppose now that α and β are in the field; we can then consider the complex number $\alpha + i\beta$, and these elements give rise to the complex extension of the field. In fact, such an element will be represented by an appropriate power series built on t where the coefficients are imaginaries. If $\alpha + i\beta$ is in the complex extension, then so will $e^{\alpha+i\beta}$ be.

Hilbert now defines an analytic geometry as follows. Points are defined by pairs of coordinates (x, y), where x and y are elements of T. Straight lines will be given by linear equations in the usual way, and so the usual incidence and order axioms will then hold, as well as the Parallel Axiom; the Archimedean Axiom, of course, will not hold. This leaves just the congruence axioms to consider, and, as one might expect, this is where the art comes. First, two segments are said to be congruent if parallel transport can shift the line segments so that they start at the coordinate origin, and then (keeping one of the segments fixed) one of them can be rotated into the other by a positive rotation through some angle θ. The two 'movements', parallel transport and rotation, are of course, properly speaking, transformations of the plane onto itself. What will be congruent to what will then depend on the definition of the transformation functions; parallel transport is trivial, so the key matter is the transformation function corresponding to rotation.

This function is based on the following trick. The coordinate (x, y) can be coded as a single complex number $\alpha = x + iy$ in the complex extension of T. The function e^α can then be used to form another complex number, which can then be decoded to form a new coordinate (x', y'). Suppose (x, y) is the coordinate of one end of a segment whose other end-point (perhaps after parallel transport) is at $(0, 0)$. Suppose the segment is positively rotated through an angle $\theta + \tau$, where θ is real (and could be 0) and τ is infinitesimal. Then the new coordinates are given by the formula

$$x' + iy' = e^{i\theta+(1+i)\tau} \cdot (x + iy)$$

Clearly $(0, 0)$ transforms into the coordinate $(0, 0)$.[37]

Segments can be assigned length acccording to the following procedure. A segment on the x-axis is said to have length l when one of its end-points lies at the origin and the other has x-coordinate $\pm l$. Any other segment has length l if one end can be shifted to the origin by parallel transport and the other end can be rotated into $(\pm l, 0)$ by the rotation function or where $(\pm l, 0)$ can be

---

[37] As Hilbert makes clear, it can be shown that given any line segment and any point in the plane, a rotation can be found so that the rotated line passes through that point.

PURITY OF METHOD IN HILBERT'S *GRUNDLAGEN* 235

rotated into it. Parallel transport does not alter length, but crucially rotation sometimes does.[38]

Hilbert now considers the point (1, 0) on the *x*-axis, and rotates this positively through the infinitesimal angle *t*. This will give a segment OA, where O is the origin and A has coordinates (x', y') determined by

$$x' + iy' = e^{(1+i)t} \cdot (x + iy) = e^{(1+i)t} \cdot x = e^{(1+i)t}$$

since $y = 0$ and $x = 1$. This is then:

$$e^{t+it} = e^t \cdot e^{it} = e^t \cdot (\cos t + i \sin t) = e^t \cos t + ie^t \sin t$$

giving the coordinates

$$x' = e^t \cos t, \quad y' = e^t \sin t$$

According to the method of determining length, segment OA necessarily has length 1.

Now we construct the reflection of A in the *x*-axis, giving A' with coordinates $(x', -y')$, and A and A' are then connected by a line perpendicular to the *x*-axis; denote by B the point where this line cuts the axis (Fig. 8.4).

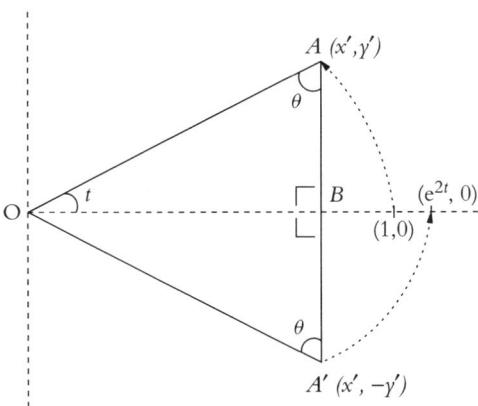

Fig. 8.4. The triangle showing the failure of the Isoceles Triangle Theorem.

---

[38] A key part of the proof is showing, of course, that the Congruence Axioms hold. Hilbert's work is not explicit about this. In his new version of Appendix II to the Seventh Edition of the *Grundlagen*, Schmidt uses one 'congruence mapping' in place of the distinct notions of parallel transport and rotation. This is essentially the rotation function described above, with 'transport' parameters added, i.e. $x' + iy' = e^{i\theta + (1+i)\tau} \cdot (x + iy) + \lambda + i\mu$, where λ, μ are taken to be elements of T. Schmidt then proceeds to give careful verifications of the Congruence Axioms.

Clearly, $AB = BA'$ and $OB = OB$; both angles $\angle OBA$ and $\angle OBA'$ are $\frac{\pi}{2}$. The question is now to compare OA and OA'. What Hilbert does is to rotate OA' positively through the angle t. This results in new coordinates (x'', y'') for the end-point, where

$$\begin{aligned} x'' + iy'' &= e^{(1+i)t} \cdot (e^t \cos t - ie^t \sin t) \\ &= e^{(1+i)t} \cdot e^t (\cos t - i \sin t) \\ &= e^{(1+i)t+t} \cdot (\cos t - i \sin t) \\ &= e^{(1+i)t+t-it} \\ &= e^{2t} \end{aligned}$$

Therefore, $x'' = e^{2t}$, and $y'' = 0$. (See Fig. 8.4.) Thus, the rotation takes the point A' to the x-axis, namely to the point $(e^{2t}, 0)$. But this point cannot coincide with $(1, 0)$, since $e^{2t} > 1$. (This is because the power series expansion of e^{2t} is $1.t^0 + 2t + \frac{4t^2}{2!} + \cdots$; since the expansion of 1 is $1.t^0 + 0 + 0 + \cdots$, the series for $e^{2t} - 1$ is $2t + \frac{4t^2}{2!} + \cdots$. The first coefficient of this is 2, hence $e^{2t} - 1 > 0$, and so $e^{2t} > 1$.) The length of OA' is therefore greater than 1, and so different from that of OA. The base angles in the extended triangle $\triangle OAA'$ are the same; the Angle-Sum Theorem holds, since the Parallel Axiom does. In consequence, we have an example of a triangle for which the base angles are equal, but in which the sides opposite these angles are not, violating the *ITT* (in its converse version).

Simple calculation in Hilbert's model also shows that $OB + BA' < OA'$, violating the Euclidean Triangle Property. As for the theory of triangle area, Hilbert shows that the Pythagorean Theorem for right-angled triangles holds in the form 'The sum of the square surface areas over the two legs of the right angle equals the square over the hypotenuse', since this depends only on the weaker Triangle Congruence Axiom. But the squares over BA' and BA are the same; since the square over OB equals itself, those over OA and OA' must be the same, too. But $OA < OA'$; hence the usual analytic conclusion from the Pythagorean Theorem ('The length of the hypotenuse is the square root of the sum of the squares of the lengths on the other two sides') fails, because we will have here sums which are the same, but squared hypotenuse lengths which are different.[39] For this reason, the geometry constructed is

---

[39] As Hilbert says in Hilbert (*1902, 125a): '...one can *no longer conclude that the sides are equal from the fact that the squares are*'. See Hallett and Majer (2004, 597).

called by Hilbert *non-Pythagorean geometry*. Moreover, since $OA < OA'$, the square over OA will fit inside the square over OA' with a non-zero area left over, so we have a square which is of equal area-content to a proper part of itself, violating one of the central pillars of the Euclidean/Hilbertian theory of triangle area-content.[40]

It is important to see that these are not just technical results, for they give us a great deal of information about what axioms are necessary for the reconstruction of classical Euclidean geometry. Hilbert sums it up in his 1905 lectures as follows:

> The result is particularly interesting again because of the way continuity is involved. In short:
>
> 1.) The theorem about the isoceles triangle and with it the Congruence Axiom in the broader sense is *not provable* from the Congruence Axiom in the narrower sense when taken with the other plane axioms I–IV [thus excluding continuity assumptions].
>
> 2.) Nevertheless, it becomes provable when one adds continuity assumptions, in particular the Archimedean Axiom.
>
> From this we see therefore that in the Euclidean system properly construed, and which allows us to dispense with continuity assumptions, the broader Congruence Axiom is a necessary component. The investigations which I have here set in train throw new light on the connections between the theorem on the isoceles triangle and many other propositions of elementary geometry, and give rise to many interesting observations [*Bemerkungen*]. Only the axiomatic method could lead to such things. (Hilbert, *1905, pp. 86–87)

This last remark is amplified by Hilbert in the direct continuation of this passage, which also ties the nature of the investigation directly to the view of geometry which we have seen emerging, namely *emancipation* from interpretation and intuition without losing contact with what underlies the axioms:

> When one enquires as to the status within the whole system of an old familiar theorem like that of the equality of the base angles in a triangle, then naturally one must liberate oneself completely from intuition and the origin of the theorem, and apply only logically arrived at conclusions from the axioms being assumed. In order to be certain of this, the proposal has often been made to avoid the usual names for things, because they can lead one astray through the numerous associations with the facts of intuition. Thus, it was proposed to introduce into the axiom system new names for point, straight line and plane etc., names which will recall only what has been set down in the axioms. It has even been proposed

---

[40] Hilbert states the conclusion as follows: '*The theory of surface content depends essentially on the theorem concerning the base angles of an isoceles triangle; it is thus not a consequence of the theory of proportions on its own*' (Hilbert (*1902, 125a), or Hallett and Majer (2004, 597)).

that words like equal, greater, smaller be replaced by arbitrary word formations, like *a*-ish, *b*-ish, *a*-ing, *b*-ing. That is indeed a good pedagogical means for showing that an axiom system only concerns itself with the properties laid down in the axioms and with nothing else. However, from a practical point of view this procedure is not advantageous, and also not even really justified. In fact, one should always be guided by intuition when laying things down axiomatically, and one always has intuition before oneself as a goal [*Zielpunkt*]. Therefore, it is no defect if the names always recall, and even make it easier to recall, the content of the axioms, the more so as one can avoid very easily any involvement of intuition in the logical investigations, at least with some care and practice. (Hilbert, *1905, pp. 87–88)

The emancipation from intuition and the 'origin' of the theorem concerned is clearly indicated in this passage, but Hilbert's remark that 'one always has intuition before oneself as a goal' is also very important. Although the *achievement* of the results necessarily requires the deliberate suspension of intuition, crucially they yield important information *about* intuition. One thing which Hilbert stresses in the passage quoted on p. 237 is that adopting the full Triangle Congruence Axiom allows us to avoid any continuity assumption in demonstrating the *ITT*. That is certainly correct, and this fact belongs alongside others concerning congruence and continuity, a prime example being Hilbert's reconstruction of the Euclidean theories of proportion and surface area without invoking continuity. But there are other subtle conclusions to be drawn. For example, as explained above, the (apparently planar) full Triangle Congruence Axiom appears to contain some hidden spatial assumption, since it licenses 'flipping' arguments. But now Hilbert's independence results seem to show that we *can* get the result without the spatial assumption, and thus with genuinely planar congruence axioms, *if* we accept some modest continuity assumptions about the plane. So one has a choice between spatial assumptions in the planar part of geometry, or continuity assumptions. However, what makes this a rather more complicated matter is that the main continuity assumption involved, the Archimedean Axiom, is quasi-numerical, something one might think should be avoided as far as possible in a purely 'geometrical' axiomatization.

But whatever the right conclusion to be drawn about our geometric intuitions, two very important things seem to follow from Hilbert's analysis. On the one hand, the results seem to strengthen Hilbert's holism about geometry, at least in the sense that the axioms are very intricately involved with each other, and that there might be more than one way to achieve many important results. Secondly, the adumbration of the intuitive picture here, and perhaps its correction or adjustment, follows from the high-level logico-mathematico investigation which Hilbert engages in. The information is obtained only

PURITY OF METHOD IN HILBERT'S *GRUNDLAGEN* 239

by using very sophisticated mathematical analysis: complex analysis over a non-Archimedean field. In other words, the higher mathematical/logical analysis based on complex numerical structures instructs and informs lower-level geometrical intuition. This, note, is just such a case where 'function theory' is used to show something about elementary geometry. (See the quotation from Hilbert on p. 201 above.)

In short, this example illustrates Hilbert's view that one is *guided* by geometrical intuition, one asks questions *suggested* by intuition, but in the end it is higher mathematics which *instructs* intuition, not the other way around. Thus, while Hilbert does carry out a kind of 'purity of method' investigation, it is much more focused, as he puts it, on the 'analysis of intuition'. One of the reasons why Hilbert thinks that intuition *requires* analysis is that it is not, for him, a *certain* source of geometrical knowledge, and certainly not a *final* source. Thus the analysis, which is designed to throw light on the question: what is one committed to exactly when one adopts certain principles, among them principles suggested by intuition?

8.4.3 The Three Chord Theorem

The third example considered here concerns another fairly elementary geometrical theorem, which says that the three chords generated by three mutually intersecting circles (lying in the same plane) always meet at a common point. Call this the Three Chord Theorem (*TCT*). (See Fig. 8.5.)

The theorem is not an ancient one, but was apparently first discovered by Monge in the middle of the 19th century. It has an interesting generalization, much studied by 19th century geometers, concerning the lines of *equal power*

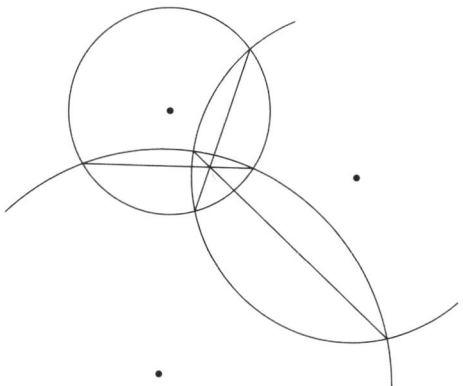

Fig. 8.5. Diagram of the Three Chord Theorem, adapted from the *Ausarbeitung* of Hilbert's 1898/1899 lectures, p.61, (Hallet and Majer (2004, 335)).

between two circles. The power of a point O with respect to a circle $\mathcal{C}$ in the same plane can be defined thus: Consider any straight line OPQ through O which cuts $\mathcal{C}$ in the two (not necessarily distinct) points P and Q. The power of O with respect to $\mathcal{C}$ is now the product $OP \cdot OQ$. This is a constant for O and $\mathcal{C}$, since this ratio is the same wherever P (respectively Q) happens to lie on the circle. (See Euclid's *Elements*, Book III, Proposition 35.) If O is inside the circle, the power is negative; if it is outside, and P and Q coincide, then OPQ is a tangent, and the power is OP^2. Given two circles, we can consider the points which have the same power with respect to both circles; these lie on a straight line perpendicular to the line joining the centres of the circles, a line which is called the *radical axis* or the *line of equal power* (in German, a '*Chordal*') between the two circles. (See Coolidge, 1916, Theorem 167.) What was often called the *Hauptsatz* of the theory of these lines is: Given three circles in a plane, intersecting or not, the three lines of equal power that the three pairs of circles generate must intersect at a single point. (*Ibid.*, Theorem 168.) Clearly, the *TCT* is just the case where each pair of circles intersect each other. There is another interesting special case. Consider a triangle, and consider three circles in the plane of the triangle centred respectively on the three vertices. As the radii of the three circles tend to zero, the three lines of equal power will tend to the perpendicular bisectors of the three bases. It follows that these three bisectors have a common point of intersection.

Hilbert gives a direct proof of the *TCT* in his 1898/1899 notes. The proof, which is clearer in the *Ausarbeitung*, pp. 61–64 (Hallett and Majer, 2004, pp. 335–337), assumes that the three circles in question, all in the plane α, are the equators of three spheres which have just two points P and Q of mutual intersection; it is then shown fairly easily (*loc. cit.*, p. 61) that the three chords of the circles, which lie, like those circles, in α, must all intersect the line PQ. But since neither P nor Q lies in α, PQ has only one point in common with α, and this point is therefore common to all three chords. Thus the simple proof depends on the assumption that three mutually intersecting spheres intersect in exactly two points, in other words, that these points *exist*. Hilbert's focus is then on this assumption, no longer on the *TCT* itself.

Hilbert's next step is to connect this with the theorem that a triangle can be constructed from any three line segments which are such that any two of the segments taken together are greater than the third. Call this the Triangle Inequality Property. In Euclid, this is proved in I, 22 (see Heath, 1925, Volume 1, pp. 292–293); call this the Triangle Inequality Theorem. The question Hilbert asks is: On what assumption is the proof of *this* proposition based? In his lecture notes (p. 64), Hilbert states this triangle property in the same breath with another, namely that, given a straight line and a circle in the same plane, if the line has both a point in the interior of the circle and a point outside it, then it must intersect it, and in exactly two points.[41] Call this the *line-circle property*. Similar to this is what can be called the *circle-circle property*, namely a circle with points both inside and outside another circle has exactly two points of intersection with it.[42] The connection to the Triangle Inequality Theorem is not surprising, for the circle-circle property is precisely what Euclid implicitly relies on in the proof of I, 22.

It is important to see Hilbert's question as one concerning 'purity of method'. It is well known that similar questions arise with the very first proposition of Euclid's *Elements* (I, 1) which shows how to construct an equilateral triangle on a given base AB. Euclid's construction takes the two circles whose centres are the endpoints of the base and whose radii are equal to the base; either of the two points of intersection of the circles can be taken as the third apex, C. But *do* the circles actually intersect? A standard objection is that there is no guarantee of this. Heath notes:

> It is a commonplace that Euclid has no right to assume, without premissing some postulate, that the two circles *will* meet in a point C. To supply what is wanted we must invoke the Principle of Continuity. (Heath, 1925, Volume 1, p. 242)

And by a 'Principle of Continuity', Heath means something like Dedekind continuity (*op. cit.*, pp. 237–238). Heath also cites one of Hilbert's contemporaries, Killing, as invoking continuity to show the line-circle property (Killing, 1898, p. 43). And many commentators on Euclid (for example, Simson in the 1700s) raised this point with respect to the proof of the Triangle Inequality Property, while at the same time stating that it is 'obvious' that the circles intersect, and that Euclid was right not to make any explicit assumption. What Hilbert investigates is what formal property of space corresponds to the implicit underlying existence/construction assumption.

Hilbert constructs a model of his geometry (i.e. Axiom Groups I–III) in which the existence assumption *fails*, and where the Triangle Inequality Theorem (Euclid's I, 22) also fails. Thus, the necessary conditions for Hilbert's proof of the Three Chord Theorem are not present in a geometry based solely on I–III. Moreover, since the Euclidean proof of I, 22 is based on a simple straightedge and compass construction, Hilbert's result is tantamount to saying that his axiom system does not have enough existential 'weight' to match this particular construction. This result is interesting, because, beginning with an

---

[41] Note that this has to be assumed for the power of point to be defined for all points and all circles.
[42] For similar properties, see pp. 62, 65 of Hilbert's lecture notes, and pp. 63–64 of the *Ausarbeitung*, i.e. Hallett and Majer (2004, 335–337).

intuitively inspired 'purity of method' question, it issues in a result on the abstract, conceptual level; not just this, but the result also shows why the assumption behind Euclid's I, 1 *is* justified in an elementary way.

Hilbert begins to consider this metamathematical problem on p. 65 of his lecture notes, pp. 64ff. of the *Ausarbeitung*. He constructs in effect the smallest Pythagorean sub-field of the reals which contains 1 and π, which yields a countable model of the axioms I–III (indeed of I–V, the whole system of the 1898/1899 lectures and the first edition of the *Grundlagen*) when the usual analytic geometry is constructed from pairs of its elements taken as coordinates. However, the number $\sqrt{1-(\pi/4)^2}$ does not exist in this field.[43] Since 1, 1, and $\frac{\pi}{2}$ are in the field, the model will possess three lines which satisfy the triangle inequality, but from which no triangle can be constructed. This is well illustrated by the diagrams in the *Ausarbeitung*, pp. 66 and 68, here Figs. 8.6 and 8.7.

One can see from Fig. 8.7 that the upper apex of the triangle depicted in Fig. 8.6 ought to have coordinates $(\pi/4, \sqrt{1-(\pi/4)^2})$; but this is not a point in the model. The same example shows that there can be a line partially within and partially without a circle but which intersects the circle nowhere: take the vertical line $x = \frac{\pi}{4}$ in the diagrams; this ought to meet the circle $x^2 + y^2 = 1$ at points with y-coordinates $\pm\sqrt{1-(\pi/4)^2}$, but again these points are missing.

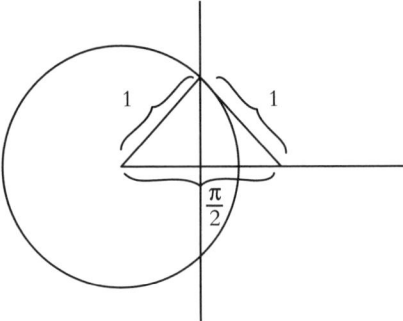

Fig. 8.6. Model of the Failure of the Triangle Inequality Property.

---

[43] Hilbert's quick sketch of the argument is as follows. Suppose $\sqrt{1-(\pi/4)^2}$ were in the Pythagorean field constructed, then, since this field is minimal, it would be represented by an expression formed from π and 1 by the five operations allowed; Hilbert denotes this expresssion by $A(1, \pi)$. But then, as he points out, $A(1, t)$ must represent the corresponding element $\sqrt{1-(t/4)^2}$ of the corresponding minimal field constructed from 1 and the real number t, whatever t is. However, while $A(1, t)$ is always real, it is obvious that $\sqrt{1-(t/4)^2}$ will be imaginary for t sufficiently large t. Hence, $A(1, t)$ will not always represent it.

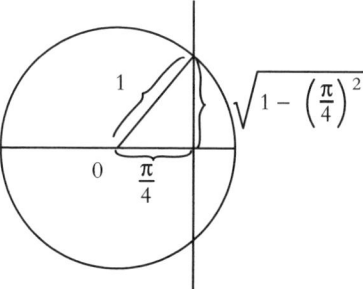

Fig. 8.7. Model of the Failure of the Triangle Inequality Property, continued.

In his review of Hilbert's *Grundlagen*, Sommer remarks that one cannot prove the line-circle property in Hilbert's system (Sommer, 1899/1900, p. 291). What Sommer observes already follows from what Hilbert shows in his lectures, and what he says there explicitly (e.g. pp. 64–65). Sommer, of course, certainly knew that Hilbert had shown this. For one thing, this is the same Julius Sommer, the 'friend', who, along with Minkowski, is thanked by Hilbert at the end of the first edition of the *Grundlagen* for help with proofreading. For another, Sommer refers to Hilbert's lectures course for 1898/1899 directly, in a different context (*loc. cit.*, p. 292). Furthermore, the mathematical core of the result is in fact *present* in the *Grundlagen*, if only implicitly and in a more abstract form. Let us turn to this now, and then return to Sommer's remark.

The model described above really presents a result about the abstract conceptual structures which mature axiom systems represent, for what Hilbert has in effect shown, in modern algebraic terms, is that not every Pythagorean field is Euclidean. An ordered field is said to be *Pythagorean* when the Pythagorean operation holds, that is, when $\sqrt{x^2 + y^2}$ is in the field whenever x and y are. (Note that, for each real r, there is a minimal, and countable, Pythagorean sub-field of the reals containing the rationals and r, a fact which Hilbert frequently employs in his independence proofs.) An ordered field K is said to be *Euclidean* when for any non-negative element $x \in K$, we also have $\sqrt{x} \in K$. It is obvious that every ordered Euclidean field is Pythagorean, but Hilbert shows here that the converse fails, for his model is formed from a field which is Pythagorean; $\frac{\pi}{4}$ is an element, given that π is, and so are $(\frac{\pi}{4})^2$ and $1 - (\frac{\pi}{4})^2$; but, as we have seen, $\sqrt{1 - (\pi/4)^2}$ is not, and so the field cannot be Euclidean.[44] The key point is summed up in the following result: In an analytic geometry whose coordinates are given by an

---

[44] That Hilbert had in fact shown this was first pointed out to me by Helmut Karzel.

ordered Pythagorean field, one can always construct a triangle from three sides satisfying the triangle inequality if and only if the underlying coordinate field is also Euclidean. Indeed, for an analytic geometry based on an ordered field, the Euclidean field property is equivalent to the line-circle property, and this is in turn equivalent to the circle-circle property, the property directly relevant to Hilbert's proof of the Three Chord Theorem (Hartshorne, 2000, pp. 144–146). Given this, the connection between the Euclidean field property and the formation of a triangle from any three lines satisfying the triangle inequality is obvious.

To repeat, the result is a thoroughly abstract one, a result about fields. The inspiration is again intuitive, but this time the major fruit is a new theorem in abstract mathematics. (The result can be found in the *Grundlagen der Geometrie*, though the background work concerns solely the algebraic equivalents to elementary constructions.)

This algebraic result is strongly hinted at in the *Ferienkurs* Hilbert gave in 1898. (See Hallett and Majer, 2004, Chapter 4; see pp. 22–23 of Hilbert's course.) Hilbert poses the question of whether, given a segment product $c \cdot d$, there is a segment x such that $x^2 = c \cdot d$, i.e. a square root for $c \cdot d$. His comment suggests that he thinks this is not always the case, and this is precisely what the counterexample outlined above confirms. For example, in the second diagram given, consider the horizontal and vertical products formed by the four segments arising from the intersection of the horizontal and vertical chords. The horizontal segment product (our $c \cdot d$) is $(1 + \pi/4) \cdot (1 - \pi/4)$, which equals the vertical segment product $\sqrt{(1 - (\pi/4)^2)} \cdot \sqrt{(1 - (\pi/4)^2)}$; thus, the x sought is $\sqrt{(1 - (\pi/4)^2)}$, which does not exist in the model, as we have seen. Thus, a question from the 1898 *Ferienkurs* is answered.

Hilbert notes that the problem exhibited here does not just arise because of the involvement of the transcendental number π; he gives an example of an elementary number which will be in any Euclidean field over the rationals, but which is not in the minimal Pythagorean field, namely $\sqrt{1 + \sqrt{2}}$. (See Hilbert's own lecture notes, p. 67, and also p. 67 of the *Ausarbeitung*. Hartshorne (2000, p. 147) gives further details of the counterexample.) Another example is given in the *Grundlagen* itself, to which we will come in a moment.

What is now interesting is how this *abstract* result is used to yield more information at the intuitive level, at the level of synthetic, Euclidean geometry rooted in elementary constructions. In the 1898/1899 lectures, Hilbert himself seems to suggest that the problem might have to do with a continuity assumption. On p. 64 of the *Ausarbeitung*, he says when assuming either the line-circle or circle-circle properties, one is actually assuming that 'the circle is a closed figure'.⁴⁵ Moreover, it is precisely in the context of the failure of continuity in Hilbert's system that Sommer makes his remark about the line-circle property, adding that 'it remains undecided whether or not the circle is a closed figure' (Sommer, 1899/1900, p. 291), thus adopting Hilbert's terminology from the lectures.

But continuity is not necessary to close this particular gap; just assuming the Euclideanness of the underlying field will do, and results in Hilbert's *Grundlagen* make this quite clear. In his lectures, having pointed out the problem with constructing the triangle described above, Hilbert says:

> We will return to these considerations later, after we have built up geometry completely, and when we investigate the means which can serve in construction. We will then become acquainted with the fine distinction which arises when one is allowed to use a pair of compasses [*der Zirkel*] in an unrestricted way, or whether it can only be used for measuring off segments and angles (the right-angle suffices). (Hilbert (*1898/1899, p. 67), p. 260 in Hallett and Majer (2004))⁴⁶

Hilbert does not return to these matters in his own lecture notes, although there is a section in the *Ausarbeitung*, pp. 170–173 which takes up directly the question of which geometrical constructions are performable in his axiom system.⁴⁷ This discussion is generalized in Chapter VII of the first edition of the *Grundlagen* (Hallett and Majer, 2004, Chapter 5), and it is this which we will consider here, although none of the discussion is motivated, as it is in the preceding lecture notes, by the original consideration of the *TCT*. Hilbert proves two results: (1) Any constructions carried out and justified on the basis of Axioms I–V are necessarily constructions using just a straightedge and a what he calls a 'segment mover [*Streckenübertrager*]' (for which a pair of dividers would serve), the first for drawing straight lines, and the second for measuring off segments. (2) The algebraic equivalent to these constructions is the Pythagorean field. The use of the '*Streckenübertrager*' corresponds to the *restricted* use of the pair of compasses in constructions, i.e. marking off given

---

⁴⁵ A related remark is made in the original lecture notes, p. 64, and here Hilbert adds that Euclid has 'a similar sounding axiom'. There is, however, no such assumption in the *Elements*, either in the Postulates or under the Common Notions. Hilbert may have been referring indirectly to the Euclidean Definitions. For Euclid, a circle is a certain kind of figure, and a figure is 'that which is contained by any boundary or boundaries' (Definition 14); see Heath (1925, Volume 1, p. 153). Perhaps the somewhat vague 'contained by' and 'boundary' suggest 'closed', and perhaps that the circle has no 'gaps'.

⁴⁶ On p. 68 of the *Ausarbeitung* of these notes (Hallett and Majer, 2004, 339–340), Hilbert writes:

> We will discover among other things that it makes an essential difference, whether one is allowed the unrestricted use of a pair of compasses or only allowed to use it for the measuring off of segments and angles.

⁴⁷ See also Hilbert's 1898 *Ferienkurs*, pp. 12–14; (Hallett and Majer, 2004, Chapter 3).

radii using the compass as, say, a pair of dividers. What this means is that the constructions licensed by Hilbert's geometrical system of the 1890s can be carried out (i.e. the existence of the points constructed justified) in any analytic geometry whose coordinates form a Pythagorean field, even the minimal Pythagorean field built over the rationals. (These are Theorems 40 and 41 of the first edition of the *Grundlagen*, pp. 79–81.)[48] Hilbert then remarks that:

> From this Theorem [41], we can see immediately that not every problem solvable by use of a pair of compasses can be solved by means of a ruler and segment mover [*Streckenübertrager*] only. (Hilbert, *Grundlagen*, p. 81, i.e. p. 516 in Hallett and Majer (2004))

In other words, the *unrestricted* use of the pair of compasses in constructions is *not* justified in his system.

To show this, Hilbert first gives an example of a real number which cannot be in the minimal Pythagorean field built over the rationals, namely $\sqrt{2|\sqrt{2}|-2}$, despite the fact 1 and $|\sqrt{2}|-1$ are both in the field. (The example is thus slightly different from that given in the lectures.) It follows that we cannot construct by means of 'Lineal und Streckenübertrager' a right-angled triangle with sides of length 1 (hypothenuse), $|\sqrt{2}|-1$ and $\sqrt{2|\sqrt{2}|-2}$, since the latter length cannot correspond to an element in the minimal Pythagorean field; hence the construction problem is not soluble in Hilbert's geometry. But, as Hilbert remarks (*Grundlagen*, p. 82), the problem *is* immediately soluble by a compass construction; the number Hilbert specifies ($\sqrt{2|\sqrt{2}|-2}$) is, of course, in any Euclidean ordered field built over the rationals.

The central point is now this: If K is the set of all real numbers obtained from the rationals by the operations of addition, multiplication, subtraction, and division, and such that K contains square roots for all of its positive elements, then K is Euclidean and is the smallest field over which straightedge and compass constructions can be carried out. (See Hartshorne, 2000, p. 147.) The salient point is even clearer in Hilbert's Theorem 44 (p. 86), which deals with the problem of characterizing which straightedge and compass constructions can be carried out in his geometry (i.e. reduced to constructions by straightedge and *Streckenübertrager*). In the statement of the condition (see *Grundlagen*, Theorem 44, p. 86), Hilbert quite clearly expresses the fact that the algebraic condition corresponding to the compass construction is that each number in

---

[48] Kürschák showed that the *Streckenübertrager* can be dispensed with in favour of an *Eichmaß*, i.e. a device which measures off a single fixed segment. (See Kürschák 1902.) Hilbert makes a corresponding adjustment, with acknowledgement to Kürschák's work, in the Second Edition of the *Grundlagen* (Hilbert, 1903, 74, 77).

the field of coordinates has a square root in the field, i.e. is Euclidean.[49] Thus, the problem is not to do with a failure of continuity, as Sommer suggests, very possibly leaning on Hilbert's original remark, but rather with the failure of a very much weaker field property, and it seems that Hilbert would certainly have been fully aware of this, by the time of the composition of the *Grundlagen* if not earlier. However, having said this, it is not entirely clear what principle should be added to the axiom system to guarantee Euclideanness; adding the circle-circle property itself as an axiom might appear somewhat *ad hoc*.

Finally, to come back to the constructions involved in Euclid's proofs of I, 1 and I, 22, although both apparently involve ruler and circle constructions, an adequate construction for the *first* case *can* be given using Pythagorean operations alone, thus, uses the compass only in the 'restricted' sense. An equilateral triangle can be constructed by Pythagorean operations just in case $\sqrt{3}$ is in the underlying coordinate field. But $\sqrt{3} = \sqrt{1 + (\sqrt{2})^2}$, and $\sqrt{2} = \sqrt{1 + 1^2}$. Hilbert shows this: see the *Ausarbeitung*, p. 173. Hence, the equilateral triangle can be constructed in Hilbert's axiom system. (The actual construction is given in the 1898 *Ferienkurs*, p. 15: see Hallett and Majer (2004, p. 169).) Thus, Euclid's construction here does not in the least assume continuity, or even 'Euclideanness'.

In sum, what we have here is another investigation which begins with a 'purity of method' question, which then employs higher mathematics in its pursuit, achieves an abstract result, and also uses the knowledge gained to inform us and instruct us about elementary geometry, the geometry closest to intuition.

8.5 Conclusion

Let us draw some general conclusions from these examples.

The concern with 'purity of method' usually focuses on some general consideration of 'appropriateness'; this at least is the way that Hilbert casts

---

[49] In the 1902 lectures, for the purposes of showing the independence of the *Vollständigkeitsaxiom*, Hilbert contructs a (countable) model based on a minimal Pythagorean field. He adds that this geometry (i.e. model) is particularly interesting, for

> ...it contains only points and lines which can be found solely by measuring off segments and angles.
> As we have shown in our *Grundlagen*, page 81ff., not every segment can be constructed by means of measuring off segments alone. Take as an example the segment $\sqrt{2\sqrt{2} - 2}$. (Hilbert (*1902, 89–90); Hallett and Majer (2004, 581)).

it in the passages quoted in the Introduction. And at *some* level, this was a concern of Hilbert's. There are many examples in Hilbert's 1898/1899 lectures on Euclidean geometry where Hilbert is concerned to show that some implicit assumption made by Euclid or his successors is in fact dispensable; some of these were cited in Section 8.2. But the examples we have concentrated on in Section 8.4 show that the focus on eliminating 'inappropriate' assumptions was only one facet of Hilbert's work on geometry; the cases examined are all ones where the appropriateness of an assumption is initially questioned, but where it is shown that it is indeed *required*. Among other things, this forces a revision of what is taken to be 'appropriate'. It is part of the lesson taught by the examples that in general this cannot be left to intuitive or informal assessment, for instance, the intuitive assessment of the complexity of the concepts used in the assumption in question.

Furthermore, full investigation of geometry requires its axiomatization, and proper examination of this requires that it be cut loose from its natural epistemological roots, or, at the very least, no longer immovably tied to them. According to Hilbert's new conception of mathematics, an important part of geometrical knowledge is knowledge which is quite independent of interpretation, knowledge of the logical relationships between the various parts of the theory, the way the axioms combine to prove theorems, the reverse relationships between the theorems and the axioms, and so on, all components we have seen in the examples. And in garnering this sort of geometrical knowledge, there is *not* the restriction to the 'appropriate' which we see in the 'Euclidean' part of Hilbert's concerns. What is invoked in pursuing this knowledge might be some highly elaborate theory, as it is in the analysis of the Isoceles Triangle Theorem, a theory far removed from the 'appropriate' intuitive roots of geometry. Even in the cases of the fairly simple models of the analytic plane used to demonstrate the failure of Desargues's Planar Theorem, the models though visualizable are far from straightforwardly 'intuitive'.

One might be tempted to say that the knowledge so achieved is not geometrical knowledge, but rather purely formal logical knowledge or (as it would be usually put now) *meta*-geometrical knowledge. But although this designation is convenient in some respects, it is undoubtedly misleading. As we have seen, the 'meta'-geometrical results have a direct bearing on what is taken to be geometrical knowledge of the most basic intuitive kind; in particular it can reveal a great deal about the *content* of intuitive geometrical knowledge. In short, it effects an alteration in geometrical knowledge, and must therefore be considered to be a *source* of geometrical knowledge. To repeat: for Hilbert, meta-mathematical investigation of a theory is as much a part of the study of a theory as is working out its consequences, or examining its foundations in the way that Frege, for instance, does.

Thus, for Hilbert's investigations in geometry, 'purity of method' analysis in the standard sense is elaborated into the 'analysis of intuition'. This resolves into two separate investigations, one at the intuitive level, and one at the abstract level, levels which frequently interact and instruct each other. Furthermore, extracting this information often itself involves a detour into the abstract. One example is given by the investigation of Desargues's Planar Theorem, where the use of the structure of segment fields requires first an abstract axiomatization of ordered fields. In particular, as the examples treated here make abundantly clear, higher mathematics is used to instruct or adumbrate intuition, or at the very least to instruct us about it and what it entails.

The second conclusion concerns the notion of the 'elementary' or 'primitive' with respect to a domain of knowledge. The examples we have considered show that often we have to adopt a *non-elementary* point of view in order to achieve results about apparently elementary theorems. Hilbert often stressed, just as Klein did, that elementary mathematics must be studied from an advanced standpoint. Certainly as far as Hilbert's work is concerned, this is much more than a pedagogical point, although Hilbert often stressed this, too.[50] For one thing, the examples considered show that genuine knowledge concerning the elementary domain can flow from such investigations, and they also show that apparently elementary propositions contain within themselves non-elementary consequences, often in a coded form.[51] Furthermore, consideration of the intuitive and elementary is used to generate results at the *abstract* level; one example of this was the result about the abstract theory of fields sketched in Section 8.4.3.

But there is a further question about appropriateness. Investigation of independence inevitably involves mathematics as broadly construed as possible, since it involves the construction of models, indeed, requires the precise description which is only afforded by mathematical models. Given this, one might ask whether there is an appropriate limit on the mathematics which can be used for the analysis of the intuitive. There is an obvious practical limitation: in constructing models, one naturally uses those branches of mathematics which are most familiar, and which will afford the finest control over the models we construct. In Hilbert's case, resort to higher analysis is especially natural, given the extensive theoretical development of analytic geometry in the 19th

---

[50] See e.g. the introductory remarks in Hilbert's *Ferienkurs* for 1896 (Hilbert, *1896), in Hallett and Majer (2004, Chapter 3).

[51] There is surely here more than an analogy with the 'hidden higher-order content' stressed by Isaacson in connection with the Gödel incompleteness phenomena for arithmetic. See Isaacson (1987).

century, which among other things produced the intricate analytic descriptions of non-Euclidean geometry (models), and also involved the treatment and extension of intuitively based geometrical ideas by highly unintuitive means. It is also worth noting that the reason behind the very development of analytic geometry was to be able to solve problems posed by synthetic geometry in an analytic way, and to construe the solutions synthetically. There is a clear sense that this is also what much of Hilbert's work with analytic models does.

But there is a philosophical reason which goes along with this. Part of the point of Hilbert's axiomatization of geometry is to remove it to an abstract sphere 'at the conceptual level' where it is indeed divorced from its 'natural' interpretation in the imperfectly understood structure of space, becoming in the process a self-standing theory. The ideal for Hilbert in this respect was the theory of numbers, and, by extension, analysis. The important philosophical point about the theory of numbers is that for Hilbert it was entirely 'a product of the mind'; it is not in origin an empirical, or empirically inspired, theory like geometry, and can be considered in some sense as already on the 'conceptual level'. Thus, Hilbert shared Gauss' view of the difference in status of geometry and number theory. In short, these geometrical investigations and the consequent extension of geometrical knowledge presuppose arithmetic at some level.

That elementary arithmetic is *explicitly* presupposed is made clear at the beginning of Hilbert's 1898/1899 lectures:

> It is of importance to fix precisely the starting point of our investigation: *We consider as given the laws of pure logic and full arithmetic.* (On the relationship between logic and arithmetic, cf. *Dedekind*, 'Was sind und was sollen die Zahlen?' [i.e. Dedekind (1888)].) Our question will then be: *What propositions must we 'adjoin' to the domain defined in order to obtain Euclidean geometry?* (*Ausarbeitung*, p. 2, p. 303 in Hallett and Majer (2004).)[52]

This use of number theory and analysis reflects two important philosophical positions which Hilbert held at that time concerning arithmetic and analysis. First, just prior to this (e.g. *circa* 1896; see the *Ferienkurs* mentioned in n.50), he seems to have held a version of the 'Dirichlet thesis' that all of higher analysis will in some sense 'reduce' to the theory of natural numbers, a thesis which is stated without challenge in the *Vorwort* to Dedekind's 1888 monograph. Secondly, there is clear indication that he thought of arithmetic as conceptually prior to geometry. This is illustrated in his 1905 lectures.

---

[52] There is no corresponding passage in Hilbert's own lecture notes, suggesting perhaps that Hilbert became aware a little later that something ought to be said about the foundation for the mathematics presupposed in the investigation of geometry.

Hilbert remarks that there are in principle three different ways in which one might provide the basis of the theory of number: 'genetically' as was common in the 19th century; through axiomatization, Hilbert's preferred method whenever possible; or *geometrically*. With respect to the latter, after sketching how such a reduction might in principle proceed, Hilbert says the following:

> The objectionable and troublesome [*mißliche*] thing in this can be seen immediately: it consists in the essential use of geometrical intuitions and geometrical propositions, while geometry and its foundation are nevertheless less simple than arithmetic and its foundations. One must also note that to lay out a foundation for geometry, we already frequently use the numbers. Thus, here the simpler would be reduced to the more complicated, or in any case to more than is necessary for the foundation. (Hilbert, *1905, p. 9)

But while the foundational investigation of geometry presupposes arithmetic (and analysis), there was at this time no similar foundational investigation of arithmetic, and no investigation of the conceptual connection between more elementary parts of arithmetic and higher arithmetic and analysis.[53] In particular, this complex of theory was not subject to axiomatic analysis and indeed not even axiomatized. In the course of Hilbert's work on geometry, he does axiomatize an important part of it, namely the theory of ordered fields, mainly for the purpose of revealing certain analytic structure in the geometry of segments in the analysis of Desargues's Theorem, giving rise to the system of complete, ordered fields published in Hilbert (1900*a*). Nevertheless, the theory of natural numbers was not treated axiomatically by Hilbert until very much later. And the important extensions of Archimedean and non-Archimedean analysis involving (say) complex function theory were never treated by Hilbert as axiomatic theories. Of course, what is in question in the work examined here are certain aspects of the foundations of geometry. Nevertheless, Hilbert was well aware that the results garnered are in a certain strong sense relative, and that the foundational investigation of geometry must be part of a wider foundational programme. Indeed, Hilbert's famous lecture on mathematical problems from 1900 sets out (as Problem 2) precisely the problem of investigating the axiom system for the real numbers, in particular showing the mathematical existence of the real numbers, where there is no recourse to a natural theoretical companion such as is possessed by synthetic geometry. Thus, a limited foundational investigation gives birth to another more general one.

---

[53] In the 1920s, Hilbert stated decisively his rejection of the Dirichlet thesis, though it is not clear when he abandoned it.

Bibliography

BERNAYS, Paul (1922), 'Über Hilberts Gedanken zur Grundlegung der Arithmetik', *Jahresbericht der deutschen Mathematiker-Vereinigung*, 31, 10–19.

BOOLOS, George (1990), 'The Standard of Equality of Numbers', in George Boolos (ed.), *Meaning and Method: Essays in Honor of Hilary Putnam* (Cambridge: Cambridge University Press) 261–277. Reprinted in Boolos (1998, 202–219). Page numbers refer to this reprinting.

—— (1998), *Logic, Logic and Logic* (Cambridge, MA: Harvard University Press).

BROWDER, Felix (1976), 'Mathematical Developments Arising from the Hilbert Problems', *American Mathematical Society, Proceedings of Symposia in Pure Mathematics*, 28 (parts 1 and 2).

COOLIDGE, Julian Lowell (1916), *A Treatise on the Circle and the Sphere* (Oxford: Clarendon Press). Reprinted by Chelsea Publishing Company, New York, 1971.

COXETER, H. M. S. (1974), *Projective Geometry*, 2nd edn (Toronto: University of Toronto Press). Reprinted in 1987 by Springer-Verlag, Berlin, Heidelberg, New York.

DEDEKIND, Richard (1872), *Stetigkeit und irrationale Zahlen* (Braunschweig: Vieweg und Sohn). Latest reprint, 1969. Also reprinted in Dedekind (1932, 315–332); English trans. in Ewald (1996, 765–779).

—— (1888), *Was sind und was sollen die Zahlen?* (Braunschweig: Vieweg und Sohn). Latest reprint, 1969. Also reprinted in Dedekind (1932, 335–391); English trans. in Ewald (1996, 787–833).

—— (1932), *Gesammelte mathematische Werke, Band 3. Herausgegeben von Robert Fricke, Emmy Noether and Öystein Ore* (Braunschweig: Friedrich Vieweg und Sohn). Reprinted with some omissions by Chelsea Publishing Co., New York, 1969.

DEHN, Max (1900), 'Die Legendre'sche Sätze über die Winkelsumme im Dreieck', *Mathematische Annalen*, 53, 404–439.

EWALD, William (ed.) (1996), *From Kant to Hilbert*. 2 vols. (Oxford: Oxford University Press).

EWALD, William and SIEG, Wilfried (eds.) (2008), *David Hilbert's Lectures on the Foundations of Logic and Arithmetic, 1917–1933* (Heidelberg, Berlin, New York: Springer). *David Hilbert's Foundational Lectures, Vol. 3*.

EWALD, William, HALLETT, Michael, and SIEG, Wilfried (eds.) (Forthcoming), *David Hilbert's Lectures on the Foundations of Logic and Arithmetic, 1894–1917* (Heidelberg, Berlin, New York: Springer). *David Hilbert's Foundational Lectures, Vol. 2*.

FREGE, Gottlob (1884), *Die Grundlagen der Arithmetik* (Wilhelm Koebner). Reprinted by Felix Meiner Verlag, Hamburg, 1986, 1988. English trans. by J. L. Austin as *The Foundations of Arithmetic*, 2nd edn (Oxford: Basil Blackwell, 1953). The Austin edition is a bilingual one, and the page numbers are the same for both the German and English texts.

―――― (1906), 'Über die Grundlagen der Geometrie', *Jahresbericht der deutschen Mathematiker-Vereinigung*, 15, 293–309, 377–403, 423–430. Reprinted in Frege (1967, 262–272).

―――― (1967), *Kleine Schriften*, ed. I. Angelelli (Hildesheim: Georg Olms Verlag).

HALLETT, Michael (1979a), 'Towards a Theory of Mathematical Research Programmes (I)', *British Journal for the Philosophy of Science*, 30, 1–25.

―――― (1979b), 'Towards a Theory of Mathematical Research Programmes (II)', *British Journal for the Philosophy of Science*, 30, 135–159.

HALLETT, Michael and MAJER, Ulrich (eds.) (2004), *David Hilbert's Lectures on the Foundations of Geometry, 1891–1902* (Heidelberg, Berlin, New York: Springer). David Hilbert's Foundational Lectures, Vol. 1.

HART, W. D. (1996), *The Philosophy of Mathematics* (Oxford: Oxford University Press).

HARTSHORNE, Robin (2000), *Geometry: Euclid and Beyond* (New York, Berlin, Heidelberg: Springer).

HEATH, Thomas L. (1925), *The Thirteen Books of Euclid's Elements*. 3 vols. 2nd edn (Cambridge: Cambridge University Press).

HILBERT, David (*1891), '*Projektive Geometrie*', Lecture notes for a course held in the Wintersemester of 1891 at the University of Königsberg. Niedersächsische Staats- und Universitätsbiblithek, Göttingen. First published in partial form in Hallett and Majer (2004, 21–64).

―――― (*1893/1894), '*Grundlagen der Geometrie*', Lecture notes for a course to have been held in the Wintersemester of 1893/1894 at the University of Königsberg. Niedersächsische Staats- und Universitätsbiblithek, Göttingen. First published in Hallett and Majer (2004, 72–144).

―――― (1895), 'Über die gerade Linie als kürzeste Verbindung zweier Punkte', *Mathematische Annalen*, 46, 91–96. Reprinted as Anhang I in Hilbert (1903).

―――― (*1896), '*Feriencurs*', Notes for a short course for *Oberlehre*, given at the University of Göttingen during the Easter holidays, 1896; Niedersächsische Staats- und Universitätsbiblithek, Göttingen. First published in Hallett and Majer (2004, 152–158).

―――― (*1898), '*Feriencurs*', Notes for a short course for *Oberlehre*, given at the University of Göttingen during the Easter holidays, 1898; Niedersächsische Staats- und Universitätsbiblithek, Göttingen. First published in Hallett and Majer (2004, 160–184).

―――― (*1898/1899), '*Grundlagen der Euklidischen Geometrie*', Lecture notes for a course held in the Wintersemester of 1898/1899 at the Georg-August Universität, Göttingen. Niedersächsische Staats- und Universitätsbiblithek, Göttingen. First published in Hallett and Majer (2004, 221–301).

―――― (*1899), '*Elemente der Euklidischen Geometrie*', *Ausarbeitung* by Hans von Schaper of the lecture notes Hilbert (*1898/1899). Niedersächsische Staats- und Universitätsbiblithek, Göttingen, and the Mathemtisches Institut of the Georg-August Universität, Göttingen. First published in Hallett and Majer (2004, 302–406).

―――― (1899), 'Grundlagen der Geometrie', in *Festschrift zur Feier der Enthüllung des Gauss-Weber-Denkmals in Göttingen* (Leipzig: B. G. Teubner). Republished in chapter 5 of Hallett and Majer (2004 436–525).

―――― (1900a), 'Über den Zahlbegriff', *Jahresbericht der deutschen Mathematiker-Vereinigung*, 8, 180–185. Reprinted (with small modifications) in 2nd to 7th editions of Hilbert (1899).

―――― (1900b), 'Les principes fondamentaux de la géométrie', *Annales scientifiques de la École Normale Supérieur*, 103–209. French trans. by L. Laugel of Hilbert (1899), with additions by Hilbert. Additions republished in Hallett and Majer (2004, 526–529).

―――― (1900c), 'Mathematische Probleme', *Nachrichten von der königlichen Gesellschaft der Wissenschaften zu Göttingen, mathematisch-physikalische Klasse*, 253–296. English trans. by Mary Winston Newson in *Bulletin of the American Mathematical Society*, (2)**8** (1902), 437–479, and in Browder (1976), 1–34. Partial English trans. in Ewald (1996), Vol. 2, 1096–1105.

―――― (*1902), '*Grundlagen der Geometrie*', *Ausarbeitung* by August Adler for lectures in the Sommersemester of 1902 at the Georg-August Universität, Göttingen. Library of the Mathematisches Institut. First published in chapter 6 of Hallett and Majer (2004, 540–606).

―――― (1902/1903), 'Über den Satz von der Gleichheit der Basiswinkel im gleichschenklichen Dreieck', *Proceedings of the London Mathematical Society*, 35, 50–67. Republished as Anhang II in Hilbert (1903), and in revised form in subsequent editions.

―――― (1903), *Grundlagen der Geometrie. Zweite Auflage* (Leipzig and Berlin: B. G. Teubner). Revised edition, also contains the sections added to Hilbert (1900b).

―――― (*1905), '*Logische Principien des mathematischen Denkens*', Lecture notes for a course held in the Sommersemester of 1905 at the Georg-August Universität, Göttingen. Library of the Mathematisches Institut. To appear in Ewald, Hallett and Sieg (Forthcoming).

―――― (*1917/1918), '*Prinzipien der Mathematik*', *Ausarbeitung* by Paul Bernays of notes for lectures in the Wintersemester 1917/1918, Mathematisches Institut of the Georg-August Universität, Göttingen. Published in Ewald and Sieg (2008).

―――― (1918), 'Axiomatisches Denken', *Mathematische Annalen*, 78, 405–415. Reprinted in Hilbert (1935, 146–156); English trans. in Ewald (1996), Vol. 2, 1105–1115.

―――― (1919), *Natur und mathematisches Erkennen* (Basil, Berlin, Boston: Birkhäuser Verlag). Lecture notes for a course held in the Wintersemester of 1921/1922 at the Georg-August Universität, Göttingen, and prepared by Paul Bernays. Library of the Mathematisches Institut. First published in 1992 in an edition edited by David Rowe.

―――― (*1921/1922), '*Grundlagen der Mathematik*', Lecture notes for a course held in the Wintersemester of 1921/1922 at the Georg-August Universität, Göttingen, and prepared by Paul Bernays. Library of the Mathematisches Institut. First published in Ewald and Sieg (2008).

―― (1923), *Grundlagen der Geometrie. Sechste Auflage* (Leipzig and Berlin: B. G. Teubner). Revised edn, with extra notes.

―― (1930), *Grundlagen der Geometrie. Siebente umgearbeitete und vermehrte Auflage* (Leipzig and Berlin: B. G. Teubner). New edn, extensively revised by Arnold Schmidt.

―― (1935), *Gesammelte Abhandlungen, Band 3* (Berlin: Julius Springer).

HILBERT, David and COHN-VOSSEN, Stefan (1932), *Anschauliche Geometrie* (Berlin: Julius Springer). 2nd edn published by Springer-Verlag in 1996. English trans. *Geometry and the Imagination* by P. Nemenyi (Chelsea Publishing Company, 1952).

ISAACSON, Daniel (1987), 'Arithmetical Truth and Hidden Higher-order Concepts', in *Logic Colloquium '85* (Amsterdam: North-Holland Publishing Co.). Reprinted in Hart (1996, 203–224).

KILLING, Wilhelm (1898), *Einführung in die Grundlagen der Geometrie. Band II* (Paderborn: Druck und Verlag von Ferdinand Schöningh).

KIRCHHOFF, Gustav (1877), *Vorlesungen über mathematische Physik. Band I: Mechanik* (Leipzig: J. Barth).

KÜRSCHÁK, Josef (1902), 'Das Streckenabtragen', *Mathematische Annalen*, 55, 597–598.

PASCH, Moritz (1882), *Vorlesungen über neuere Geometrie* (Leipzig: B. G. Teubner).

POINCARÉ, Henri (1891), 'Les géometries non-euclidiennes', *Revue générale des sciences pures et appliquées*, 2, 769–776. Reprinted with alterations in Poincaré (1902).

―― (1902), *La Science et l'Hypothése* (Paris: Ernst Flammarion). English translation as *Science and Hypothesis*, Walter Scott Publishing Co., London, reprinted by Dover Publications, New York, 1952.

SAMUEL, Pierre (1988), *Projective Geometry* (Berlin, Heidelberg, New York: Springer-Verlag). English trans. by Silvio Levy of *Géometrie Projective* (Paris: Presses Universitaires de France, 1986).

SOMMER, Julius (1899/1900), 'Hilbert's Foundations of Geometry; Review of David Hilbert: *Grundlagen der Geometrie*', *Bulletin of the American Mathematical Society*, 6 (2nd ser.), 287–289.

STÄCKEL, Paul and ENGEL, Friedrich (eds.) (1895), *Theorie der Parallellinien von Euklid bis auf Gauss, eine Urkundensammlung zur vorgeschichte der nichteuklidischen Geometrie* (Leipzig: B. G. Teubner).

9

Mathematical Concepts and Definitions

JAMIE TAPPENDEN

> These are some of the rules of classification and definition. But although nothing is more important in science than classifying and defining well, we need say no more about it here, because it depends much more on our knowledge of the subject matter being discussed than on the rules of logic.
>
> (Arnauld and Nicole, 1683, p. 128)

9.1 Definition and mathematical practice

The basic observation structuring this survey is that mathematicians often set finding the 'right'/'proper'/'correct'/'natural' definition as a research objective, and success—finding 'the proper' definition—can be counted as a significant advance in knowledge. Remarks like these from a retrospective on 20th century algebraic geometry are common:

> ... the thesis presented here [is that] the progress of algebraic geometry is reflected as much in its definitions as in its theorems (Harris, 1992, p. 99)

I am indebted to many people. Thanks to Paolo Mancosu both for comments on an unwieldy first draft and for bringing together the volume. Colin McClarty and Ian Proops gave detailed and illuminating comments. Colin also alerted me to a classic treatment by Emmy Noether (1921, pp. 25–29) of the multiple meanings of 'prime'. An exciting conversation with Steven Menn about quadratic reciprocity set me on a path that led to some of the core examples in this paper. Early versions of this material were presented at Wayne State University and Berkeley. I'm grateful to both audiences, especially Eric Hiddleston, Susan Vineberg, and Robert Bruner. Thanks to the members of my philosophy of mathematics seminar for discussing this material, especially Lina Jansson and Michael Docherty for conversation about the 'bad company' objection. As usual, I would have been lost without friendly and patient guidance of the U. of M. mathematicians, especially (on this topic) Jim Milne (and his class notes on class field theory, algebraic number theory, and Galois theory, posted at http://www.jmilne.org/math/) and Brian Conrad.

Similarly, from a popular advanced undergraduate textbook:

> Stokes' theorem shares three important attributes with many fully evolved major theorems:
>
> a) It is trivial
> b) It is trivial because the terms appearing in it have been properly defined
> c) It has significant consequences (Spivak, 1965, p. 104)

Harris is speaking of the stipulative introduction of a new expression. Spivak's words are most naturally interpreted as speaking of an improved definition of an established expression. I will address both stipulative introduction and later refinement here, as similar issues matter to both.

What interest should epistemology take in the role of definition in mathematics? Taking the question broadly, of course, since the discovery of a proper definition is rightly regarded in practice as a significant contribution to mathematical knowledge, our epistemology should incorporate and address this fact, since epistemology is the (ahem) *theory* of (ahem) *knowledge*. A perfectly good question and answer, I think, but to persuade a general philosophical audience of the importance and interest of mathematical definitions it will be more effective, and an instructive intellectual exercise, to take 'epistemology' and 'metaphysics' to be fixed by presently salient issues: what connection can research on mathematical definition have to current debates?

9.2 Mathematical definition and natural properties

Both stipulative definitions of new expressions and redefinitions of established ones are sometimes described as 'natural'. This way of talking suggests a connection to metaphysical debates on distinctions between natural and artificial properties or kinds. Questions relevant to 'naturalness' of mathematical functions and classifications overlap with the corresponding questions about properties generally in paradigm cases. We unreflectively distinguish 'grue' from 'green' on the grounds that one is artificial and the other isn't, and we distinguish 'is divisible by 2' from 'is π or a Riemann surface of genus 7 or the Stone-Čech compactification of ω' on the same ground.[1]

---

[1] For those unfamiliar with the philosophical background, 'grue' is an intentionally artificial predicate coined by Nelson Goodman. 'x is grue if x is observed before t and found to be green or x is observed after t and found to be blue.' See the collection Stalker (1994) for discussion and an extensive annotated bibliography.

A particularly influential presentation of the issues appears in the writings of David Lewis.[2] It is useful here less for the positive account (on which Lewis is noncommittal) than for its catalogue of the work the natural/nonnatural distinction does. Most entries on his list (underwriting the intuitive natural/nonnatural distinction in clear cases, founding judgements of similarity and simplicity, supporting assignments of content, singling out 'intended' interpretations in cases of underdeterminacy…) are not different for mathematical properties and others.[3] In at least one case (the 'Kripkenstein' problem) naturalness will not help unless some mathematical functions are counted as natural. In another—the distinction between laws of nature and accidentally true generalizations—it is hard to imagine how an account of natural properties could help unless at least some mathematical properties, functions, and relations are included. The criteria in practice for lawlikeness and simplicity of laws often pertain to mathematical form: can the relation be formulated as a partial differential equation? Is it first or second order? Is it linear? The role of natural properties in inductive reasoning may mark a disanalogy, but as I indicate below this is not so clear. Of course, the use of 'natural' properties to support analyses of causal relations is one point at which mathematics seems out of place, though again as we'll see the issue is complicated. In short, an account of the natural/nonnatural distinction is incomplete without a treatment of mathematical properties.

Obviously the prospects of a smooth fit between the background account of mathematical naturalness and the treatment of physical properties will depend on the broader metaphysical picture. If it takes as basic the shape of our classifying activity, as in Taylor (1993), or considerations of reflective equilibrium, as in Elgin (1999) there is no reason to expect a deep disanalogy. Accounts positing objective, causally active universals could present greater challenges to any effort to harmonize the mathematical and non-mathematical cases. However, though the issues are complicated, they principally boil down to two questions: first, what difference, if any, devolves from the fact that properties in the physical world interact through contingent causal relations and mathematical properties don't? Second: to what extent is it plausible to set aside the distinctions between natural and nonnatural that

---

[2] See for example Lewis (1986, especially pp. 59–69). A helpful critical overview of Lewis' articles on natural properties is Taylor (1993); Taylor proposes what he calls a 'vegetarian' conception based on principles of classification rather than objective properties.

[3] Sometimes less grand distinctions than 'natural–nonnatural' are at issue. In Lewis' treatment of intrinsic properties the only work the natural–nonnatural distinction seems to do is secure a distinction between disjunctive and non-disjunctive properties. Someone might regard the latter distinction as viable while rejecting the former. If so, the Legendre symbol example of §3 illustrates the intricacy of even the more modest distinction.

arise in mathematical practice as somehow 'merely pragmatic' questions of 'mathematical convenience'?[4] Here too we can't evaluate the importance of mathematical practice for the metaphysical questions unless we get a better sense of just what theoretical choices are involved. To make progress we need illustrations with enough meat on them to make clear how rich and intricate judgements of naturalness can be in practice. The next two sections sketch two examples.

9.3 Fruitfulness and stipulative definition: the Legendre symbol

Spivak's remark suggests that one of the criteria identifying 'properly defined' terms is that they are fruitful, in that they support 'trivial' results with 'significant consequences'. It is an important part of the picture that the consequences are 'significant'. (Any theorem will have infinitely many consequences, from trivial inferences like $A \vdash A \& A$.) So what makes a consequence 'significant'? I won't consider everything here, but one will be especially relevant in the sequel: a consequence is held in practice to be significant if it contributes to addressing salient 'why?' questions. Evaluations of the explanatoriness of arguments (theories, principles, etc.) and evaluations of the fruitfulness of definitions (theories, principles, etc.) interact in ways that make them hard to surgically separate. I'm not suggesting that considerations about explanation *exhaust* the considerations relevant to assessing whether or not a consequence is significant or a concept fruitful because it has significant consequences. I'm just making the mild observation that explanation is easier to nail down and better explored than other contributors to assessments of significance, so it is helpful as a benchmark. As a contrast, it is also common for proofs and principles to be preferred because they are viewed as more natural.[5] However,

---

[4] This is a pivotal argumentative support in Sider (1996) to cite just one example. Discussion of the arguments would go beyond this survey, so I'll leave it for the companion article.

[5] For example, many number theorists count the cyclotomic proof as particularly natural. (Frölich and Taylor, 1991, p. 204) opine that this proof is 'most natural' (or rather: ' "most natural" '). Similarly, in the expository account of Artin's life by Lenstra and Stevenhagen (2000, p. 48) we read: 'Artin's reciprocity law over $\mathbb{Q}$ generalizes the quadratic reciprocity law and it may be thought that its mysteries lie deeper. Quite the opposite is true: the added generality is the first step on the way to a natural proof. It depends on the study of cyclotomic extensions.' (p. 48). Gauss, on the other hand, though one of his proofs exploits cyclotomy, preferred a more direct argument using what is now called 'Gauss's lemma'. Of other proofs he wrote: 'Although these proofs leave nothing to be desired as regards rigor, they are derived from sources much too remote...I do not hesitate to say that until now a *natural* proof has not been produced.' (Gauss, 1808, p. 1). Gauss might have

the relevant idea of 'natural proof' is uncharted and poorly understood; it would hardly clarify 'mathematically natural property' to analyze it in terms of 'mathematically natural proof'. On the other hand, though the study of mathematical explanation is still in early adolescence, we have learned enough about it to use it for orientation.

An illustration of the quest for explanation in mathematics is the often re-proved theorem of quadratic reciprocity:[6] If p and q are odd primes, then $x^2 \equiv p \pmod{q}$ is solveable exactly when $x^2 \equiv q \pmod{p}$ is, except when $p \equiv q \equiv 3 \pmod 4$.[7] In that case, $x^2 \equiv p \pmod{q}$ is solveable exactly when $x^2 \equiv q \pmod{p}$ isn't. Gauss famously found eight proofs and many more have been devised[8]. One reason that it attracts attention is that it cries out for explanation, even with several proofs already known. As Harold Edwards puts it:

> The reason that the law of quadratic reciprocity has held such fascination for so many great mathematicians should be apparent. On the face of it there is absolutely no relation between the questions 'is p a square mod λ?' and 'is λ a square mod p?' yet here is a theorem which shows that they are practically the same question. Surely the most fascinating theorems in mathematics are those in which the premises bear the least obvious relation to the conclusions, and the law of quadratic reciprocity is an example *par excellence*. ... [Many] great mathematicians have taken up the challenge presented by this theorem to find a natural proof or to find a more comprehensive 'reciprocity' phenomenon of which this theorem is a special case. (Edwards, 1977, p. 177)

A similar expression of amazement, and a demand for explanation and understanding, appears in a review of a book on reciprocity laws:

> We typically learn (and teach!) the law of quadratic reciprocity in courses on Elementary Number Theory. In that context, it seems like something of a miracle. Why should the question of whether p is a square modulo q have any relation to the question of whether q is a square modulo p? After all, the modulo p world

---

revised his opinion were he to have seen subsequent research, given his often expressed view of the 'fruitfulness' of the study of cyclotomic extensions. On Gauss on cyclotomy and reciprocity, see Weil (1974).

[6] The basic facts are available in many textbooks. A particularly appealing, historically minded one is Goldman (1998). Cox (1989) is an engagingly written, historically minded essay on a range of topics in the area. Chapter 1 is a clear presentation of the basic number theory and history accessible to anyone with one or two university math courses. The presuppositions jump up significantly for Chapter 2 (covering class field theory and Artin reciprocity). Jeremy Avigad is doing penetrating work exploring philosophical ramifications of algebraic number theory. See Avigad (2006) and elsewhere.

[7] $a \equiv b \pmod{c}$ means $(\exists n)$ $a = nc + b$, or as we put it in school arithmetic 'a divided by c has remainder b'. When $(\exists x)\, x^2 \equiv p \pmod{q}$ we say p is a *quadratic residue* mod q.

[8] 221 proofs using significantly different techniques are listed at http://www.rzuser.uni-heidelberg.de/hb3/fchrono.html. A hardcopy is in Lemmermeyer (2000, pp. 413–417).

and the modulo q world seem completely independent of each other... The proofs in the elementary textbooks don't help much. They prove the theorem all right, but they do not really tell us why the theorem is true. So it all seems rather mysterious... and we are left with a feeling that we are missing something. What we are missing is what Franz Lemmermeyer's book is about. ... he makes the point that even the quadratic reciprocity law should be understood in terms of algebraic number theory, and from then on he leads us on a wild ride through some very deep mathematics indeed as he surveys the attempts to understand and to extend the reciprocity law.[9]

The search for more proofs aims at more than just explaining a striking curiosity. Gauss regarded what he called 'the fundamental theorem' as exemplifying the fruitfulness of seeking 'natural' proofs for known theorems.[10] His instinct was astonishingly accurate. The pursuit of general reciprocity proved to be among the richest veins mined in the last two centuries. Nearly one hundred years after Gauss perceived the richness of quadratic reciprocity, Hilbert ratified the judgement by setting the 'proof of the most general law of reciprocity in any number field' as ninth on his list of central problems. The solution (the Artin reciprocity law) is viewed as a major landmark.[11]

Gauss recognized another key point: the quest for mathematically natural (or, indeed, *any*) proofs of higher reciprocity laws forces extensions of the original domain of numbers.[12] (Once quadratic reciprocity is recognized, it is natural to explore higher degree equations. Are there cubic reciprocity

---

[9] Review of Lemmermeyer (2000) by F. Gouvêa at: <http://www.maa.org/ reviews/brief_jun00.html>.

[10] A typical expression of his attitude is:

> It is characteristic of higher arithmetic that many of its most beautiful theorems can be discovered by induction with the greatest of ease but have proofs that lie anywhere but near at hand and are often found only after many fruitless investigations with the aid of deep analysis and lucky combinations. This significant phenomenon arises from the wonderful concatenation of different teachings of this branch of mathematics, and from this it often happens that many theorems, whose proof for years was sought in vain, are later proved in many different ways. As a new result is discovered by induction, one must consider as the first requirement the finding of a proof by *any possible* means. But after such good fortune, one must not in higher arithmetic consider the investigation closed or view the search for other proofs as a superfluous luxury. For sometimes one does not at first come upon the most beautiful and simplest proof, and then it is just the insight into the wonderful concatenation of truth in higher arithmetic that is the chief attraction for study and often leads to the discovery of new truths. For these reasons the finding of new proofs for known truths is often at least as important as the discovery itself. (Gauss, 1817, pp. 159–60. Translation by May (1972, p. 299) emphasis in original)

[11] See Tate (1976). As Tate notes, the richness of the facts incorporated in quadratic reciprocity has not run out even after two centuries of intense exploration. A 2002 Fields Medal was awarded for work on the Langlands program, an even more ambitious generalization.

[12] See Gauss (1828). Weil (1974a, p. 105) observes that for Gauss, even conjecturing the right laws wasn't possible without extending the domain.

laws? Seventeen-th?) To crack biquadratic reciprocity, Gauss extended the integers to the Gaussian integers $\mathbb{Z}[i] = \{a + bi \mid a, b \in \mathbb{Z}\}$. Definitions that are interchangeable in the original context can come apart in the expanded one. This can prompt an analysis that reshapes our view of the definitions in the original environment, when we use the extended context to tell us which of the original definitions was really doing the heavy lifting.[13] This pressure to understand the original phenomena generated 'class field theory': the study of (Abelian) extension fields. A central 20th century figure put it this way:

> ... by its form, [quadratic reciprocity] belongs to the theory of rational numbers ... however its contents point beyond the domain of rational numbers ... Gauss ... recognized that Legendre's reciprocity law represents a special case of a more general and much more encompassing law. For this reason he and many other mathematicians have looked again and again for new proofs whose essential ideas carry over to other number domains in the hope of coming closer to a general law ... The development of algebraic number theory has now actually shown that the content of the quadratic reciprocity law only becomes understandable if one passes to general algebraic numbers and that a proof appropriate to the nature of the problem can best be carried out with these higher methods.[14] (Hecke, 1981, p. 52–53)

This provides the background for our central example of a mathematically natural stipulative definition: the Legendre symbol. The statement of the law of quadratic reciprocity given above was broken into cases, and it's good intellectual hygiene to find a uniform statement. For this purpose we define the Legendre symbol $\left(\frac{n}{p}\right)$ (p an odd prime):

$$\left(\frac{n}{p}\right) =_{def} \begin{cases} 1, & \text{if } x^2 \equiv n \pmod{p} \text{ has a solution and } n \not\equiv 0 \pmod{p} \\ -1, & \text{if } x^2 \equiv n \pmod{p} \text{ has no solution and } n \not\equiv 0 \pmod{p} \\ 0 & \text{if } n \equiv 0 \pmod{p} \end{cases}$$

---

[13] For another example, the definition of 'integer' requires serious thought. Say we begin with the normal $\mathbb{Z}$ (a ring) in $\mathbb{Q}$ (a field). What ring is going to count as 'the integers' if we extend the field to $\mathbb{Q}[\alpha]$? (That is: when we toss in α and close under $+$, $\times$ and inverses.) The obvious answer is $\mathbb{Z}[\alpha]$; that's what Gauss and Kummer used as 'the integers', and in the specific cases they addressed it happened to do the job. But in general this won't work, and it becomes a genuine problem to identify the natural ring of integers for a general algebraic number field. The question was analyzed and answered in what remains the accepted way by Dedekind in 1871. The basic details and core references are in Goldman, 1998, pp. 250–252).

[14] Jim Milne has drawn my attention to a remark of Chevalley, echoing Hecke's point about 'higher methods' with a puckish parody of Hegel: 'The object of class field theory is to show how the Abelian extensions of an algebraic number field K can be determined by objects drawn from our knowledge of K itself; or, if one wants to present things in dialectical terms, how a field contains within itself the elements of its own transcending (and this without any internal contradictions!).' (Chevalley, 1940, p. 394, my translation)

The Legendre symbol supports a single-case statement of quadratic reciprocity:

$$\text{for odd prime } p, q : \left(\frac{p}{q}\right)\left(\frac{q}{p}\right) = (-1)^{\frac{(p-1)(q-1)}{4}}$$

The Legendre symbol doesn't just give this one-off compression of a statement. It also streamlines proofs, as in Dirichlet's (1858) reformulation of Gauss's first proof of quadratic reciprocity. Proof 1 used nothing fancier than mathematical induction. Dirichlet pointed out that economy of machinery was traded for fragmentation: Gauss proves eight separate cases. Using (a minor generalization of) the Legendre symbol Dirichlet trims the cases to two.

The value of this kind of unification has been discussed in the philosophical literature on explanation.[15] On first pass, it might seem that this is a textbook case of a definition effecting a valuable unification: The new statement of the theorem is brief and seemingly uniform, and a proof that the definition supports reduces many cases to just a couple. However, a recognized problem is that only some unifications are genuine advances. When the unification is effected by appeal to predicates that are 'gerrymandered' the unification may be unilluminating and spurious.[16] We gain nothing if we produce a 'unified theory' of milkshakes and Saturn by rigging concepts like 'x is a milkshake or Saturn'.

So: Does the Legendre symbol reflect a natural mathematical function or an artifice? To appreciate the force of the question, note that one's first impression is likely to be that the definition is an *obvious* gerrymander. We apparently artificially simplify the statement of the theorem by packing all the intricacy and case distinctions into the multi-case Legendre symbol definition. The question is especially pressing because discussions of the metaphysics of properties often view 'disjunctive' predicates as *prima facie* artificial, or unnatural, or 'gruesome', or 'Cambridge', or [insert favourite label for 'metaphysically fishy'].[17] In short, our problem is to find principled reasons to go beyond this impasse:

Thesis: The Legendre symbol is a useful stipulation that contributes to mathematical knowledge. It allows for one-line statements of theorems that had required several clauses, and it supports streamlined proofs by unifying a

---

[15] For more, see the chapters by Paolo Mancosu in this volume and Tappenden (2005).
[16] Kitcher (1989) recognizes this as a limitation on his account of explanation as unification and he accordingly considers only patterns based on projectible predicates. I suggest in Tappenden (2005) that this limits the account to applications where the concept of 'projectibility' makes sense, and mathematics seems to be a place where it doesn't. But as I note in the research paper, I may have been too hasty. Perhaps mathematical practices of conjecture and verification afford more of a basis for a distinction like that between inductively projectible and inductively nonprojectible predicates than I appreciated.
[17] On 'disjunctiveness' of properties see Sanford (1994) and Kim (1992).

variety of cases into two. This supports the verdict that it is mathematically natural.

Antithesis: The symbol is paradigmatic as a definition that is valuable only for limited reasons pertaining to accidental facts of human psychology. That it is a hack, not a conceptual advance is displayed right in the syntax of its definition.

The foothold allowing us to progress is that the function corresponding to the Legendre symbol is itself the object of mathematical investigation.[18] It is a mathematical question whether the Legendre symbol carves mathematical reality at the joints and the verdict is unequivocally yes. There is so much mathematical detail that I don't have nearly enough space to recount it all. To convey the point, it will have to suffice here to gesture at some of the mathematics, with the assurance that there is much more where this came from and a few references to get curious readers started learning more for themselves.

The Legendre symbol (restricted to p relatively prime to the argument on top) is a function from numbers to $\{1, -1\}$. For fixed p, the function is completely multiplicative (for any m and n, $(\frac{mn}{p}) = (\frac{m}{p})(\frac{n}{p})$). It is periodic mod p $((\frac{m}{p}) = (\frac{m+p}{p}))$. Under multiplication, $\{1, -1\}$ and $\{1, 2, , \ldots, p-1\}$ (aka $(\mathbb{Z}/p\mathbb{Z})^*$) are groups, so the function restricted to $(\mathbb{Z}/p\mathbb{Z})^*$ is a surjective homomorphism. These are all mathematically good things, in that they are experience-tested indicators that a function is well-behaved. Another is that multiplication on $\{1, -1\}$ is commutative. (That is: the group is *Abelian*.) This is usually handy; a century of research revealed it be especially pivotal here.[19] Many more such indicators could be listed, but this list says enough to clarify some of the simpler criteria for 'mathematical naturalness' at issue.

What of higher-powered considerations? It would require too much technical background to do more than glance, but some of the flavour of the mathematics is needed to convey what reasons can be given. To crack Hilbert's ninth problem we need to properly generalize in many directions, and to do this we need to reformulate the question. (In contrast to the picture one might get from meditating too long on first-order logic, generalizing is much more than just picking constants and replacing them with variables. Articulating the right structures, which then can be generalized, is an incredibly

---

[18] Some years ago, in a early effort to articulate some of these ideas, I called this the 'self-consciousness' of mathematical investigation. (See Tappenden, 1995a, b).

[19] Class field theory and the Langlands programme differ precisely on whether the Galois group is Abelian. The fact that seems trivial for $\{\{1, -1\}, x\}$ casts a long shadow.

involved process and it is hard to get it right.)[20] The issue in quadratic reciprocity for primes p and q can be rethought as the circumstances under which we can split (i.e. factor into linear factors) an equation in a field extending $\mathbb{Q}$. This can in turn be seen as a question about the relations between $\mathbb{Q}$ and the field $\mathbb{Q}[\sqrt{q*}]$ ($q* = q$ or $-q$ depending on q). Field extensions K/F are sorted by the smallest degree of a polynomial from F that splits in K. In the case of $\mathbb{Q}[\sqrt{q*}]/\mathbb{Q}$ the degree is 2. The basic fact of Galois theory is: If K is the splitting field of a polynomial $\theta(x)$ over a field F, we can associate with K/F a group Gal(K/F) (the *Galois group*) encoding key information about K/F and $\theta(x)$.

These statements have been fruitfully generalized in many ways: we can consider not just degree 2 polynomials but other degrees and their Galois groups. We can consider not just prime numbers but more general structures sharing the basic property of primes. Considering other fields besides $\mathbb{Q}$ induces the need to generalize 'integers (of the field)'. The Galois group is a group of functions, and we can define other useful functions in terms of those functions..., and more. After this and more reformulating and generalizing, lasting nearly 200 years (if the clock starts with Euler) we arrive at the Artin reciprocity law. It has quadratic reciprocity, cubic reciprocity, ..., seventeen-ic reciprocity... as special cases. The core is a function called the Artin symbol fixed by several parameters (base and extension field with induced general integers, given generalized prime, ...). The punch line for us is that when you plug in the values for quadratic reciprocity, (fields: $\mathbb{Q}$ and $\mathbb{Q}[\sqrt{q*}]$, generalized integers: the ordinary $\mathbb{Z}$, generalized prime: $p \in \mathbb{Z}$, ...) the Legendre symbol appears![21]

Now of course given any function in any theorem, we can always jigger up *many* more general versions with the given function and theorem as special cases. Most will be completely uninteresting. The generalizations we've been discussing are mathematically natural, as measured in many ways, some subtle and some obvious. The core fact is the one Gauss noted: Investigations of quadratic reciprocity and its generalizations reveal deep connections among (and support proofs in) an astonishing range of fields that initially seem to have little in common. (Not just elementary arithmetic, but circle-division, quadratic forms, general algebra of several kinds, complex analysis, elliptic

---

[20] Wyman (1972) is a superb exposition of a few of the reformulations and re-reformulations needed to generalize quadratic reciprocity.

[21] Cox (1989, pp. 97–108) describes how the Legendre symbol falls immediately out of the Artin symbol as a special case. Samuel (1970, esp. pp. 92–93) conveys with beautiful economy the basic idea in a less general form with fewer mathematical prerequisites. (Think 'Artin symbol' when Samuel refers to the 'Frobenius automorphism'.)

functions...) The upshot of the general investigation is a collection of general theories regarded by mathematicians (and hobbyists) of a range of different backgrounds and subspecialties as *explaining* the astonishing connection between arbitrary odd primes. The judgement that the Legendre symbol carves at a joint interacts with a delicate range of mathematical facts and judgements: verified conjectures, the practice of seeking explanations and understanding, efforts to resolve more general versions of known truths (and the evaluations of 'proper' generalizations that support this practice), judgements of similarity and difference, judgements about what facts would be antecedently expected (quadratic reciprocity not among them), and more.

The history of quadratic reciprocity also illustrates the importance of inductive reasoning in mathematics. Euler conjectured the law decades before Gauss proved it, on the basis of enumerative induction from cases.[22] This issue will be revisited in the research article, so I'll just note the key point.[23] In many cases, the natural/artificial distinction is linked to projectibility: Natural properties ('green') support correct predictions and artificial ones ('grue') don't. It is common, as in the influential work of Sydney Shoemaker (1980a, b), to connect this with a thesis about causality. Simplifying Shoemaker's picture drastically: natural properties are those that enter into causal relations, and it is because of this that only natural properties support induction properly. Euler and quadratic reciprocity reveal a limit to this analysis: induction, as a pattern of reasoning, does not depend for its correctness on physical causation. The properties supporting Euler's correct inductive reasoning have the same claim to naturalness deriving from projectibility that 'green' has. This is consistent with the observation that mathematical properties don't participate in causation as Shoemaker understands it. Even though there is much more about inductive reasoning in mathematics that we need to understand better, and we should have due regard for the differences between mathematical and empirical judgements, we shouldn't underestimate the affinities.

Delicate issues of identity and difference of content also arise, as in Dedekind's proof of quadratic reciprocity in (Dedekind, 1877/1996). Dedekind describes himself as presenting 'essentially the same as the celebrated sixth proof of Gauss' (from Gauss, 1817). The derivation recasts the treatment of cyclotomic extensions in section 356 of (Gauss, 1801). Dedekind is plausibly described as presenting the same argument in a conceptual form avoiding most of the

---

[22] Edwards (1983), Cox (1989, pp. 9–20), and Weil (1984) are excellent treatments of the inductive reasoning that led Euler to his conjecture. There is a particularly beginner-friendly discussion at the online Euler archive <http://www.maa.org/news/howeulerdidit.html>.

[23] For more on plausible reasoning in mathematics see Jeremy Avigad's first article.

calculations. The challenge for the analyst of reasoning is to find some way of characterizing what kind of success it was that Dedekind achieved.

Returning to the main line, we can draw a moral about the division of mathematical concepts into artificial and non-artificial. Whether or not a concept is 'disjunctive' in a way that gives it a grue-like artificiality is not something we can just read off the syntax of the definition. This example (or rather, this example plus the rest of the mathematics whose surface I've only scratched) illustrates how inextricably judgements as to the naturalness and proper definition of basic concepts are embedded in mathematical method.

Let's remind ourselves of the orienting question: (i) is the Legendre symbol artificial or natural and (ii) how can we tell? The answer is (i) natural, and (ii) that is a mathematical question, to which a substantial answer can be given. The richness and depth of the mathematical rationale provides the basic answer to the suggestion that the mathematical naturalness of a property, function, or relation can be shrugged off as 'pragmatic/mere mathematical convenience'. It is hard to see how this assessment of one category as better than others is different than cognate assessments in the natural sciences or philosophy. This response won't work against someone who maintains that *all* such distinctions of natural and nonnatural are 'merely pragmatic' matters of 'philosophical/physical/chemical/biological/etc. convenience' but that is a separate topic. Our point is that mathematics is on an equal footing.

9.4 Prime numbers: real and nominal definition revisited

We've observed that a common pattern in mathematics is the discovery of the proper definition of a word with an already established meaning. The Legendre symbol can be redefined in light of later knowledge, to reflect the best explanations of the facts. This may appear to be in tension with a longstanding philosophical presumption that the definition introducing an expression is somehow privileged as a matter of meaning. But the unfamiliarity of the Legendre symbol may make the point seem abstruse. It will be worthwhile to look at a comfortingly familiar example: 'prime number'. We learn the original definition in elementary school: $n \neq 1$ is prime if it is evenly divided by only 1 and n. Over $\mathbb{N}$ the familiar definition is equivalent to: $a \neq 1$ is prime if, whenever a divides a product bc (written $a \mid bc$) then $a \mid b$ or $a \mid c$. In extended contexts, the equivalence can break down. For example, in $\mathbb{Z}[\sqrt{5}i] = \{a + b\sqrt{5}i \mid a, b \in \mathbb{Z}\}$, 2 is prime in the original sense, but not in the second, since $6 = (1 - \sqrt{5}i)(1 + \sqrt{5}i)$; 2 divides 6 but not $(1 - \sqrt{5}i)$ or $(1 + \sqrt{5}i)$.

Once the two options are recognized, we need to say which will be canonical. The choice will be context-sensitive, so I'll take algebraic number theory as background. The word *'prime'* is given the *second* definition.[24] (*'Irreducible'* is used for the first.) The reason for counting the second definition as the proper one is straightforward: The most significant facts about prime numbers turn out to depend on it. As with quadratic reciprocity, this is complicated, so I'll just observe that ongoing efforts to explain facts about the structure of natural numbers justify choosing the novel definition of prime.[25] Say we describe the situation this way: investigations into the structure of numbers discovered what prime numbers really are. The familiar school definition only captures an accidental property; the essential property is: $a \mid bc \rightarrow (a \mid b$ or $a \mid c)$. These descriptions of the situation might seem philosophically off. First, it might be maintained that it is *obviously* wrong to say that in $\mathbb{Z}[\sqrt{5}i]$, 2 is only divisible by itself and 1, but isn't a prime number, since that is *analytically* false, given what '2 is a prime number' *means*. Such objections would be unnecessary distractions, so I'll reformulate the point in extensional terms. The original definition of prime number fixes a set $\{2, 3, 5, 7, ...\}$ with interesting properties. In the original domain $\mathbb{N}$ it can be picked out by either definition. The new definition is the more important, explanatory property, so it is the natural one.

The situation is the mathematical analogue of an example in Hilary Putnam's classic 'The analytic and the synthetic' (Putnam, 1962). In the 19th century, one could hold 'Kinetic energy is $\frac{1}{2}mv^2$' to be true by definition. With relativity, the new situation can be described in two ways. We could say 'Kinetic energy is $\frac{1}{2}mv^2$' is not true by definition and also we have learned kinetic energy is not $\frac{1}{2}mv^2$, or that it *was* true by definition but Einstein changed the definition of 'kinetic energy' for theoretical convenience. Neither of these options aptly captures what went on. Saying that we have just changed our mind about the properties of kinetic energy doesn't respect the definitional character of the equation, but saying we have embraced a more convenient definition while retaining the same words fails to do justice to the depth of the reasons for

---

[24] This point is addressed in many places; one is Stewart and Tall (2002, pp. 73–76).

[25] There also also deep reasons for excluding 1. Widespread folklore sees the rationale to be a clean statement of the uniqueness of prime factorization; this makes excluding 1 appear to be a matter of minor convenience in the statement of theorems. As against this, an algebraic number theorist I know remarked in correspondence: '... with the advent of modern algebra and the recognition of the different concepts of unit and prime in a general commutative ring it became clear that units are not to be considered as prime.' The statement of unique factorization, on the other hand, was seen as 'not particularly compelling'.

the change, and so fails to capture how the new definition was a genuine advance in knowledge. As with 'kinetic energy', so with 'prime': it is of less consequence whether we say that word meaning changed or opinions about objects changed. What matters is that the change was an advance in knowledge, and we need a philosophical niche with which to conceptualize such advances.

The idea that discovering the proper definition can be a significant advance in knowledge has overtones of a classical distinction between 'real' and 'nominal' definition. 'Real definition' has fallen on hard times in recent decades. Enriques' *The Historic Development of Logic* (Enriques, 1929), addresses the topic throughout, in a whiggish recounting of the emergence of the idea that 'real definition' is empty and that all definitions are nominal. That seems to be where things stand now. We might need to rework too many entrenched presuppositions to revive the distinction in its traditional form, but it would help to reconstruct a minimal doctrine to support the distinctions we want to draw and connections we want to make. The core motivation is that in mathematics (and elsewhere) finding the proper principles of classification can be an advance in knowledge. We appear to have enough of a grip on the real/nominal distinction that it might be useful in this connection, and we don't have to accept everything that comes bundled in the package. For instance, one concession to changing attitudes seems sensible. Since we are no longer sure what to make of defining a thing as opposed to a word let's stick with the contemporary view that definitions are stipulations of meaning for expressions.

One reason for carefully delineating what should be involved with a prospective concept of real definition is indicated by Richard Robinson in the classic *Definition* (Robinson, 1950). The concept has been called on to play too many different roles: Robinson lists twelve importantly different ones (1950, pp. 189–90). Robinson's recommendation is that talk of 'real definition' be dropped, and that each of the components be treated separately. But this might be an overreaction, giving up the prospect of salvaging valuable insights from the tradition. The activities did not come to be associated by accident, and even if no single concept can do *everything*, we can identify a subset of roles that can usefully be welded together. From Robinson's list, these include: 'Searching for a key that will explain a mass of facts', 'improvement of concepts', plus (if these can be construed neutrally) 'searching for essences' and 'searching for causes'.

Of course, to speak of 'improvement of concepts' we'll need to specify a criterion for improvement. Putting together a useful, metaphysically uncontentious doctrine of real definition seems promising if we take the relevant improvement to partially involve finding 'a key that will explain a mass of facts'. That is, a concept is improved is if it is made to fit better into explanations (or into arguments that are good in some other way). The connection between singling out specific definitions as real and the practice of giving explanations is also indicated in the recent revival of 'essence', if we understand 'essence' in terms of real definition and role in argument. Since 'essence' is seen as metaphysically charged we'll take 'real definition' to be basic, and essences, to the extent we want to use that idea, will be what real definitions define.[26]

The idea that real definition fits with 'cause' draws on the observation that, for Aristotle, some mathematical facts cause others. This is not how we use 'cause' today, so it might be better to appeal to conceptual or metaphysical dependence. Once more this can be understood in terms of explanatory argument structure.[27] This need not make mathematical explanation fundamentally different from ordinary explanation. We could opt for an account of explanation and causation as sketched in Kim (1994), taking dependence as basic. (We'll need an account of dependence anyway, to spell out what it is for facts about primes to depend on the second definition, for example.) Causation, as one kind of dependence, supports explanation, but it is not the only kind that does.

Our theme has been that theorizing has to be driven by the data, so we should look to see how the cases we've just considered appear in this light. Taking real definition and essence to be functions of explanation (and other, harder to pin down properties of argument) promises to give a plausible accounting of why the algebraic definition of 'prime' counts as picking out the essential property of the class of primes in the context of algebraic number theory. Relative to an explanation of quadratic reciprocity via Artin reciprocity, the 'real definition' of the Legendre symbol becomes its specification as a special case of the Artin symbol rather than the multicase initial definition. As with 'prime number', the definition with the central role in the central arguments is not the one that

---

[26] This is the approach of Fine (1994), though one would not call the view 'metaphysically innocuous'. Our search for a neutral account is not driven by any sense that a meaty view like Fine's isn't right, but only by the goal of avoiding extra debate while charting a framework to explore mathematical cases in parallel with cases in science and common life.

[27] More on this cluster of Aristotelean ideas can be found in Mic Detlefsen's contribution. The core idea, that there is a prospectively illuminating, broadly Aristotelean, way of understanding 'essence' in terms of explanation has been noted for some time. Unpublished work by my colleague Boris Kment promises to be illuminating on this point. In talks, Fine (Eastern meetings of the American Philosophical Association, December 2004) has sketched an account of essence as connected fundamentally to explanation; he tells me (correspondence, May 2006) that though he has not yet published anything on the explanation–essence connection, he has explored it in unpublished work.

introduced the symbol. Here too further investigation unearthed and justified something that it makes sense to call the 'real definition'.

9.5 Prospective connections

(a) 'Fruitful concepts'

Frege, in various writings, addressed the issue of valuable concepts and definitions. Though he does not develop the idea systematically, there are enough tantalizing hints to spur efforts to reconstruct the doctrines behind them.[28] The importance of 'fruitfulness' in theory choice has long been recognized. Kuhn (1977) lists it among the theoretical values in science, remarking that despite its importance the idea was little studied. This situation changed little in subsequent decades, though an essay by Daniel Nolan (1999) takes the first steps toward analyzing what is virtuous about this theoretical virtue. A prerequisite for real progress is a more detailed analysis of cases that have unfolded in practice.

(b) Burge on 'partial grasp' of senses; Peacocke on 'implicit conceptions'

Euler, Gauss, and others had true thoughts about prime numbers before the concept was properly defined. It seems right to say that Dedekind's presentation of Gauss' third proof of quadratic reciprocity really does present 'essentially' what is going on, while omitting the calculations that Gauss himself (for all we know) may have thought essential. Such cases are examples looking for a theory (specifically, a theory of partially grasped content and sameness of content). Burge (1990) has articulated a theory of meaning that builds upon the idea of 'partial grasp'. Following Burge, Peacocke (1998a, b) explores the relation of the 19th century $\delta - \epsilon$ definitions to the Newton/Leibniz presentations of calculus, plausibly suggesting that the later definitions are implicit in the earlier. As critics made clear, many complications remain to be sorted out.[29] In particular, Rey (1998) notes that Newton and Leibniz may have had a conception explicitly in mind contradicting the $\delta - \epsilon$ treatment. On the other hand, the commitments of researchers are complex, and could involve methodology that, when consistently followed out, might undercut other aspects of the researchers' view of what they are doing. (Here one remembers the amusing episodes in the development of the axiom of choice, where

---

[28] See Tappenden (1995a).
[29] The volume containing Peacocke's papers has several more devoted to criticism.

opponents aggressively rejected the axiom while finding themselves repeatedly applying it tacitly in proofs.) Peacocke has identified a deep problem that we have only begun to see clearly. Here too, philosophical analysis and study of rich cases from practice will have to go hand in hand.

(c) Wright/Hale on the Caesar problem

The insight inspiring neo-logicism is that arithmetic can be derived from an abstraction principle. This suggests that we revisit Frege's reservation: How can we conclude, from this definition, that Julius Caesar is not a number? One appealing strategy is to spell out and defend the naive reaction: Caesar isn't a number because he just isn't the *kind of thing* that *can* be. Pursuing this line, Hale and Wright (2001) argue that in sortal classification some categories are inherently disjoint. They concentrate their brief discussion on abstract reflections on individuation and criteria of identity; my impression is that most readers have been left wanting more. A good way to go beyond foot-stomping staredowns where each side just reaffirms favoured intuitions is to assemble more information about a range of cases. The neo-logicist programme may need clarification of the structure of natural classification in mathematics for a further reason: the so-called 'bad company objection'. Some abstraction principles have evidently false consequences, but might be excluded if a viable artificial/natural distinction ('Cambridge'/'non-Cambridge' in Wright/Hale's parlance) could be articulated and defended. Not all the troublesome abstracts are plausibly ruled unnatural, so this is not a cure-all. But we shouldn't assume that all problematic abstracts are problematic for the same reason.

9.6 Summing up: the *Port Royal* principle

The examples we've seen are enough to conclude that mathematical defining is a more intricate activity, with deeper connections to explanation, fruitfulness of research, etc. than is sometimes realized. We need a philosophical framework to help us keep the books while we sort it out, and to guide us as to what consequences if any this can have for other areas of philosophy. The philosophical stance implicit in the above has affinities to the position that Penelope Maddy ((1997) and elsewhere) has called 'mathematical naturalism'. Though I differ with Maddy on some details, we agree that rich analysis of mathematical practice is a *sine qua non* for judgements in the philosophy of mathematics. The lines that open this paper, from the *Port Royal Logic*'s treatment of real definition, prompt the name '*Port Royal* principle' for the kind of 'naturalism' at issue: 'nothing is more important in science than classifying and defining well ... [though] it depends much more on our knowledge of the subject matter being discussed than on the rules of logic.'

Bibliography

ARNAULD, Antoine and NICOLE, Pierre (1683), *La Logique, ou L'art du Penser*, 5th edn. Ed. and trans. by Jill Buroker as *Logic, or the Art of Thinking* (Cambridge: Cambridge University Press, 1996). Page references to the translation.

AVIGAD, Jeremy (2006), 'Methodology and Metaphysics in the Development of Dedekind's Theory of Ideals' in J. Ferrirós and J. Gray (eds.) *The Architecture of Modern Mathematics* (Oxford: Oxford University Press), pp. 159–186.

BURGE, Tyler (1990), 'Frege on Sense and Linguistic Meaning', in D. Bell and N. Cooper (eds.), *The Analytic Tradition* (Oxford: Blackwell) pp. 30–60.

CHEVALLEY, Claude (1940), 'La Theorie du Corps de Classes', *Annals of Mathematics*, 41, 394–418.

Cox, David (1989), *Primes of the Form $x^2 + ny^2$* (New York: Wiley).

DEDEKIND, Richard (1877), 'Sur la Théorie des nombres entiers algébrique' (Gauthier-Villars: Paris). Ed. and trans. by J. Stillwell as *Theory of Algebraic Integers* (Cambridge: Cambridge University Press, 1996). Page references to the translation.

DIRICHLET, Pierre (1854), 'Über den ersten der von Gauss gegebenen Beweise des Reciprocitätsgesetzes in der Theorie der quadratischen Reste', *Journal für die reine und angewandte Mathematik*, 47, 139–50 (= *Werke* II: 121–138).

EDWARDS, Harold (1977), *Fermat's Last Theorem* (Berlin: Springer).

____ (1983) 'Euler and Quadratic Reciprocity', *Mathematics Magazine*, 56, 285–291.

ELGIN, Catherine (1999), *Considered Judgement* (Princeton: Princeton University Press).

ENRIQUES, Federigo (1929), *The Historic Development of Logic* (New York: Holt and Co.).

FINE, Kit (1994), 'Essence and Modality', *Philosophical Perspectives*, 8, 1–16.

FRÖLICH Albrecht and TAYLOR, Martin J. (1991), *Algebraic Number Theory* (Cambridge: Cambridge University Press).

GAUSS, Carl F. (1801), *Disquisitiones Arithmeticae* (Leipzig: G. Fleischer) (= *Werke* I) Trans. with a preface by A. Clarke (New Haven: Yale University Press, 1966). Page references to the translation.

____ (1808), 'Theorematis arithmetici demonstratio nova', *Commentationes Societatis Regiæ Scientiarum Göttingensis*, 16 (= *Werke* II 1–8).

____ (1817), 'Theorematis fundamentalis in doctrina de residuis quadraticis demonstrationes et amplicationes novae', *Werke* II 159–164.

____ (1828), 'Theoria Residuorum Biquadraticorum Commentatio Prima', *Werke* II 165–68.

GOLDMAN, Jay (1998), *The Queen of Mathematics: A Historically Motivated Guide to Number Theory* (Wellesley, Mass.: A. K. Peters).

HALE, Bob and WRIGHT, Crispin (2001), 'To Bury Caesar...' in B. Hale and C. Wright, *The Reason's Proper Study* (Oxford: Oxford University Press), pp. 335–397.

HARRIS, Joe (1992), 'Developments in Algebraic Geometry', *Proceedings of the AMS Centennial Symposium* (Providence: A. M. S. Publications).

HECKE, Erich (1981), *Lectures on the Theory of Algebraic Numbers*, trans. G. Brauer, J. Goldman and R. Kotzen (New York: Springer).

KIM, Jaegwon (1992), 'Multiple Realization and the Metaphysics of Reduction', *Philosophy and Phenomenological Research*, 12, 1–26.

—— (1994), 'Explanatory Knowledge and Metaphysical Dependence' in *Truth and Rationality Philosophical Issues*, 5, 52–69.

KITCHER, Philip (1989), 'Explanatory Unification and the Causal Structure of the World' in P. Kitcher and W. Salmon, (eds.) *Scientific Explanation* (Minneapolis: University of Minnesota Press), pp. 410–505.

KUHN, Thomas (1977), 'Objectivity, Value Judgement and Theory Choice' in *The Essential Tension* (Chicago: University of Chicago Press).

LEMMERMEYER, Franz (2000), *Reciprocity Laws From Euler to Eisenstein* (Berlin: Springer).

LENSTRA, Hendrik and STEVENHAGEN, Peter (2000), 'Artin Reciprocity and Mersenne Primes', *Nieuw Archief voor Wiskunde*, 5, 44–54.

LEWIS, David (1986), *On the Plurality of Worlds* (New York: Blackwell).

MADDY, Penelope (1997), *Naturalism in Mathematics* (Oxford: Clarendon Press).

MAY, Kenneth (1972), 'Gauss' in C. Gillispie (ed.), *Dictionary of Scientific Biography* 5 (New York: Charles Scribner's Sons), pp. 298–310.

NOETHER, Emmy (1921), 'Idealtheorie in Ringbereichen', *Mathematishe Annalen*, 83, 24–66.

NOLAN, Daniel (1999), 'Is Fertility Virtuous in its Own Right?', *British Journal for the Philosophy of Science*, 50, 265–282.

ROBINSON, Richard (1950), *Definition* (Oxford: Clarendon Press).

PEACOCKE, Chris (1998a), 'Implicit Conceptions, Understanding and Rationality', *Philosophical Issues*, 9, 43–88.

—— (1998b), 'Implicit Conceptions, the A Priori, and the Identity of Concepts', *Philosophical Issues*, 9, 121–148.

PUTNAM, Hilary (1962), 'The Analytic and the Synthetic' in H. Feigl and G. Maxwell (eds.) *Minnesota Studies in the Philosophy of Science* 3 (Minneapolis: University of Minnesota Press), pp. 358–397.

REY, George (1998), 'What Implicit Conceptions are Unlikely to Do', *Philosophical Issues*, 9, 93–104.

SAMUEL, Pierre (1970), *Algebraic Theory of Numbers* trans. A. Silleberger (Paris: Hermann).

SANFORD, David (1994), 'A Grue Thought in a Bleen Shade: 'Grue' as a Disjunctive Predicate' in D. Stalker (ed.), *Grue: the New Riddle of Induction* (Chicago: Open Court), pp. 173–192.

SHOEMAKER, Sydney (1980a), 'Causality and Properties' in Peter van Inwagen (ed.), *Time and Cause* (Dordrecht: Reidel), pp. 109–135.

—— (1980b), 'Properties, Causation, and Projectibility' in L. J. Cohen and M. Hesse (eds.), *Applications of Inductive Logic* (Oxford: Clarendon Press), pp. 291–312.

SIDER, Ted (1996), 'Naturalness and Arbitrariness', *Philosophical Studies*, 81, 283–301.

SPIVAK, Michael (1965), *Calculus on Manifolds* (Reading, Mass.: Addison-Wesley).

STALKER, Douglas (1994), *Grue!: The New Riddle of Induction* (LaSalle, Ill.: Open Court).

STEWART, Ian and TALL, David (2002), *Algebraic Number Theory and Fermat's Last Theorem*, 3rd edn (Natik, Mass.: A. K. Peters).

TATE, John (1976), 'The General Reciprocity Law', *Mathematical Developments Arising from Hilbert Problems* (Providence: American Mathematical Society Publications), pp. 311–322.

TAPPENDEN, Jamie (1995a), 'Extending Knowledge and "Fruitful Concepts": Fregean Themes in the Foundations of Mathematics', *Noûs*, 29, 427–467.

—— (1995b), 'Geometry and Generality in Frege's Philosophy of Arithmetic', *Synthèse*, 102, 319–361.

—— (2005), 'Proofstyle and Understanding in Mathematics I: Visualization, Unification and Axiom Choice' in P. Mancosu, K. Jørgensen and S. Pedersen (eds.), *Visualization, Explanation and Reasoning Styles in Mathematics* (Berlin: Springer), pp. 147–214.

—— (Forthcoming), *Philosophy and the Origins of Contemporary Mathematics: Frege and his Mathematical Context* (Oxford: Oxford University Press).

TAYLOR, Brian (1993), 'On Natural Properties in Metaphysics', *Mind*, 102, 81–100.

WEIL André (1974a), 'La Cyclotomie, Jadis et Naguère', *Enseignements Mathématiques*, 20, 247–263.

—— (1974b), 'Two Lectures on Number Theory: Past and Present', *Enseignements Mathématiques*, 20, 87–110.

—— (1984), *Number Theory: An Approach through History; From Hammurapi to Legendre* (Boston: Birkhäuser).

WYMAN, Bostwick (1972), 'What is a Reciprocity Law?', *American Mathematical Monthly*, 79, 571–586.

10

Mathematical Concepts: Fruitfulness and Naturalness

JAMIE TAPPENDEN

[Speaking of a hypothetical mineralogist who rejects color as a basis for classification, preferring crystalline structure or chemical composition:] The introduction of such a concept as a *motif* for the arrangement of the system is, as it were, a hypothesis which one puts to the inner nature of the science; only in further development does the science answer; the greater or lesser effectiveness of a concept determines its worth or worthlessness. (Dedekind, 1854, p. 429)

10.1 Introduction

The preceding chapter urged the 'Port Royal Principle': 'nothing is more important in science than classifying and defining well... [though] it depends much more on our knowledge of the subject matter being discussed than on the rules of logic'. This chapter will present, in broad outline, a position on mathematical 'classification and definition' that is true to the principle and that is interesting both as history and as independently defensible philosophy. An orienting reference is a seemingly innocent passage from an 1899 textbook on elliptic functions:

> ... the peculiarities of Riemann's treatment lie first in the abundant use of geometrical presentations, which bring out in a flexible way the essential properties

I'm grateful to Paolo Mancosu for comments and patience. Some of this material, in an early form, was discussed in a talk at the Kansas State University philosophy department, and I'm grateful to the participants for helpful discussion. The influence of Hilary Putnam's paper 'What is Mathematical Truth?' (Putnam, 1975) runs throughout the discussion of evidence for conjectures. Some of this material was presented to a conference in Princeton in honor of Paul Benacerraf; I'm grateful to the participants—especially Paul Benacerraf, Steven Wagner, John Collins, John Balfe, and Hilary Putnam—for comments.

of the elliptic functions and at the same time immediately throw light on the fundamental values and the true relations of the functions and integrals which are also particularly important for applications. Second, in the synthetic treatment of analytic problems which builds up the expression for the functions and integrals solely *on the basis of their characteristic properties* and *nearly without computing* from the given element and thereby guarantees a multifaceted view into the nature of the problem and the variety of its solutions. (Stahl, 1899, p. III; italics mine)

We expect to encounter such seemingly generic motivational writing in the prefaces of textbooks, and we tend to flip past to get to the meat. But these words are worth pausing over: through the topic (Riemann's lectures), the means of expression (especially the phrases in italics, which were recognized, red-letter clichés), and the orientation toward 'essential properties' gained through the 'synthetic presentation' the author displays his allegiance to a methodology and tacit metaphysics of concepts that had been percolating and crystalizing for fifty years, beginning with the revolutionary techniques of Riemann, and winding through the work of his followers (notably Dedekind).[1] These principles of method were widely enough known that Hilbert needed no further explication for every mathematician to know what he meant when he said that in his approach to higher reciprocity laws: 'I have tried to avoid Kummer's elaborate computational machinery so that here too Riemann's principle may be realized and the proofs driven not by calculations but by thought alone.' (Hilbert, 1998, p. X)

Though the occasional methodological asides are helpful as benchmarks, it is really in the mathematics itself that the power and scope of the revolution in method is manifest. The core idea was that the decision about what to regard as a fundamental property should be seen as part of the problem to be solved rather than as antecedentently evident. (For example, it might be held as evident that + and × are fundamental concepts and others should be reduced to them. (This was basically Weierstrass' view.) By contrast on the revolutionary view it is possible, and indeed in many cases was concluded, that other concepts should be taken to be basic and representations in terms of + and × treated as relatively accidental.) The evidence that the selection of basic categories was correct was grounded in the fruitfulness of the relevant formulations for subsequent research. The theoretical outlook and the resulting

---

[1] Riemann's basic point of view was interpreted quite differently by different followers. Inspired by Riemann, Felix Klein/Sophus Lie, Alfred Clebsch/Ludwig Brill/Max Noether, and Dedekind initiated distinct traditions of mathematics, with different implicit conceptions of method. I believe that the features of Riemann's understanding that I'm describing here are common coin among Riemann's followers, but I will restrict attention here to Riemann as Dedekind understood him, postponing some scholarly subtleties for another time.

278　JAMIE TAPPENDEN

mathematics have had a profound effect on the emergence of the styles of mathematical reasoning that evolved in the subsequent century.

　My goal here is to describe the Riemann–Dedekind approach to "essential characteristic properties" and indicate some of the mathematics that gives it substance. Along the way, I'll spell out why I regard this as a promising and philosophically profound strategy for arriving at an account of naturalness in mathematical classification that can coalesce with an account of properties and definitions in general. To set the stage, I'll first discuss the current situation in the general metaphysics of properties as it pertains to the naturalness of mathematical properties. The point of the context-setting will be to explain why the way things currently stand—especially the role of metaphysical intuitions, and the stock of examples used as reference points (plus the potential examples that are not used)—make it difficult to address mathematical properties in an illuminating way.

10.2 Analytic metaphysics: the 'rules of the game' and the method of intuitions

Even if we acknowledge the Port Royal principle of Chapter 9, and ensure that our account of mathematical 'classification and definition' pays due heed to 'the subject matter being discussed', we should work toward a synoptic treatment of mathematical and non-mathematical cases. It's unlikely that mathematical and non-mathematical reasoning are so disjoint as to exclude interesting points of overlap. In recent decades there has been a revival of old-fashioned metaphysical debates about the reality of universals, the artificial/natural distinction, and cognate topics. It might seem initially promising to draw on these debates to illuminate the questions appearing in the survey essay. In this section I'll illustrate why the methods accepted as defining the 'rules of the game' in the relevant areas of contemporary analytic metaphysics are unlikely to help us as things currently stand. This will force a different perspective on the problem; I'll explore one possibility that centers on inductive practices of conjecture and verification in the subsequent section.

　Consider again the example from the survey essay: In algebraic number theory, the definition '$a \neq 1$ is prime if, whenever $a \mid bc$ then $a \mid b$ or $a \mid c$' is, in an important way, the 'correct', or 'proper' definition of 'prime number', and the school definition '$n \neq 1$ is prime if it is evenly divided by only 1 and n'. is comparatively accidental. As in the suggested comparison—the change in the definition of kinetic energy made necessary by relativity theory—we need to explain what kind of advance in knowledge it is to replace one definition with another for reasons of the relevant kind. Similarly, we need to clarify what we learn when we arrive at a proper choice of basic categories. Two hundred years of research into number theory reveal that the Legendre symbol determines a central number-theoretic function, carving the subject at its joints, despite the disjunctiveness of the definition that introduces it. But just what is it we are learning when we learn this?

Implicit in the suggestion that finding the right definition or arriving at the proper choice of basic categories can be an advance in *knowledge*, is the rider that the choice of definition (or of primitive concepts to be defined) can be in some sense 'objectively' correct. We are really learning something *about the subject we are studying*—not just something about ourselves and our cognitive peculiarities—when we learn the 'best' definition of 'prime', or when we learn that the concept of scheme forms the basis of 'the language of algebraic geometry'.[2] What we are learning may be difficult to spell out with precision, of course. To spell it out we may need in addition to clarify some of the other ideas this volume is devoted to (such as mathematical understanding or explanation or purity of method). But the suggestion is that there is some tenable idea of objective correctness to be spelled out.

Of course, this sets a stumbling block: to clarify the suggestion of the last paragraph, we need to flesh out what we *mean* by 'objective'. And to be sure, both in the present narrow context and in general, this is not easy to do. Much of the discussion of the objectivity of classifications seems to draw more on some unanalyzed intuitive idea of what 'objectivity' must involve, and less on a clearly worked out doctrine or principled analysis. I am in no position to give anything like a principled, general account of a useful concept of 'objectivity' here, so my goal will be more modest: to come up with some tangible sufficient conditions for judging a classification to be 'objectively correct' in a way that will allow mathematical and non-mathematical cases to be treated uniformly.

For orientation, we'll need to look to the ways that the intuitive idea is unpacked in general metaphysics. The work of David Lewis is useful at this point because it helps bring out what is at stake. Lewis famously adopted a striking change in direction in the early 1980s when he argued that we need to accept a class of properties distinguished as universals.[3] Lewis's mode of argument is characteristic of his style: he points to the amount of *work* that the recognition of distinguished properties does, which would be threatened if

---

[2] Here I am echoing a common way of speaking of schemes, as incorporated for example in the title of the Eisenbud and Harris monograph *Schemes: The Language of Modern Algebraic Geometry*.

[3] See Lewis (1983, 1984).

they were to be given up. This work includes the entries in this list (of course, many of these are close relatives):

1. Underwriting the intuitive natural/artificial distinction in clear cases.
2. Founding judgements of simplicity and similarity (which in turn informs the selection of referents of variables for Ramsey sentences).
3. Evaluations of general truths as lawlike or not.
4. Supporting assignments of content (for example in 'Kripkenstein' cases).
5. Underwriting a distinction between intrinsic and non-intrinsic properties.
6. Singling out 'intended' interpretations in cases of underdeterminacy (for example in connection with Löwenheim–Skolem indeterminacy arguments).
7. Distinguishing correct from incorrect inductive predictions in 'grue' type examples.

As noted in the introductory essay, there don't appear to be significant differences in the listed respects between selecting distinguished properties in mathematics and in areas dealing with contingent properties of contingently existing things. Two of the cases Lewis treats (Löwenheim–Skolem indeterminacy and 'Kripkenstein') directly address mathematical examples. The 'Kripkenstein' problem is, in a simple form, the problem of explaining how we come to mean the regular + function when we say 'plus', since the specific computations a speaker has performed in a lifetime will always also be compatible with infinitely many other bizarre 'quus' functions from pairs of numbers to numbers. Lewis's suggestion is that an answer can begin with the simple observation that interpretations tend to pick out natural candidates rather than wacky ones. The case of inductive prediction might appear to indicate a disanalogy, but as I noted in the introduction, this isn't so. We'll revisit induction in more detail later in this paper. Causation, of course, is a potential spoiler, but it's complicated for reasons indicated in Chapter 9, and so we'll set it aside here.

It is worth making a remark about terminology, to avoid the appearance of prejudging any issues, since 'natural' has an unfortunate dual connotation. A choice of categories can be 'natural' if it possesses a certain kind of appropriateness or correctness (as when one says 'ϕ is the natural map', or 'this interpretation is the natural way to understand Kant's remarks on page 17'), or if it pertains to the physical world (as when one speaks of reducing talk of mental properties to talk of natural properties). Discussion of 'natural properties' in metaphysics seem to me most, well, naturally understood as drawing on the former meaning in the cases that are relevant here (such as when 'plus' is counted as more natural than 'quus', or 'grue' less natural than 'green') and so it would be natural to speak of 'natural properties' in mathematics as well. But it might be objected that speaking of 'natural' mathematical functions, definitions, domains, proofs, generalizations, etc. gains rhetorical effect misleadingly from the 'pertaining to nature' meanings.[4] This is a fair complaint, but here I'll have to simply note it as something to keep in mind. It would be difficult to adequately introduce new terminology to disambiguate until we have a clearer sense of what we want the terminology to do. So at this early stage, I'll stick with the common use of the word 'natural' and its cognates—ambiguities and all—while keeping alert to the potential for fallacies of equivocation.

Of course, part of what gives force to the Lewis treatment is the sheer intuitive *umph* of some simple common-sense cases. It is hard to deny the *prima facie* pull of the suggestion that *however* we understand 'objective' it is an *objective* fact that two electrons A and B are more like one another than either of them is like the moon, or the Eiffel Tower, or a moose. This gives urgency to the quest to find an account to either underwrite this intuitive judgement or explain it away. Something like this immediate intuitive force attends the suggestion that 'plus' is somehow objectively natural or simple in a way that 'quus' isn't. However, such an appeal to the intuitive force of a judgement can hardly suffice for more than preliminary orientation. This is true not just because it is inadequate, in general, to rely solely on brute intuition in philosophical argument, but because the force of some of the better mathematical examples of prospective natural categories (like 'genus' or 'scheme') and definitions ('prime') requires training to appreciate. If we can make sense of the idea that categories can be 'objectively mathematically natural' in a way that relates to ongoing mathematical investigation we'll need to make room for the fact that we can *discover* that a category is in fact natural even if it seemed to lack intuitive naturalness at the outset. Indeed, in some cases (like the Legendre symbol) the *prima facie* impression may be that the definition is an obvious disjunctive gerrymander.

Some of the presuppositions about mathematical activity that appear to frame the discussion in the general metaphysics literature present a systematic obstacle to incorporating the relevant mathematical details. It is difficult to arrive at a compelling diagnosis with just vague handwaving about nameless authors. For illustration, I'll look at one representative treatment by Ted Sider (1996) in

---

[4] I'm grateful to conversations with David Hilbert (the Chicago philosopher, not the Göttingen mathematician) and Hartry Field for helping me see that the ambiguities in the word 'natural' could become a distraction in this context.

which mathematical examples are taken to support metaphysical conclusions about the naturalness of properties. Of course, restricting attention to one article gains concreteness but has the potential to sacrifice scope. However, it seems fair to take Sider's paper as a paradigm. Sider is recognized as a significant researcher in the field, the paper appears in a major journal and is widely cited, and my impression is that the intuitions it appeals to are on the whole regarded as acceptable moves by contributors to these debates. I'll leave much of the intricacy of Sider's careful article unmentioned since my concern here is just to use certain aspects of his arguments to illustrate ways in which the presuppositions and accepted moves of current analytic metaphysics make it difficult to address mathematical judgements informatively. In particular, I want to bring out that a general problem with the method of appealing to intuitions about the objectivity of judgements and theories—the extreme context-sensitivity of these intuitions—is especially acute here because of the isolation of metaphysical speculation about mathematics from ongoing mathematics.

Sider sets out to refine an argument of Armstrong (1986, 1989) and Forrest (1986) against Lewis. The resulting objection speaks less to the idea of naturalness and more to a further position (called 'class nominalism') Lewis endorsed: properties, functions, and relations are constructed out of sets. I have reservations about class nominalism, but for orientation it will be useful to accept it and follow out Sider's refinements.[5]

The basic Armstrong–Forrest argument runs as follows: Begin (for the sake of *reductio ad absurdum*) with the assumption that some relations are natural, and choose one natural relation R. The class nominalist reduces relations to sets of ordered pairs, and ordered pairs to sets. There are different ways to do this. The reductions considered explicitly are Kuratowski's definition $<x, y> = \{x, \{x, y\}\}$ and Weiner's $<x, y> = \{\{x, \emptyset\}, \{y\}\}$. If R is natural, then the collections of sets corresponding to R should be natural. The objection is: It is *arbitrary* which reduction of ordered pairs is adopted, which conflicts with the sought-after objective, non-arbitrary character of the assignment of naturalness to the set that represents R. The point, as so distilled, needs fleshing out and strengthening, but the core idea is clear enough. Sider's paper suggests a sequence of such fleshings out and strengthenings, followed by his criticisms of these proposals, before arriving at what he takes to be tenable final results.

Sider considers this reply to Armstrong–Forrest: perhaps one of the reductions should be counted as the *right* one. Sider suggests that there couldn't be a

---

[5] Sider suggests that he is not addressing positions that take functions and ordered pairs to be unreduced, *sui generis* entities. However, Sider's argument against the variation suggested by Phillip Bricker appeals to a premise sufficiently broad as to apply also to any view that takes (unreduced) functions to be distinguished at least in part because of their centrality for mathematical practice.

reason for preferring one over the other, which seems plausible in the specific case of these reductions of ordered pairs. It's less clear what to say about another example Sider considers: the well-known 'multiple reduction problem' first broached in Benacerraf's 'What Numbers Could Not Be' (Benacerraf, 1965). It is observed that the natural numbers can be reduced to either of two series: Zermelo's {∅, {∅}, {{∅}}, ...} where each member of the series is the unit set of the previous one, or von Neumann's {∅, {∅}, {∅, {∅}}, ...} where each member of the series is the set of its predecessors. The suggestion that one reduction or the other might be correct is rejected on the grounds that a reason to prefer one over the other is inconceivable: 'Perhaps one line of our thought here is that we don't see what could possibly count as a reason' (Sider, 1996, p. 289) and 'The only features that distinguish one method from another involve mathematical convenience, and so seem irrelevant to the existence of an ontologically distinguished method' (Sider, 1996, p. 289).

Sider addresses another variation on the Armstrong–Forrest point. Say that instead of arguing for a single distinguished method, we accept that there will be *many*, and maintain that *all* of them are equally natural. Sider rejects this on the (reasonable) ground that *some* methods are *obviously* unnatural, and he produces a grotesquely gerrymandered case to witness the point. Thus we come to what seems to me the best option (given an antecedent acceptance of class nominalism): Say that there is a class C containing more than one method, such that no method is more natural than any member of C, every member of C is equally natural, and every method as natural as a member of C is in C. Say also that there is at least one method that is less natural than the members of C. That is, C is the class of methods such that nothing is any better than a member of C. Why shouldn't we just pick an arbitrary member of C and stick with it, so long as we avoid the less natural methods not in C?[6] The fact that there are several winners in equal first place doesn't take away our ability to distinguish more and less natural.[7]

Sider's only grounds to reject this option is to amplify the claim noted above: 'While a class of distinguished methods may be less implausible than the single distinguished method ... it is still implausible. [This] theory seems to

---

[6] The situation would be like the one we face when choosing a coordinate scheme for Euclidean space. Of course, there is no uniquely reasonable choice. Does this mean that every choice is equally good? Of course not: some assignments of algebraic objects to geometric ones are terrible. Say, for instance that we have Cartesian coordinates and polar coordinates in alternating octants. Then the mathematical description of (say) a sphere centered at the origin would be wild indeed despite the mathematical simplicity of the figure described, and that would be the fault of the choice of unnatural coordinates. Here it seems right to say that there are a class of distinguished coordinatizations, all more or less equally acceptable, and all superior to a range of unacceptable ones.

[7] Sider attributes this alternative to Phillip Bricker.

mistake a pragmatic distinction (Kuratowski's method is more mathematically convenient than method X) for an ontological one. Thus, I continue to reject the idea that any method of constructing ordered pairs is ontologically more distinguished from any other' (Sider, 1996, p. 292). Given that the 'method X's at issue include the obviously gerrymandered case whose intuitive unnaturalness Sider appealed to, and indeed any reduction however artificial and clumsy, Sider's claim will only do the work he needs if it is understood broadly enough that *any* distinction between natural and unnatural mathematical reductions, if made on the basis of an assessment of mathematical naturalness, is counted as based on merely 'mathematical convenience', which is to be counted as merely 'pragmatic' and not an objective guide to the way things are.

Two related suggestions underwrite Sider's stance: the purported inconceivability of any reason to prefer one reduction (of numbers, and by implication of ordered pairs), and the claim that any reason that could be adduced could only be a matter of 'mathematical convenience', and consequently a 'pragmatic' distinction rather than an 'objective'/'ontological' one. Many theoretical choices made in the course of successful mathematical reasoning would be thereby shrugged onto the 'merely pragmatic' scrapheap, so we should pause and take stock. The plausibility of these claims rests significantly on the examples taken as paradigmatic. If we restrict ourselves to ordered pairs, it does indeed seem inconceivable that there could be a substantial, non-pragmatic reason to prefer Kuratowski over Weiner. But it is a mistake to generalize this impression, because of the mathematical insignificance of the example. If 'mathematical convenience' means 'convenience for the practice of mathematics' then the reduction of ordered pairs to sets is at best valueless. In mathematical practice, reduction for the *sole* purpose of paring down the number of basic entities is regarded with indifference or even distaste. (Reductions can be valued, of course, but only if they bring some kind of mathematical benefit, as measured by, for example, improved understanding or enhanced potential for discovering proofs, or 'purity of method' considerations of the sort considered in Mic Detlefsen's contribution.) The reduction of ordered pairs to sets has been used by analytic metaphysicians as a paradigm of mathematics since Quine singled it out as one of his canonical examples. But it is not a paradigm of *mathematics*. In fact, the reduction of ordered pairs to sets has become a stock example in philosophical discussions because it fits with widely shared and well-entrenched *philosophical* assumptions. Quine took it as a paradigm of ontological economy, and ontological economy was proposed as a core methodological objective. Through an unnoticed philosophical metamorphosis, ontological economy came to be counted as a measure of simplicity, with simplicity of the relevant kind counted as a virtue of scientific theories.

Of course, even those who hold that some preferences for one representation over another are based on reasons of more than 'mere convenience' will grant that *some* such preferences *are* just based on relatively trivial pragmatic grounds. Recall that, as noted in Chapter 9, we can cite several reasons for not counting 1 as a prime number. Some of these involve deep facts about the behavior of units in general fields, others can reasonably be seen as minor improvements in streamlined bookkeeping, such as the availability of a compact statement of the prime decomposition theorem. It's plausible that any reason for preferring the Weiner reduction over Kuratowski or conversely could only cite bookkeeping advantages. But the tacit suggestion that this comparison is paradigmatic of mathematical choices of this type masks just what a range of distinct, subtle considerations will get shrugged onto the 'pragmatic' heap.

The point comes out more pressingly in connection with Sider's example of the natural numbers. Though the accepted philosophical folklore holds otherwise, there *are* mathematically important differences between the von Neumann and Zermelo representations. Thanks to these differences, von Neumann's series has long been accepted by set theorists as the most natural, canonical representation of $\mathbb{N}$ and Zermelo's is a forgotten historical curiosity. It is important to recognize that substantive reasons can be cited for this; it is not just a contingent historical accident. (The von Neumann ordinals are not merely VHS to Zermelo's Betamax.) It would be distracting to dissect all the reasons (some of them are rather involved), so I'll concentrate on just a particularly clear and simple one: The principle generating the finite von Neumann ordinals generalizes naturally to infinite numbers while the one generating Zermelo's ordinals doesn't. Take ω, the ordinal of the series $\{0, 1, 2, 3, ...\}$. A von Neumann ordinal is the set of its von Neumann predecessors, so the principle generating the finite ordinals gives us immediately the representation $\omega = \{\emptyset, \{\emptyset\}, \{\emptyset, \{\emptyset\}\}, ...\}$. A Zermelo ordinal is the unit set containing its predecessor, which for ω is... well, ω has no predecessor. So there is no Zermelo ω.

We'd need to jigger up something else to regard the Zermelo ordinals as the finite initial segment of all the ordinals. Of course, it isn't too hard to come up with *something*. For example, we could make every successor ordinal the unit set of its predecessor, and every limit ordinal the set of all its predecessors. But this would evidently be an artificial patch job, and a uniform account would be preferable. Though the von Neumann/Zermelo hybrid might occasionally be clumsy to work with, the problem with it is not *inconvenience* but rather unnaturalness. A proper account of the ordinals should display finite and infinite ordinal numbers as the same kind of thing, with ordinals in general as a natural generalization of the finite ordinals. Whether or not finite and infinite numbers should be generated by the same or different principles is the sort of thing that should be regarded as a fundamental question, contributing to an at least potentially objective case for one representation over the other.

This serves up a striking example of the context-sensitivity of the metaphysical intuitions at work in this debate. If the debate is carried out in isolation from the ongoing mathematical research that is ostensibly its subject, we should expect situations like this to arise: The non-existence, and indeed the *inconceivability* of a kind of mathematical argument is put forward as a central datum, when such arguments are not only possible but are widely, if tacitly, acknowledged in practice. In addition, with the reduction of ordered pairs taken as a *mathematical* paradigm, the student of mathematical method is put in a difficult dialectical position: the reduction that is put forward as fulfilling mathematical desiderata does indeed seem arbitrary and artificial. It is indeed hard to see how either reduction of ordered pairs could contribute to our view of what reality really is like. But from the point of view of mathematical method, it is *appropriate* to think this: the reductions *are* mathematically artificial, whatever their philosophical virtues may be.

One reason that informal mathematical reasoning of this type has stayed off of philosophy's radar is hinted at in a remark Sider quotes as sympathetic to his position:

> In awaiting enlightenment on the true identity of 3 we are not awaiting a proof of some deep theorem. Having gotten as far as we have without settling the identity of 3, we can go no farther. We do not know what a proof of that *could* look like. The notion of 'correct account' is breaking loose from its moorings if we admit of the possible existence of unjustifiable but correct answers to questions such as this. (Benacerraf, 1965, p. 58)

Certainly proving theorems is the canonical means of obtaining knowledge of mathematical facts. If we narrow our picture of the cognitive activity informing mathematical reasoning to include just deductively valid arguments from indubitable premises, then we won't have any access to the sorts of reasons that we are interested in here. The arguments that can be given to justify counting one definition as 'correct' or one reduction as 'natural' are not going to be *theorems*. (Of course, sometimes central theorems can be crucial to making a case for the naturalness of one definition or function over another. This was the case for the Legendre symbol: its claim to the status of natural rather than artificial was buttressed when the Artin Reciprocity Theorem was proven, and it turned out to be the special case of a more general fundamental concept.) But the fact that this reasoning is 'softer' than what we are accustomed to finding in mathematics textbooks doesn't mean it is an insignificant contributor to mathematical knowledge. The principles provide reasons for mathematical expectations, guide conjectures, and inform problem-solving strategies. The reasons can be debated, and those debates can be resolved by further investigation. Why shouldn't we count this as rational activity producing objective knowledge?

At its crucial points, Sider's argument seems to amount to simply appealing to the perceived plausibility of the suggestion that mathematical judgements of more or less natural can never be more than judgements of 'mathematical convenience' that cannot be reasonably taken as a guide to the way things are objectively. Part of what gives initial appeal to this claim is the example of the reduction of ordered pairs, which I've suggested is misleading as a guide to the judgements informing mathematical practice. A mathematically insignificant case is taken as paradigmatic, which colors the intuitions about the other cases it is meant to illustrate. Of course, this observation puts the ball in my court: I'll need to lay out other examples which will turn the intuitions around. The examples developed in the first essay can take some steps in the right direction. Given the extensiveness, intricacy, and multiplicity of connections exhibited by the reasoning at issue it doesn't seem nearly so obvious that we can devalue as a judgement of 'mere mathematical convenience' the claim that the proof of quadratic reciprocity using cyclotomic extensions is especially natural because it provides a conceptual stepping stone to more general Artin reciprocity, or that the Legendre symbol, despite the artificial-seeming initial definition, can be seen to carve things at the joints when it is recognized as a special case of the Artin symbol.

Of course, the topic of context-sensitivity is a double-edged sword here. It is open to the defender of the thesis that judgements of mathematical naturalness are all ('merely') pragmatic to suggest that familiarity with mathematical details distorts intuitions as well. Immersion in the details of any rich theory with a range of systematic interconnections can foster a sense that certain facts that may seem soft from the outside are hard and inescapable. This is true for systematic theology, chess theory, and the theory of what Harry Potter would do under hypothetical circumstances consistent with, but not occurring in, the Harry Potter books. Enough immersion in the Potter books, discussion groups with other fans, fan fiction groups on the internet, etc. might give a fan the sense that it is an objective fact that one continuation of the story is natural and another is artificial, and the fan might be able to provide an extensive and ingenious rationale. A well developed theory of the Potter stories could provide *explanations* of the actions of the characters, and could be assessed for simplicity, elegance, and Potter-theoretic analogues of 'purity of method' considerations, to mention just a few cognitive virtues. But a skeptic could maintain, quite plausibly, that this doesn't make the naturalness of one storyline versus another anything but an artifact of the cognitive peculiarities of the enthusiastic fan, and perhaps of human cognitive particularities generally. Not unreasonably, the defender of the thesis that judgements of mathematical naturalness are all pragmatic could point to such analogies, to argue that it is just irrelevant that Sider considers a trivial example instead of richer cases with complex details.

At bottom, this reply could run, the preference for one mathematical formulation over another, in principle logically equivalent, one is a pragmatic matter, *however* detailed the rationale, *however* intricately the preferences may be bound up with attributions of mathematical understanding or explanation or simplicity or [insert your favorite virtue of theories here], and *however* subtly the preferences may interact with problem-solving potential. The mere fact that the judgements are deeply embedded in the practice of mathematics does not change their *nature*. It may make the pragmatic judgements more complicated, and they may 'feel' differently to people familiar with the subject, but they remain ultimately pragmatic, just as artistic judgements about musical works are not changed into something other than aesthetic judgements if they are complex and involved.

However, as other strands of the Sider article bring out, the context-sensitivity of intuitions about objectivity is a universal concern. Following out a different cluster of options, Sider takes up the possibility that the idea of naturalness might admit of many subgenres, depending on what reduction is accepted. (So, there could be 'Kuratowski-natural' relations, 'Weiner-natural' relations, and so on, but no generally definable class of natural relations.) The sole grounds given for rejecting this option are appeals to intuitions about complexity and the aesthetics of theories: '[This is not] a knockdown objection, but [it is] a forceful one nonetheless: [The theory] would require an infinitude of primitive properties, with no hope of subsumption under a single formula or explanation. If [the theory] requires such an unlovely menagerie to do its work, we'd do better to look elsewhere for a theory of naturalness' (1996, p. 293). This objection is redescribed later as an appeal to 'theoretical economy' (1996, p. 295) and later still as an appeal to 'prohibitive cost in complication of theory' (1996, p. 299). The obvious question presents itself: why are offhand intuitive assessments of simplicity and aesthetic appeal reliable guides to the way things really are for philosophical theories but pragmatic matters of 'mere convenience' for mathematical discriminations? It is hard to see any principled basis motivating the distinction in the cases at hand.

This draws us to a deep problem to be sure, but it is a problem for *everyone*: why should theoretical virtues serve as a guide to the way things are? How can we justify an appeal to relative simplicity of one account over another, or the fact that one account explains or confers understanding better than another, or some other cognitive advantage, when there would seem to be no *a priori* reason to expect that objective reality must be so considerate as to conform to the particularities of human cognitive makeup. There is no generally accepted response to the suggestion that *every* choice of a theory—in mathematics, philosophy, or any other field—on the grounds of theoretical virtues like simplicity or apparent naturalness or 'loveliness' is 'merely pragmatic'. So for any particular, local appeal to such theoretical reasons, the charge that these reasons are 'merely pragmatic' is available. If we want to maintain that even so, *some* apparently pragmatic judgements are guides to the way things objectively are and others are not, we need a solid basis for distinguishing the cases. The problem with the method of appealing to intuitions about what is and isn't 'pragmatic' is that these intuitions are simply too context-sensitive to afford an informative distinction between cases where a preference motivated by theoretical virtues *is* 'objective' and when it isn't.

In the end, it is hard to make any progress beyond a wheel-spinning clash of intuitions without more information about what 'objective' means, or some harder criteria of objectivity. But if our intuitions are too fluid to adjudicate whether or not some discrimination is objective, what else is there?

10.3 Conjecture and verification: a foothold on objectivity

As I've said, I am in no position to propose a general theory of, or criterion for, the objectivity of a judgement. Rather, I'll borrow a practical strategy from mathematical research: if you can't solve the general problem, find a tractable special case and solve that. The foothold will be the interaction between finding the proper definitions or concepts and the practice of successfully verifying conjectures. This won't be the whole story but it can be the beginning of one. In particular, I'll concentrate on one orienting mark: it is a *prima facie* sign that a judgement is objective if it has consequences that can be confirmed or refuted and whose truth or falsehood is independent of the original conjecture. In the simple cases that we know from discussions of the 'grue' paradox, these consequences are successful empirical predictions. As noted in the introductory essay, such predictions are made in mathematics as well, as witness the inductive reasoning Euler exploited to arrive at quadratic reciprocity. Euler's argument for quadratic reciprocity forms a pleasingly clear-cut analogy to Goodman's New Riddle of Induction. In each case we are dealing with simple enumerative induction, and the correctness of the prediction depends upon hitting on the proper category. The fact that one is a prediction about a contingent fact and the other a necessary truth about prime numbers makes no difference to the reasoning itself.

One of the reasons that the Goodman's 'New Riddle' is such a compelling philosophical set-piece is that it links—in a simple and clear way—the question of whether or not a category is homogeneous with a judgement whose rightness or wrongness is *outside the control* of the judger. No amount of reclassifying or pragmatic revision of conceptual schemes will make the next emerald examined be (what is called in the current vocabulary) blue. The color of an emerald is a contingent empirical fact, but as noted the reasoning doesn't turn on contingency. Euler's records of the computations that prompted his conjecture of the law of quadratic reciprocity display the same pattern.[8] He examined cases, made conjectures and then tested them against an expanding series of examples. As with induction about the physical world, Euler's analyses of the data were, of course, not unreflectively mechanical. It took considerable ingenuity and pattern-spotting before Euler recognized that (what amounted to) quadratic reciprocity was the pattern to project. After he hit on the proper way to classify the cases he had examined, he was able to predict the correct theorem, which was then verified through examination of further specific values. It is remarkable how well the patterns of computation, and conjecture, refute conjecture with more computations, refine concepts, and make new conjecture,...look like textbook examples of enumerative induction. (Find some green rocks. Conjecture all rocks are green. Find some non-green rocks, re-analyze situation, note that the green rocks are emeralds. Conjecture all emeralds are green...) To the extent that the naturalness of a property plays a role in the correctness of a projection, it has to play a role here, which would lead us to the conclusion that the properties like 'x is a quadratic residue' supporting Euler's correct inductive reasoning should have the same claim to 'naturalness' deriving from projectibility that 'green' has. The elementary inductive pattern is in no way altered by the fact that the theorem could also be directly proven, or that the relevant facts about numbers aren't contingent, and the pattern is perfectly consistent with the observation that the concept of causation doesn't have any obvious application in connection to mathematical properties. The correctness or otherwise of the categories Euler used was tested by a hard, objective criterion: will the values still to be computed be the

---

[8] There are several excellent treatments of the inductive reasoning that led Euler to his conjecture. Edwards (1983), Cox (1988), Cox (1989, pp. 9–20), and Weil (1984) are all worth consulting. Sandifer (2005) is a particularly beginner-friendly discussion at the online Euler archive.

predicted ones? And further down the road: if the conjecture is resolved by a proof, will it be proven true or false?

Of course, the Euler case is unusual, in that elementary enumerative induction of the form 'a_1 is P, a_2 is, P, ..., a_n is P, therefore a_{n+1} will be P' is as rare in mathematics as it is in ordinary reasoning about the physical world: both ordinary and mathematical conjectures are typically more subtle.[9] In both empirical and mathematical cases, the simple enumerative induction is an artificially simplified touchstone that distills certain essential features of more formless kinds of prediction and conjecture that inform ongoing mathematical and empirical thinking.

The practice of conjecturing and then striving to ascertain whether or not the conjecture is correct is ubiquitous in mathematical practice.[10] Even without the formality of explicitly announcing a conjecture, the practice of forming expectations about what will turn out to be correct, grounded in prior discoveries, is an indispensable component of mathematical reasoning. These expectations guide the choice of directions to search for proofs, for example. The concepts that are marked out as mathematically natural support further inductive practice: on the basis of similarities that are evaluated using these concepts, conjectures are made about what further investigations will discover. If it is discoved that these conjectures are true, it reinforces the judgement that the concepts regarded as mathematically natural *really are* mathematically natural. In at least some cases the expectations are sufficiently well-grounded in computations or other quasi-empirical data or plausibility considerations that it seems right to say that the theorems are *known* to be true even without a proof. (This seems to be a reasonable thing to say about Euler's justified belief in quadratic reciprocity, for example.)

Inductive reasoning doesn't just appear in the direct form of conjectures or specific expectations that are then ideally refuted or verified. Broader methodological virtues like 'significance' and 'fruitfulness' have a kind of quasi-empirical inductive character.[11] Singling out a property as central tacitly makes a prediction that it will reappear in unexpected contexts, and serve as

---

[9] Though there are significant examples in which an approach is initially validated through its ability to make verifiable predictions. Perhaps the most famous is the Schubert calculus for counting intersections of curves. Long before the system could be rigorously formulated and demonstrated, Schubert was making astonishing predictions of intersection numbers. See Kleiman and Laksov (1972).

[10] A good recent discussion of conjecture as an aspect of mathematical reasoning is in Mazur (1997). The classic investigations of inductive reasoning in mathematics are of course Pólya (1968) and Lakatós (1976). Putnam (1975) is an illuminating reflection on the philosophical angles. Corfield (2003, Chapters 2–6) contains some valuable discussion.

[11] 'Quasi-empirical' is the expression coined by Hilary Putnam for this sort of mathematical evidential support in his 'What is Mathematical Truth?' (Putnam, 1975).

support to unanticipated proofs. It need not be obvious at the outset what shape the proofs will take or what the 'unexpected contexts' will be for this to be a prediction that can succeed or fail. This supports a response to the suggestion that these evaluations are 'merely pragmatic' or 'subjective' or 'merely psychological' or otherwise lacking objectivity in a way that makes them uninteresting to epistemology. Just as with more narrowly circumscribed conjectures, the predictive character of an evaluation of 'fruitfulness' brings a degree of independence from the attitudes of the person making the judgement. To judge that a program, theorem, or definition is 'fruitful' is to make a prediction that solutions to given problems will in fact be found by means of the program, theorem, or definition. Such judgements involve previsions that can be found to be mistaken in light of subsequent investigation. Not all predictions are as unequivocal and precise as Euler predicting quadratic reciprocity, but they can still be confirmed or disconfirmed. There is, of course, a social component to the prediction that some approach will yield research deemed to be valuable and informative. But a prediction like 'The Weil conjectures will be solved using the concept of scheme and not with the concepts available earlier', is not just a forecast of social facts. It turns both on what mathematicians will choose to do and on what proofs are objectively there to be found.

In stating that inductive and plausible arguments in mathematics can secure knowledge, I am not echoing the suggestion one occasionally encounters that the mathematical community ought to relax its emphasis on rigorous proof. Something can be known without being established to the degree required for the mathematical community to count it as officially in the bank. (Analogously, I would not be suggesting that criminal trial procedure should relax the standard of 'proof beyond a reasonable doubt' if I were to point out that we know certain former defendants to be guilty even though they were ruled not guilty at their trial.) I am suggesting no revisions at all to regular practice. I am rather urging that we extend the range of mathematical reasoning that we take to generate knowledge and to be material for epistemological study. Forming expectations on the grounds of plausible reasoning and using these expectations as a guide is an integral part of mathematical practice. My point has been that if we expand our epistemological horizons to include these varieties of mathematical knowledge, it brings with it an additional bonus: A foothold on the study of the naturalness of mathematical concepts and definitions.

I should also make clear that in urging the importance and quasi-empirical character of the tacit predictions informing mathematical reasoning, I am not suggesting that our study of mathematical concepts must be linked to direct applications in physical science and common-sense reasoning about the physical world. (Of course, in some cases there will be close connections, but this isn't necessary to the point.)[12] Whether or not the conjectures in question have any obvious connection to any direct applications makes no difference to our assessment of the patterns of reasoning informing the making and supporting of conjectures. It may well be that in an ideal epistemology, the status of mathematics as knowledge must *ultimately* appeal to physical applications, or it may not; we can be neutral on that point here.

10.4 Fruitfulness, conjecture, and Riemannian mathematics

The Riemann–Dedekind methodology mentioned in the introduction incorporates exactly the connection between identification of core properties and mathematical fecundity discussed in the previous section. In this section I'll say a bit more about the Riemann–Dedekind approach. First consider these Riemannian remarks. They seem innocuous and even dull, but understood in context they are the first shot in a revolution.

> Previous methods of treating [complex] functions always based the definition of the function on an expression that yields its value for each value of the argument. Our study shows that, because of the general nature of a function of a complex variable, a part of the determination through a definition of this kind yields the rest... This essentially simplifies the discussion...
>
> A theory of these functions on the basis provided here would determine the presentation of a function (i.e. its value for every argument) independently of its mode of determination by operations on magnitudes, because one would add to the general concept of a function of a variable complex quantity just the attributes necessary for the determination of the function, and only then would one go over to the different expressions the function is fit for. The common character of a class of functions formed in a similar way by operations on quantities, is then represented in the form of boundary conditions and discontinuity conditions imposed on them. (Riemann, 1851, p. 38)

Riemann's point is simple. We are wondering how to characterize a well-behaved complex function or a class of them. How should we proceed? One

---

[12] Though I'm not exploring the point here, a closeness to physical applications was in fact recognized as a feature of Riemann's mathematics. (This is touched on in the opening quotation from Stahl.) For example, Helmholtz found Riemann's complex analysis congenial, presumably in part because Riemann's classification of functions by their singularities fit smoothly with Helmholtz's approach to potential flow (cf. Darrigol, 2005, p. 164).

possibility would be to appeal to a formula that would explicitly display the relation between every argument and every value in terms of elementary operations like plus or times. For example, one could write down a polynomial or power series. But this uses more information than necessary: the function can be uniquely determined (up to a constant multiple) by just a fragment of this information (the 'boundary conditions and discontinuity conditions').

I'll illustrate with an example where the conditions are particularly straightforward: elliptic functions, the subject of the Stahl textbook quoted in the introduction.[13] These are (defined today as) well-behaved (i.e. meromorphic) functions Φ satisfying the condition that there are complex numbers ω_1 and ω_2 such that for any $z \in \mathbb{C}$ and $n, m \in \mathbb{Z}$, $\Phi(z) = \Phi(z + n\omega_1 + m\omega_2)$ (i.e. Φ is doubly periodic.) Given periods ω_1 and ω_2, how should we single out specific functions? We could give an explicit rule for computing every value for every argument, but we have a sparer alternative. Fixing the natural domain (period lattice) corresponding to the periods ω_1 and ω_2, given any two functions f and g on this domain with the same zeros and poles (counted with the same multiplicities) there is a constant c such that for every z, $f(z) = cg(z)$. That is, the function is essentially fixed by its zeros and poles; the remaining values need be pulled up only if necessary for some specific purpose. Though I am restricting attention to a simple case to avoid discussion of Riemann surfaces, the fact holds more generally: we lose no important information about a well-behaved function if we just know the zeros and certain singularities.[14]

With such results in hand, Riemann opts to characterize functions by their singularities. There are too many ramifications of this preference to discuss here. I'll restrict attention to the one that is directly relevant to our present topic: getting the concepts right allows you to see results more easily.[15] For example, concerning one application of his methods, Riemann deploys the language that gave rise to the clichés echoed by Stahl: Riemann's reconceptualizations

---

[13] Here I'll only be able to gesture at some of the relevant information. I explore elliptic functions in connection with Riemann's conception of mathematical method in more detail in Tappenden (forthcoming).

[14] Given any compact Riemann surface R, a meromorphic function defined on R is characterized up to a constant by its zeros and poles (with multiplicities).

[15] Among the other considerations Riemann cites are purity of method considerations arising from his sense that it isn't proper for language-independent properties of objects to be characterized in ways that essentially involve language:

> By one of the theorems quoted above, this property of single-valuedness in a function is equivalent to that of its developability in a series... However, it seems inappropriate to express properties independent of the mode of representation by criteria based on a particular expression for the function. (Riemann, 1857b)

make it easy to recognize at a glance what had previously required effort and laborious computations.

> [My method yields] almost immediately from the definition results obtained earlier, partly by rather tedious computations. (Riemann, 1857b, p. 67)
>
> [My method is one] by means of which all the results found earlier can be obtained virtually without computation. (Riemann, 1857a, p. 85)

Not only does the reconceptualization allow you to see easily what had previously been established with difficulty, but also as Dedekind explicitly notes, the temporal order can be reversed. The formulation 'based immediately on fundamental characteristics' supports the effective *anticipation* of the results of as yet unfinished computations. Dedekind sees 'almost all areas of mathematics' as requiring the choice between essential ('internal') and accidental ('external'):

> It is preferable, as in the modern theory of [complex] functions to seek proofs based immediately on fundamental characteristics, rather than on calculation, and indeed to construct the theory in such a way that it is able to predict the results of calculation... Such is the goal I shall pursue in the chapters of this memoir that follow. (Dedekind, 1877, p. 102)
>
> [Gauss remarks in the *Disquisitiones Arithmeticae*]: "But neither [Waring nor Wilson] was able to prove the theorem, and Waring confessed that the demonstration was made more difficult by the fact that no notation can be devised to express a prime number. But in our opinion truths of this kind ought to be drawn out of notions not out of notations." In these last words lies, if they are taken in the most general sense, the statement of a great scientific thought: the decision for the internal in contrast to the external. This contrast also recurs in mathematics in almost all areas; [For example, complex] function theory, and Riemann's definition of functions through internal characteristic properties, from which the external forms of representation flow with necessity. (Dedekind, 1895, p. 54–55)

In some cases, such as the zeros and poles of a complex function, the 'internal characteristic properties' at issue are already familiar, so the cognitive success involved in identifying them is just the recognition of their importance. In other cases, it requires substantial reformulation and analysis even to be in a position to notice and define the properties. Thus, for example, Riemann recognized the importance of the property we now call the *genus* of a surface in connection with complex functions. (In the simplest cases, the genus is the 'number of holes' in the surface.) Riemann showed how to associate with each well-behaved complex function a surface (now called a Riemann surface) serving as the natural domain of definition for the function. The genus is then defined in terms of the topological properties of this surface. This sets the context for a core result proven partly by Riemann and partly by his student

Roch (called, reasonably, the Riemann–Roch theorem) that had incalculable importance in the mathematics of the 20th century. Here again I can hardly give more than a vague impression, so I'll just state the punchline for this chapter: The result links together the number of linearly independent functions with given collections of poles defined on a Riemann surface with the genus of the surface. The subsequent importance of the theorem, and the fact that it links together the properties that would have been understood as the kinds of things being seen as 'inner characteristic properties' makes it one of the many examples that give mathematical content and force to Riemann's and Dedekind's methodological asides. The characterization of poles and genus as essential properties is tacitly a prediction that the Riemann–Roch theorem, among others, will prove fruitful in solving additional problems and supporting new conjectures. This is a chancy prediction, since the theorem might have led nowhere. Had the research jumping off from core theorems about genus and poles of functions petered out without interesting consequences, this would have forced a reassessment of the original evaluation of these properties as 'internal characteristic' ones.

The most salient example of his own work that Dedekind is alluding to in the above remarks is the concept of an ideal. Once again there is much more to discuss than I can address here, so I'll just stick to the immediately relevant punchlines.[16] It will suffice to think of an ideal as a kind of generalized number that secures the possibility of factoring otherwise irreducible numbers. Among the discoveries that gave impetus to Dedekind's introduction was an analogy between fields of numbers and fields of functions unveiled in the classic paper by Dedekind and Weber (1892).[17] (Riemann's characterization of functions by their zeros and poles was a crucial building block.) Among other things, Dedekind and Weber interpreted Riemann surfaces as algebraic objects and proved core results such as the Riemann–Roch theorem in this environment. Speaking informally we might say that Dedekind and Weber revealed that a family of structures were genuinely, rather than just superficially, similar. The analogy between fields of numbers and fields of functions is very deep.

It is possible to lay out an extensive rationale for the judgement of similarity.[18] As just one illustration of how layered the judgement of similarity can be,

---

[16] Fortunately for any readers hungry for more, the relevant details are well covered by historians, in particular Harold Edwards. See for example Edwards (1980). Recently philosophers have begun to reflect on the methodology described in Dedekind's remarks and displayed in his mathematics. See for example Avigad (2006) and Tappenden (2005).

[17] This analogy, with specific reference to ramification, is explored with a philosophical eye in Corfield (2003, p. 90–96)

[18] An illustration can be found in a letter of 1940 from André Weil to his sister Simone, containing a strikingly detailed discussion of the many consequences of this analogy, what matters in the analogy,

consider the property of *ramification* for complex functions and generalized numbers.[19] Among the singularities Riemann recognized to be especially important are *branch points* (a.k.a. *ramification points*). With proper stage setting, these can be understood as points where the function behaves locally like $z \mapsto z^e$, with $e > 1$. When dealing with number fields, the idea of prime decomposition can be generalized so that generalized numbers (ideals) can be written as products of generalized primes $\mathfrak{p}_i$:

$$\prod_{i=1}^{n} \mathfrak{p}_i^{e_i}$$

The generalized number is said to *ramify* if $e_i > 1$ for any i. This point of contact between complex analysis and algebraic number theory was already pointed out by Dedekind and Weber.

Of course, in one sense this particular connection is laying on the surface of the algebraic representations. In both cases you have exponents—either greater than one or equal to one—sitting right there just waiting to be noticed. But if all we know is that some exponent is greater than one, we have no reason to get excited, or judge there to be a point of similarity robust enough to warrant the introduction of new terminology. The point of similarity could be a superficial accident of the representation, or a mathematically inert coincidence. The bedrock support for the claim that this is a genuine point of deep similarity is the fact that powerful general theorems can be proven which exploit the analogy. Had it turned out that, at a more advanced point of reckoning, no interesting theorems emerged, the shared terminology would stand like the word 'elliptic' in the phrase 'elliptic curve' as a quaint reminder of a connection that once seemed to matter. As always, the aesthetic judgements, assessments of the depth of a similarity, judgements about appropriateness of techniques, evaluations of relative simplicity, etc. are subordinated to the fundamental bottom-line consideration: does it lead to genuinely interesting new results? Without the *sine qua non* of fruitfulness, the other considerations are counted as ultimately incidental.

Riemann's envisioned connection between correct definition and fruitfulness was revolutionary at the time, but now seems familiar, even banal. Indeed, it is fair to say that it is the dominant attitude among contemporary mathematicians. It is hardly profound or surprising that mathematicians typically

what is needed for the analogy to be complete, etc. (Weil also spells out connections to quadratic reciprocity and the Artin reciprocity theorem.) See Weil (1984).

[19] I'm indebted to Brian Conrad for illuminating conversations on this point.

have a hardheaded, bottom-line orientation that ties the ultimate validity of softer judgements of 'naturalness' or 'depth' to the facilitation of far-reaching general results. This piece of rudimentary empirical sociology is obvious to anyone after a few department lounge conversations. Of course, for Riemann and Dedekind, and even more for most contemporary mathematicians, the attitude is more often displayed in the practice of mathematical research than articulated as something like a philosophical theory of the natural/artificial distinction. But there is a tacit theory behind the pronouncement that (say) genus is a natural property and as philosophers we'd be wise to work it out.

10.5 Summing up

Chapter 9 made a descriptive observation and posed a problem. The descriptive observation is that mathematical practice is colored and guided by a kind of normative judgement: it can be counted as an advance in knowledge to identify a 'natural' definition (or 'deep' analogy, or 'proper' context for investigation, or 'correct' formulation of a question...). The problem is to find a way for epistemology to explore how such acts of recognition can be what mathematical practice seems to treat them as being. This paper has been an extended prospectus for a program to address the problem. On a quick pass through Lewis' reckoning of the some of the work done by discriminations of natural and non-natural properties there appeared to be enough affinity between, and overlap among mathematical and non-mathematical cases to give us hope. My first task in this paper was to illustrate, with reference to a representative instance, that appealing to metaphysical intuition (either directly or as coded into theoretical measures such as 'simplicity' or 'theoretical economy') to separate the cases where such choices of preferred 'natural' categories are 'merely pragmatic' and the cases where the choices are genuinely 'objective' is unlikely to help us realize that hope, because of the fluidity and context-sensitivity of the relevant intuitions. We need some rough ground. The second goal was to point to one domain—the interaction between concept choice and successful prediction—where some such rough ground might be secured. A key point is that induction, as a pattern of reasoning under uncertainty, does not depend for its cogency on the predicted outcomes being contingent. Hence the link between assessments of the naturalness of properties and the correctness of predictions exhibited in 'green'/'grue' cases carry over to mathematical cases. (The model of enumerative induction is just a handy orienting and simplifying device; the key foothold is the ubiquitous interaction in mathematical practice between choices of basic categories and the success of predictions (where the predictions may be clearly stated conjectures or less specific expectations of 'fruitfulness').) Finally, I sketched the Riemann–Dedekind account of 'inner characteristic properties' as a promising example of a method, and a methodology, incorporating this insight. This methodology has the additional attractive feature that in its details it resonates with the Port Royal Principle, since identifiable mathematical cases drive it and give it substance. Of course, so far I've given only the roughest sketch of the Riemann–Dedekind account and the mathematics informing it, and I've given only unargued hints about its relationship to contemporary mathematical practice. In part this has been a function of space, but in part it is a necessity arising from the texts themselves. Riemann and Dedekind doled their methodological *dicta* out frugally. As mathematicians are prone to do, they let their mathematics do most of the talking, which leaves the philosopher/scribe a lot of detail to spell out. But this shouldn't be a surprise: Arnauld and Nicole warned us that this is what to expect.

Bibliography

ARMSTRONG, David M. (1986), 'In Defence of Structural Universals', *Australasian Journal of Philosophy*, 1, 85–88.
—— (1989), *Universals: An Opinionated Introduction* (Boulder: Westview Press).
AVIGAD, Jeremy (2006), 'Methodology and Metaphysics in the Development of Dedekind's Theory of Ideals', in José Ferreirós and Jeremy Gray (eds.), *The Architecture of Modern Mathematics* (Oxford: Oxford University Press), pp. 159–186.
BENACERRAF, Paul (1965), 'What Numbers Could Not Be', *The Philosophical Review*, 74(1), 47–73.
CORFIELD, David (2003), *The Philosophy of Real Mathematics* (Cambridge: Cambridge University Press).
COX, David (1988), 'Quadratic Reciprocity: its Conjecture and Application', *The American Mathematical Monthly*, 95(5), 442–448.
—— (1989), *Primes of the Form $x^2 + ny^2$* (New York: Wiley).
DARRIGOL, Olivier (2005), *Worlds of Flow: A History of Hydrodynamics from the Bernoullis to Prandtl* (Oxford: Oxford University Press).
DEDEKIND, Richard (1854), *Über die Einfürung neuer Funktionen in der Mathematik*, Habilitation lecture, Göttingen. Reprinted in *Werke*, III, pp. 428–438.
—— (1877), *Theory of Algebraic Integers*, ed. and trans. John Stillwell (Cambridge Mathematical Library) Translation published 1996; originally published 1877.
—— (1895), 'Über die Begründung der Idealtheorie', *Nachrichten der Königlichen Gesellschaft der Wissenschaften zu Göttingen*, 106–113. Reprinted in *Werke*, II pp. 50–58.

DEDEKIND, Richard and WEBER, Heinrich (1882), 'Theorie der Algebraischen Funktionen einer Veränderlichen', *Journal für die Reine und Angewante Mathematik*, 181–290. Reprinted in Dedekind's *Werke*, I, pp. 248–349.

EDWARDS, Harold (1980), 'The Genesis of Ideal Theory', *Archive for History of Exact Sciences*, 23, 321–378.

―― (1983), 'Euler and Quadratic Reciprocity', *Mathematics Magazine*, 56, 285–291.

EISENBUD, David and HARRIS, Joe (1992), *Schemes: The Language of Modern Algebraic Geometry* (Boston: Wadsworth).

FORREST, Peter (1986), 'Neither Magic Nor Mereology: A Reply to Lewis', *Australasian Journal of Philosophy*, 64, 89–91.

HILBERT, David (1998), *Theory of Algebraic Number Fields*, trans. Ian T. Adamson (Berlin: Springer). Originally published 1897.

KLEIMAN, Steven and LAKSOV, Dan (1972), 'Schubert Calculus', *American Mathematical Monthly*, 79, 1061–1082.

LAKATOS, Imre (1976), *Proofs and Refutations: The Logic of Mathematical Discovery*, ed. John Worrall and Elie Zahar (Cambridge: Cambridge University Press).

LEWIS, David (1983), 'New Work for a Theory of Universals', *Australasian Journal of Philosophy*, 62, 343–377. Reprinted in Lewis (1999).

―― (1984), 'Putnam's Paradox', *Australasian Journal of Philosophy*, 62, 221–236. Reprinted in Lewis (1999).

―― (1999), *Papers on Metaphysics and Epistemology* (Cambridge: Cambridge University Press).

―― (2001), 'Redefining 'Intrinsic'', *Philosophy and Phenomenological Research*, 63, 381–398.

MAZUR, Barry (1997), 'Conjecture', *Synthèse*, 111, 197–210.

PÓLYA, George (1968), *Mathematics and Plausible Reasoning*, 2 vols. (Princeton: Princeton University Press).

PUTNAM, Hilary (1975), 'What is Mathematical Truth?', in *Mathematics, Matter and Method: Philosophical Papers*, Vol. 1 (Cambridge: Cambridge University Press).

RIEMANN, Bernhard (1851), '*Grundlagen für eine allgemeine Theorie der Functionen einer veränderlichen complexen Grösse*', Inauguraldissertation, Göttingen. Reprinted in *Werke*, pp. 3–45.

―― (1857a), 'Abstract of [1857b]', *Göttinger Nachrichten*, 1. Reprinted in *Werke*, pp. 84–85.

―― (1857b), 'Beiträge zur Theorie der durch die Gauss'sche Reihe $F(\alpha, \beta, \gamma, x)$ darstellbaren Functionen', *Abhandlungen der Königlichen Gesellschaft der Wissenschaften zu Göttingen*, 7. Reprinted in *Werke*, pp. 67–83.

―― (1857c), 'Theorie der Abel'schen Functionen', *Journal für die reine und angewandte Mathematik*, 54. Reprinted in *Werke*, pp. 88–142.

SANDIFER, Ed (2005), 'How Euler Did It: Factors of Forms', MAA online column accessed from index: <http://www.maa.org/news/howeulerdidit.html>.

SIDER, Ted (1996), 'Naturalness and Arbitrariness', *Philosophical Studies*, 81, 283–301.

STAHL, Hermann (ed.) (1899), *Elliptische Functionen: Vorlesungen von Bernhard Riemann* (Leipzig: Teubner).

TAPPENDEN, Jamie (2005), 'The Caesar Problem in its Historical Context: Mathematical Background', *Dialectica*, 59 (fasc. 2), 237–264.

―― (Forthcoming),'Reflections on Mathematical Explanation: Why Do Elliptic Functions Have Two Periods?'.

WEATHERSON, Brian (2006), 'Natural Quantities', *manuscript*.

WEIL, André (1940/2005) 'Letter of 1940 to Simone Weil', *Notices of the American Mathematical Society*, trans. M. Krieger, 52, 3, pp. 335–341.

―― (1984), *Number Theory: An Approach through History; From Hammurapi to Legendre* (Boston: Birkhäuser).

11

Computers in Mathematical Inquiry

JEREMY AVIGAD

11.1 Introduction

Computers are playing an increasingly central role in mathematical practice. What are we to make of the new methods of inquiry?

In Section 11.2, I survey some of the ways in which computers are used in mathematics. These raise questions that seem to have a generally epistemological character, although they do not fall squarely under a traditional philosophical purview. The goal of this article is to try to articulate some of these questions more clearly, and assess the philosophical methods that may be brought to bear. In Section 11.3, I note that most of the issues can be classified under two headings: some deal with the ability of computers to deliver appropriate 'evidence' for mathematical assertions, a notion that is explored in Section 11.4, while others deal with the ability of computers to deliver appropriate mathematical 'understanding', a notion that is considered in Section 11.5. Final thoughts are provided in Section 11.6.

11.2 Uses of computers in mathematics

Computers have had a dramatic influence on almost every arena of scientific and technological development, and large tracts of mathematics have been developed to support such applications. But this essay is not about the

---

I am grateful to Ben Jantzen and Teddy Seidenfeld for discussions of the notion of plausibility in mathematics; to Ed Dean, Steve Kieffer, and Paolo Mancosu, for comments and corrections; and to Alasdair Urquhart for pointing me to Kyburg's comments on Pólya's essay.

numerical, symbolic, and statistical methods that make it possible to use the computer effectively in scientific domains. We will be concerned, rather, with applications of computers *to* mathematics, that is, the sense in which computers can help us acquire mathematical knowledge and understanding.

Two recent books, *Mathematics by Experiment: Plausible Reasoning in the 21st Century* (Borwein and Bailey, 2004) and *Experimentation in Mathematics: Computational Paths to Discovery* (Borwein et al., 2004) provide a fine overview of the ways that computers have been used in this regard (see also the associated 'Experimental Mathematics Website', which provides additional links and resources). Mounting awareness of the importance of such methods led to the launch of a new journal, *Experimental Mathematics*, in 1992. The introduction to the first book nicely characterizes the new mode of inquiry:

> The new approach to mathematics—the utilization of advanced computing technology in mathematical research—is often called *experimental mathematics*. The computer provides the mathematician with a 'laboratory' in which he or she can perform experiments: analyzing examples, testing out new ideas, or searching for patterns... To be precise, by experimental mathematics, we mean the methodology of doing mathematics that includes the use of computations for:
>
> 1. Gaining insight and intuition.
> 2. Discovering new patterns and relationships.
> 3. Using graphical displays to suggest underlying mathematical principles.
> 4. Testing and especially falsifying conjectures.
> 5. Exploring a possible result to see if it is worth a formal proof.
> 6. Suggesting approaches for formal proof.
> 7. Replacing lengthy hand derivations with computer-based derivations.
> 8. Confirming analytically derived results.

In philosophical discourse it is common to distinguish between discovery and justification; that is, to distinguish the process of formulating definitions and conjectures from the process of justifying mathematical claims as true. Both types of activities are involved in the list above.

On the discovery side, brute calculation can be used to suggest or test general claims. Around the turn of the 19th century, Gauss conjectured the prime number theorem after calculating the density of primes among the first tens of thousands of natural numbers; such number-theoretic and combinatorial calculations can now be performed quickly and easily. One can evaluate a real-valued formula to a given precision, and then use an 'Inverse Symbolic Calculator' to check the result against extensive databases to find a simplified expression. Similarly, one can use Neil Sloan's 'On-Line Encyclopedia of Integer Sequences' to identify a sequence of integers arising from a particular calculation. These, and more refined methods along these lines, are described in Bailey and Borwein (2005). Numerical methods can also be used to simulate dynamical systems and determine their global properties, or to calculate approximate solutions to systems of differential equations where no closed-form solution is available. Graphical representations of data are often useful in helping us understand such systems.

Computers are also used to justify mathematical claims. Computational methods had been used to establish Fermat's last theorem for the first four million exponents by the time its general truth was settled in 1995. The Riemann hypothesis has been established for all complex numbers with imaginary part less than 2.4 trillion, though the general claim remains unproved. Appel and Haken's 1977 proof of the four-color theorem is a well-known example of a case in which brute force combinatorial enumeration played an essential role in settling a longstanding open problem. Thomas Hales' 1998 proof of the Kepler conjecture, which asserts that the optimal density of sphere packing is achieved by the familiar hexagonal lattice packing, has a similar character: the proof used computational methods to obtain an exhaustive database of several thousand 'tame' graphs, and then to bound nonlinear and linear optimization problems associated with these graphs. (This pattern of reducing a problem to one that can be solved by combinatorial enumeration and numerical methods is now common in discrete geometry.) Computer algebra systems like *Mathematica* or *Maple* are used, in more mundane ways, to simplify complex expressions that occur in ordinary mathematical proofs. Computers are sometimes even used to find justifications that can be checked by hand; for example, William McCune used a theorem prover named EQP to show that a certain set of equations serve to axiomatize Boolean algebras (McCune, 1997), settling a problem first posed by Tarski.

The increasing reliance on extensive computation has been one impetus in the development of methods of formal verification. It has long been understood that much of mathematics can be formalized in systems like Zermelo–Fraenkel set theory, at least in principle; in recent decades, computerized 'proof assistants' have been developed to make it possible to construct formal mathematical proofs in practice. At present, the efforts required to verify even elementary mathematical theorems are prohibitive. But the systems are showing steady improvement, and some notable successes to date suggest that, in the long run, the enterprise will become commonplace. Theorems that have been verified, to date, include Gödel's first incompleteness theorem, the prime number theorem, the four color theorem, and the Jordan curve theorem (see Wiedijk, 2006). Hales has launched a project to formally verify his proof of the Kepler conjecture, and Georges Gonthier has launched a project to verify the Feit–Thompson theorem. These are currently among the most ambitious mathematical verification efforts under way.

Thus far, I have distinguished the use of computers to suggest plausible mathematical claims from the use of computers to verify such claims. But in many cases, this distinction is blurred. For example, the Santa Fe Institute is devoted to the study of complex systems that arise in diverse contexts ranging from physics and biology to economics and the social sciences. Computational modeling and numeric simulation are central to the institute's methodology, and results of such 'experiments' are often held to be important to understanding the relevant systems, even when they do not yield precise mathematical hypotheses, let alone rigorous proofs.

Computers can also be used to provide inductive 'evidence' for precise mathematical claims, like the claim that a number is prime. For example, a probabilistic primality test due to Robert Solovay and Volker Strassen works as follows.[1] For each natural number n, there is an easily calculable predicate, $P_n(a)$, such that if n is prime then $P_n(a)$ is always true, and if n is not prime then at least half the values of a less than n make $P_n(a)$ false. Thus, one can test the primality of n by choosing test values $a_0, a_1, a_2, \ldots$ less than n at random; if $P_n(a_i)$ is true for a large number of tests, it is 'virtually certain' that n is prime.[2]

In sum, the new experimental methodology relies on explicit or implicit claims as to the utility of computational methods towards obtaining, verifying, and confirming knowledge; suggesting theorems and making conjectures plausible; and providing insight and understanding. These claims have a patent epistemological tinge, and so merit philosophical scrutiny.[3] For example, one can ask:

- In what sense do calculations and simulations provide 'evidence' for mathematical hypotheses? Is it rational to act on such evidence?
- How can computers be used to promote mathematical understanding?
- Does a proof obtained using extensive computation provide mathematical certainty? Is it really a proof?

---

[1] A probabilistic test later developed by Michael Rabin, based on a deterministic version by Gary Miller, has similar properties and is now more commonly used.

[2] This can be made mathematically precise. For example, suppose a 100-digit number is chosen at random from a uniform distribution. Number-theoretic results show that there is a non-negligible prior probability that n is prime. If one then chooses $a_0, \ldots, a_l$ at random, one can show that the probability that n is prime given that $P_n(a_i)$ holds for every i approaches 1, quickly, as l increases.

[3] The list is not exhaustive. For example, uses of computers in storing, organizing, and communicating mathematical knowledge also raise issues that merit philosophical attention.

- Is knowledge gained from the use of a probabilistic primality test any less certain or valuable than knowledge gained from a proof? What about knowledge gained from simulation of a dynamical system?
- Does formal verification yield absolute, or near absolute, certainty? Is it worth the effort?

As presented, these questions are too vague to support substantive discussion. The first philosophical challenge, then, is to formulate them in such a way that it is clear what types of analytic methods can have a bearing on the answers.

11.3 The epistemology of mathematics

A fundamental goal of the epistemology of mathematics is to determine the appropriate means of justifying a claim to mathematical knowledge. The problem has a straightforward and generally accepted solution: the proper warrant for the truth of a mathematical theorem is a mathematical proof, that is, a deductive argument, using valid inferences, from axioms that are immediately seen to be true. Much of the effort in the philosophy of mathematics has gone towards determining the appropriate inferences and axioms, or explaining why knowledge obtained in this way is worth having. These issues will not be addressed here.

There are at least two ways in which one may wish to broaden one's epistemological scope, neither of which denies the correctness or importance of the foregoing characterization. For one thing, one may want to have a philosophical account of warrants for mathematical knowledge that takes into consideration the fact that these warrants have to be recognized by physically and computationally bounded agents. A formal proof is an abstract object, albeit one that we may take to be reasonably well instantiated by symbolic tokens on a physical page. But proofs in textbooks and mathematical journals are somewhat further removed from this idealization: they are written in a regimented but nonetheless imprecise and open-ended fragment of natural language; the rules of inference are not spelled out explicitly; inferential steps are generally much larger than the usual formal idealizations; background knowledge is presupposed; and so on. Few can claim to have verified any complex theorem from first principles; when reading a proof, we accept appeals to theorems we have learned from textbooks, journal articles, and colleagues. The logician's claim is that the informal proof serves to indicate the existence of the formal idealization, but the nature of this 'indication' is never spelled out precisely. Moreover, we recognize that proofs can be mistaken, and often express degrees of faith depending on the nature of the theorem, the complexity of proof, the methods that have been used to prove it, and the reliability of the author or the authorities that are cited. Just as mathematical logic and traditional philosophy of mathematics provides us with an idealized model of a perfect, gapless deduction, we may hope to model the notion of an 'ordinary' proof and ask: when is it rational to accept an ordinary proof as indicating the existence of an idealized one?[4]

To explore this issue, one need not conflate the attempt to provide an idealized account of the proper warrants for mathematical knowledge with the attempt to provide an account of the activities we may rationally pursue in service of this ideal, given our physical and computational limitations. It is such a conflation that has led Tymozcko (1979) to characterize mathematics as a quasi-empirical science, and Fallis (1997, 2002) to wonder why mathematicians refuse to admit inductive evidence in mathematical proofs. The easy answer to Fallis' bemusement is simply that inductive evidence is not the right sort of thing to provide mathematical knowledge, as it is commonly understood. But when their remarks are taken in an appropriate context, Tymoczko and Fallis do raise the reasonable question of how (and whether) we can make sense of mathematics, more broadly, as an activity carried out by agents with bounded resources. This question should not be dismissed out of hand.

A second respect in which one may wish to broaden one's epistemological ambitions is to extend the analysis to value judgments that go beyond questions of correctness. On the traditional view, the role of a proof is to warrant the truth of the resulting theorem, in which case, all that matters is that the proof is correct. But when it comes to proofs based on extensive computation, a far more pressing concern is that they do not provide the desired mathematical insight. Indeed, the fact that proofs provide more than warrants for truth becomes clear when one considers that new proofs of a theorem are frequently judged to be important, even when prior proofs have been accepted as correct. We tend to feel that raw computation is incapable of delivering the type of insight we are after:

> ...it is common for people first starting to grapple with computers to make large-scale computations of things they might have done on a smaller scale by hand. They might print out a table of the first 10,000 primes, only to find that their printout isn't something they really wanted after all. They discover by this kind of experience that what they really want is usually not some collection of 'answers'—what they want is *understanding*. (Thurston, 1994, p. 162)

---

[4] For an overview of issues related to the 'surveyability' of proofs, see Bassler (2006).

We often have good intuitions as to the ways that mathematical developments constitute conceptual advances or further understanding. It is therefore reasonable to ask for a philosophical theory that can serve to ground such assessments, and account for the more general epistemological criteria by which such developments are commonly judged.

In sum, questions about the use of computers in mathematics that seem reasonable from a pre-theoretic perspective push us to extend the traditional philosophy of mathematics in two ways: first, to develop theories of mathematical evidence, and second, to develop theories of mathematical understanding. In the next two sections, I will consider each of these proposals, in turn.

11.4 Theories of mathematical evidence

We have seen that some issues regarding the use of computers in mathematics hinge on assessments of the 'likelihood' that a mathematical assertion is true:

- a probabilistic primality test renders it highly likely that a number is prime;
- numeric simulations can render it plausible that a hypothesis is true;
- formal verification can render it nearly certain that a theorem has a correct proof.

Since judgments like these serve to guide our actions, it is reasonable to ask for a foundational framework in which they can be evaluated. Such a framework may also have bearing on the development of computational support for mathematics; for example, systems for automated reasoning and formal verification often attempt to narrow the search space by choosing the most 'plausible' or 'promising' paths.

Probabilistic notions of likelihood, evidence, and support have long played a role in characterizing inductive reasoning in the empirical sciences, and it is tempting to carry these notions over to the mathematical setting. However, serious problems arise when one tries to do so. Roughly speaking, this is because any mathematical assertion is either true, in which case it holds with probability 1, or false, in which case it holds with probability 0, leaving no room for values in between.

Put more precisely, classical approaches to probability model 'events' as measurable subsets of a space whose elements are viewed as possible outcomes of an experiment, or possible states of affairs. The laws of probability dictate that if an event A entails an event B, in the sense that $A \subseteq B$, then the probability of *A* is less than or equal to the probability of *B*. In particular, if a property holds of all possible outcomes, the set of all possible states of affairs that satisfy that property has probability 1. So, to assign a probability other than 1 to an assertion like '5 is prime', one needs to characterize the primality of 5 as a property that may or may not hold of particular elements of a space. But 5 is prime, no matter what, and so it is difficult to imagine what type of space could reasonably model the counterfactual case. I may declare *X* to be the set $\{0, 1\}$, label 0 the state of affairs in which 5 is not prime, label 1 the state of affairs in which 5 is prime, and then assign $\{0\}$ and $\{1\}$ each a probability $1/2$. But then I have simply modeled a coin flip; the hard part is to design a space that can convincingly be argued to serve as an appropriate guide to behavior in the face of uncertainty.

It is tempting to resort to a Bayesian interpretation, and view probabilities as subjective degrees of belief. I can certainly claim to have a subjective degree of belief of $1/2$ that 5 is not prime; but such claims cannot play a role in a theory of rationality until they are somehow linked to behavior. For example, it is common to take the outward signs of a subjectively held probability to be the willingness to bet on the outcome of an experiment (or the result of determining the true state of affairs) with corresponding odds. In that case, F. P. Ramsey and Bruno de Finetti have noted that the dictates of rationality demand that, at the bare minimum, subjective assignments should conform to the laws of probability, on pain of having a clever opponent 'make book' by placing a system of bets that guarantees him or her a profit no matter what transpires. But such coherence criteria still depend, implicitly, on having a model of a space of possible outcomes, against which the possibility of book can be judged. So one has simply shifted the problem to that of locating a notion of coherence on which it is reasonable to have a less-than-perfect certainty in the fact that 5 is prime; or, at least, to develop a notion of coherence for which there is anything interesting to say about such beliefs.

The challenge of developing theories of rationality that do not assume logical omniscience is not limited to modeling mathematical beliefs; it is just that the difficulties involved in doing so are most salient in mathematical settings. But the intuitions behind ascriptions of mathematical likelihood are often so strong that some have been encouraged to overcome these difficulties. For example, Pólya (1941) discusses a claim, by Euler, that it is nearly certain that the coefficients of two analytic expressions agree, because the claim can easily be verified in a number of specific cases. Pólya then suggested that it might be possible to develop a 'qualitative' theory of mathematical plausibility to account for such claims. (See also the other articles in Pólya 1984, and Kyburg's remarks at the end of that volume.) Ian Hacking (1967), I. J. Good (1977), and, more recently, Haim Gaifman (2004) have proposed ways of making sense of probability judgments in mathematical settings. David Corfield (2003) surveys such attempts, and urges us to take them seriously.

Gaifman's proposal is essentially a variant of the trivial '5 is prime' example I described above. Like Hacking, Gaifman takes sentences (rather than events or propositions) to bear assignments of probability. He then describes ways of imposing constraints on an agent's deductive powers, and asks only that ascriptions of probability be consistent with the entailments the agent can 'see' with his or her limited means. If all I am willing to bet on is the event that 5 is prime and I am unable or unwilling to invest the effort to determine whether this is the case, then, on Gaifman's account, any assignment of probability is 'locally' consistent with my beliefs. But Gaifman's means of incorporating closure under some deductive entailments allows for limited forms of reasoning in such circumstances. For example, if I judge it unlikely that all the random values drawn to conduct a probabilistic primality test are among a relatively small number of misleading witnesses, and I use these values to perform a calculation that certifies a particular number as prime, then I am justified in concluding that it is likely that the number is prime. I may be wrong about the chosen values and hence the conclusion, but at least, according to Gaifman, there is a sense in which my beliefs are locally coherent.

But does this proposal really address the problems raised above? Without a space of possibilities or a global notion of coherent behavior, it is hard to say what the analysis does for us. Isaac Levi (1991, 2004) clarifies the issue by distinguishing between theories of *commitment* and theories of *performance*. Deductive logic provides theories of the beliefs a rational agent is ideally committed to, perhaps on the basis of other beliefs that he or she is committed to, independent of his or her ability to recognize those commitments. On that view, it seems unreasonable to say that an agent committed to believing 'A' and 'A implies B' is not committed to believing 'B', or that an agent committed to accepting the validity of basic arithmetic calculations is not committed to the consequence of those calculations.

At issue, then, are questions of performance. Given that physically and computationally bounded agents are not always capable of recognizing their doxastic commitments, we may seek general *procedures* that we can follow to approximate the ideal. For example, given bounds on the resources we are able to devote to making a certain kind of decision, we may seek procedures that provide correct judgments most of the time, and minimize errors. Can one develop such a theory of 'useful' procedures? Of course! This is exactly what theoretical computer science does. Taken at face value, the analysis of a probabilistic primality test shows that if one draws a number at random from a certain distribution, and a probabilistic primality test certifies the number as prime, then with high probability the conclusion is correct. Gaifman's theory tries to go one step further and explain why it is rational to accept the result of a test in a *specific* case where it provides a false answer. But it is not clear that this adds anything to our understanding of rationality, or provides a justification for using the test that is better than the fact that the procedure is efficient and usually reliable.

When it comes to empirical events, we have no problem taking spaces of possibilities to be implicit in informal judgments. Suppose I draw a marble blindly from an urn containing 500 black marbles and 500 white marbles, clasp it in my fist, and ask you to calculate the probability that the marble I hold is black. The question presupposes that I intend for you to view the event as the result of a draw of a ball from the urn. Without a salient background context, the question as to the probability that a marble I clasp in my fist is black is close to meaningless.

In a similar fashion, the best way to understand an ascription of likelihood to a mathematical assertion may be to interpret it as a judgment as to the likelihood that a certain manner of proceeding will, in general, yield a correct result. Returning to Pólya's example, Euler seems to be making a claim as to the probability that two types of calculation, arising in a certain way, will agree in each instance, given that they agree on sufficiently many randomly or deterministically chosen test cases. If we assign a probability distribution to a space of such calculations, there is no conceptual difficulty involved in making sense of the claim. Refined analyses may try to model the types of calculations one is 'likely' to come across in a given domain, and the outcome of such an analysis may well support our intuitive judgments. The fact that the space in question may be vague or intractable makes the problem little different from those that arise in ordinary empirical settings.[5]

---

[5] Another nice example is given by Wasserman (2004, Example 11.10), where statistical methods are used to estimate the value of an integral that is too hard to compute. As the discussion after that example suggests, the strategy of suppressing intractable information is more congenial to a classical statistician than to a Bayesian one, who would insist, rather, that all the relevant information should be reflected in one's priors. This methodological difference was often emphasized by I. J. Good, though Wasserman and Good draw opposite conclusions. Wasserman takes the classical statistician's ability to selectively ignore information to provide an advantage in certain contexts: 'To construct procedures with guaranteed long run performance, ... use frequentist methods.' In contrast, Good takes the classical statistician's need to ignore information to indicate the fragility of those methods; see the references to the 'statistician's stooge' in Good (1983). I am grateful to Teddy Seidenfeld for bringing these references to my attention.

I have already noted, above, that Good (1977) favors a Bayesian approach to assigning probabilities to outcomes that are determined by calculation. But, once again, Levi's distinction between commitment and performance is helpful: what Good seems to propose is a theory that is capable of modeling

Along the same lines, the question as to the probability of the correctness of a proof that has been obtained or verified with computational means is best understood as a question as to the reliability of the computational methods or the nature of the verification. Here, too, the modeling issues are not unlike those that arise in empirical contexts. Vendors often claim 'five-nines' performance for fault-tolerant computing systems, meaning that the systems can be expected to be up and running 99.999% of the time. Such judgments are generally based on past performance, rather than on any complex statistical modeling. That is not to say that there are not good reasons to expect that past performance is a good predictor, or that understanding the system's design can't bolster our confidence. In a similar manner, formal modeling may, pragmatically, have little bearing on our confidence in computational methods of verification.

In sum, there are two questions that arise with respect to theories of mathematical evidence: first, whether *any* philosophical theory of mathematical plausibility can be put to significant use in any of the domains in which the notions arise; and second, if so, whether a fundamentally different concept of rationality is needed. It is possible that proposals like Pólya's, Hacking's, and Gaifman's will prove useful in providing descriptive accounts of human behavior in mathematical contexts, or in designing computational systems that serve mathematical inquiry. But this is a case that needs to be made. Doing so will require, first, a clearer demarcation of the informal data that the philosophical theories are supposed to explain, and second, a better sense of what it is that we want the explanations to do.

11.5 Theories of mathematical understanding

In addition to notions of mathematical evidence, we have seen that uses of computers in mathematics also prompt evaluations that invoke notions of mathematical understanding. For example:

- results of numeric simulation can help us understand the behavior of a dynamical system;
- symbolic computation can help shed light on an algebraic structure;
- graphical representations can help us visualize complex objects and thereby grasp their properties (see Mancosu, 2005).

'reasonable' behavior in computationally complex circumstances, without providing a normative account of what such behavior is supposed to achieve.

Such notions can also underwrite negative judgments: we may feel that a proof based on extensive computation does not provide the insight we are after, or that formal verification does little to promote our understanding of a theorem. The task is to make sense of these assessments.

But the word 'understanding' is used in many ways: we may speak of understanding a theory, a problem, a solution, a conjecture, an example, a theorem, or a proof. Theories of mathematical understanding may be taken to encompass theories of explanation, analogy, visualization, heuristics, concepts, and representations. Such notions are deployed across a wide range of fields of inquiry, including mathematics, education, history of mathematics, cognitive science, psychology, and computer science. In short, the subject is a sprawling wilderness, and most, if not all, of the essays in this collection can be seen as attempts to tame it. (See also the collection Mancosu et al., 2005.)

Similar topics have received considerably more attention in the philosophy of science, but the distinct character of mathematics suggests that different approaches are called for. Some have expressed skepticism that anything philosophically interesting can be said about mathematical understanding, and there is a tradition of addressing the notion only obliquely, with hushed tones and poetic metaphor. This is unfortunate: I believe it is possible to develop fairly down-to-earth accounts of key features of mathematical practice, and that such work can serve as a model for progress where attempts in the philosophy of science have stalled. In the next essay, I will argue that philosophical theories of mathematical understanding should be cast in terms of analyses of the types of mathematical abilities that are implicit in common scientific discourse where notions of understanding are employed. Here, I will restrict myself to some brief remarks as to the ways in which recent uses of computers in mathematics can be used to develop such theories.

The influences between philosophy and computer science should run in both directions. Specific conceptual problems that arise in computer science provide effective targets for philosophical analysis, and goals like that of verifying common mathematical inferences or designing informative graphical representations provide concrete standards of success, against which the utility of an analytic framework can be evaluated. There is a large community of researchers working to design systems that can carry out mathematical reasoning effectively; and there is a smaller, but significant, community trying to automate mathematical discovery and concept formation (see e.g. Colton et al., 2000). If there is any domain of scientific inquiry for which one might expect the philosophy of mathematics to play a supporting role, this is it. The fact that the philosophy of mathematics provides virtually no practical guidance in the appropriate use of common epistemic terms may lead some to wonder what, exactly, philosophers are doing to earn their keep.

In the other direction, computational methods that are developed towards attaining specific goals can provide clues as to how one can develop a broader philosophical theory. The data structures and procedures that are effective in getting computers to exhibit the desired behavior can serve to direct our attention to features of mathematics that are important to a philosophical account.

In Avigad (2006), I addressed one small aspect of mathematical understanding, namely, the process by which we understand the text of an ordinary mathematical proof. I discussed ways in which efforts in formal verification can inform and be informed by a philosophical study of this type of understanding. In the next essay, I will expand on this proposal, by clarifying the conception of mathematical understanding that is implicit in the approach, and discussing aspects of proofs in algebra, analysis, and geometry in light of computational developments. In focusing on formal verification, I will be dealing with only one of the many ways in which computers are used in mathematics. So the effort, if successful, provides just one example of the ways that a better interaction between philosophical and computational perspectives can be beneficial to both.

11.6 Final thoughts

I have surveyed two ways in which the philosophy of mathematics may be extended to address issues that arise with respect to the use of computers in mathematical inquiry. I may, perhaps, be accused of expressing too much skepticism with respect to attempts to develop theories of mathematical evidence, and excessive optimism with respect to attempts to develop theories of mathematical understanding. Be that as it may, I would like to close here with some thoughts that are relevant to both enterprises.

First, it is a mistake to view recent uses of computers in mathematics as a source of philosophical puzzles that can be studied in isolation, or resolved by appeal to basic intuition. The types of questions raised here are only meaningful in specific mathematical and scientific contexts, and a philosophical analysis is only useful in so far as it can further such inquiry. Ask not what the use of computers in mathematics can do for philosophy; ask what philosophy can do for the use of computers in mathematics.

Second, issues regarding the use of computers in mathematics are best understood in a broader epistemological context. Although some of the topics explored here have become salient with recent computational developments, none of the core issues are specific to the use of the computer *per se*. Questions having to do with the pragmatic certainty of mathematical results, the role of computation in mathematics, and the nature of mathematical understanding have a much longer provenance, and are fundamental to making sense of mathematical inquiry. What we need now is *not* a philosophy of computers in mathematics; what we need is simply a better philosophy of mathematics.

Bibliography

AVIGAD, Jeremy (2006), 'Mathematical method and proof', *Synthese*, 153, 105–159.

BAILEY, David and BORWEIN, Jonathan (2005), 'Experimental mathematics: examples, methods and implications', *Notices of the American Mathematical Society*, 52, 502–514.

BASSLER, O. Bradley (2006), 'The surveyability of mathematical proof: A historical perspective', *Synthese*, 148, 99–133.

BORWEIN, Jonathan and BAILEY, David (2004), *Mathematics by Experiment: Plausible Reasoning in the 21st Century* (Natick, MA: A. K. Peters Ltd).

BORWEIN, Jonathan, BAILEY, David, and GIRGENSOHN, Roland (2004), *Experimentation in Mathematics: Computational Paths to Discovery* (Natick, MA: A. K. Peters Ltd).

COLTON, Simon, BUNDY, Alan, and WALSH, Toby (2000), 'On the notion of interestingness in automated mathematical discovery', *International Journal of Human–Computer Studies*, 53, 351–365.

CORFIELD, David (2003), *Towards a Philosophy of Real Mathematics* (Cambridge: Cambridge University Press).

FALLIS, Don (1997), 'The epistemic status of probabilistic proof', *Journal of Philosophy*, 94, 165–186.

—— (2002), 'What do mathematicians want?: probabilistic proofs and the epistemic goals of mathematicians', *Logique et Analyse*, 45, 373–388.

GAIFMAN, Haim (2004), 'Reasoning with limited resources and assigning probabilities to arithmetical statements', *Synthese*, 140, 97–119.

GOOD, I. J. (1977), 'Dynamic probability, computer chess, and the measurement of knowledge', in E. W. Elcock and Donald Michie (eds.), *Machine Intelligence 8* (New York: John Wiley & Sons), pp. 139–150. Reprinted in Good (1983), pp. 106–116.

—— (1983), *Good Thinking: The Foundations of Probability and its Applications* (Minneapolis: University of Minnesota Press).

HACKING, Ian (1967), 'A slightly more realistic personal probability', *Philosophy of Science*, 34, 311–325.

LEVI, Isaac (1991), *The Fixation of Belief and its Undoing* (Cambridge: Cambridge University Press).

―― (2004), 'Gaifman', *Synthese*, 140, 121–134.

MANCOSU, Paolo (2005), 'Visualization in logic and mathematics', in Mancosu *et al.* (2005).

MANCOSU, Paolo, JØRGENSEN, Klaus Frovin, and PEDERSEN, Stig Andur (2005), *Visualization, Explanation and Reasoning Styles in Mathematics* (Dordrecht: Springer-Verlag).

MCCUNE, William (1997), 'Solution of the Robbins problem', *Journal of Automated Reasoning*, 19, 263–276.

PÓLYA, George (1941), 'Heuristic reasoning and the theory of probability', *American Mathematical Monthly*, 48, 450–465.

―― (1984), *Collected papers. Vol. IV: Probability; Combinatorics; Teaching and Learning in Mathematics*, Gian-Carlo Rota, M. C. Reynolds, and R. M. Short eds. (Cambridge, MA: MIT Press).

THURSTON, William P. (1994), 'On proof and progress in mathematics', *Bulletin of the American Mathematical Society*, 30, 161–177.

TYMOZCKO, Thomas (1979), 'The four-color problem and its philosophical significance', *Journal of Philosophy*, 76, 57–83. Reprinted in Tymoczko (1998), pp. 243–266.

―― (ed.) (1998), *New Directions in the Philosophy of Mathematics*, expanded edn (Princeton, NJ: Princeton University Press).

WASSERMAN, Larry (2004), *All of Statistics* (New York: Springer-Verlag).

WIEDIJK, Freek (2006), *The Seventeen Provers of the World* (Berlin: Springer-Verlag).

12

Understanding Proofs

JEREMY AVIGAD

'Now, in calm weather, to swim in the open ocean is as easy to the practised swimmer as to ride in a spring-carriage ashore. But the awful lonesomeness is intolerable. The intense concentration of self in the middle of such a heartless immensity, my God! who can tell it? Mark, how when sailors in a dead calm bathe in the open sea—mark how closely they hug their ship and only coast along her sides.' (Herman Melville, *Moby Dick*, Chapter 94)

12.1 Introduction

What does it mean to understand mathematics? How does mathematics help us understand?

These questions are not idle. We look to mathematics for understanding, we value theoretical developments for improving our understanding, and we design our pedagogy to convey understanding to students. Our mathematical practices are routinely evaluated in such terms. It is therefore reasonable to ask just what understanding amounts to.

The issue can be addressed at different levels of generality. Most broadly, we need to come to terms with the sort of thing that understanding is, and the sort of thing that mathematics is, in order to discuss them in an appropriate manner. We can narrow our focus by noticing that the term 'understanding' is used in different ways; we can ask, for example, what it

---

Early versions of parts of this essay were presented at a conference, *La Preuve en Mathématique: Logique, Philosophie, Histoire*, in Lille, May 2005, and at a workshop organized by Ken Manders at the University of Pittsburgh in July 2005. I am grateful to the participants and many others for comments, including Andrew Arana, Mic Detlefsen, Jeremy Heis, Jukka Keranen, Paolo Mancosu, Ken Manders, John Mumma, Marco Panza, and Stewart Shapiro. I am especially grateful for sharp criticism from Clark Glymour, who still feels that Part I is a fuzzy and unnecessary framing of the otherwise promising research program surveyed in Part II.

means to understand a mathematical definition, a theorem, or a proof; or to understand a theory, a method, an algorithm, a problem, or a solution. We can, alternatively, focus our task by restricting our attention to particular types of judgments, such as historical or mathematical evaluations of theoretical developments, or pedagogical evaluations of teaching practices and lesson plans. We can be even more specific by considering individual objects of understanding and particular evaluatory judgments. For example, we can ask what it means to understand algebraic number theory, the spectral theorem for bounded linear operators, the method of least squares, or Gauss's sixth proof of the law of quadratic reciprocity; or we can try to explain how the introduction of the group concept in the 19th century advanced our understanding, or why the 'new math' initiative of the 1960s did not deliver the desired understanding to students. The way we deal with the specific examples will necessarily presuppose at least some conception of the nature of mathematical understanding; but, conversely, the things we find to say in specific cases will help us establish a more general framework.

In this chapter, I will defend the fairly simple claim that ascriptions of understanding are best understood in terms of the possession of certain abilities, and that it is an important philosophical task to try to characterize the relevant abilities in sufficiently restricted contexts in which such ascriptions are made. I will illustrate this by focusing on one particular type of understanding, in relation to one particular field of scientific search. Specifically, I will explore what it means to understand a proof, and discuss specific efforts in formal verification and automated reasoning that model such understanding.

This chapter is divided in two parts. In Part I, I will argue that the general characterization of understanding mentioned above provides a coherent epistemological framework, one that accords well with our intuitions and is capable of supporting rational inquiry in practical domains where notions of mathematical understanding arise. In Part II, I will present four brief case studies in formal verification, indicating areas where philosophical reflection can inform and be informed by contemporary research in computer science.

Part I. The nature of understanding

12.1 Initial reflections

A central goal of the epistemology of mathematics has been to identify the appropriate support for a claim to mathematical knowledge. We show that a theorem is true by exhibiting a proof; thus, it has been a primary task of the philosophy of mathematics to clarify the notion of a mathematical proof, and to explain how such proofs are capable of providing appropriate mathematical knowledge. Mathematical logic and the theory of formal axiomatic systems have done a remarkably good job of addressing the first task, providing idealized accounts of the standards by which a proof is judged to be correct. This, in and of itself, does not address more difficult questions as to what ultimately justifies a particular choice of axiomatic framework. But, as far as it goes, the modern theory of deductive proof accords well with both intuition and mathematical practice, and has had practical applications in both mathematics and computer science, to boot.

But in mathematics one finds many types of judgment that extend beyond evaluations of correctness. For example, we seem to feel that there is a difference between *knowing that* a mathematical claim is true, and *understanding why* such a claim is true. Similarly, we may be able to convince ourselves that a proof is correct by checking each inference carefully, and yet still feel as though we do not fully understand it. The words 'definition' and 'concept' seem to have different connotations: someone may know the definition of a group, without having fully understood the group concept. The fact that there is a gap between knowledge and understanding is made pointedly clear by the fact that one often finds dozens of published proofs of a theorem in the literature, all of which are deemed important contributions, even after the first one has been accepted as correct. Later proofs do not add to our knowledge that the resulting theorem is correct, but they somehow augment our understanding. Our task is to make sense of this type of contribution.

The observation that modern logic fails to account for important classes of judgments traces back to the early days of modern logic itself. Poincaré wrote in *Science et Méthode* (1908):

> Does understanding the demonstration of a theorem consist in examining each of the syllogisms of which it is composed in succession, and being convinced that it is correct and conforms to the rules of the game? In the same way, does understanding a definition consist simply in recognizing that the meaning of all the terms employed is already known, and being convinced that it involves no contradiction?
>
> ... Almost all are more exacting; they want to know not only whether all the syllogisms of a demonstration are correct, but why they are linked together in one order rather than in another. As long as they appear to them engendered by caprice, and not by an intelligence constantly conscious of the end to be attained, they do not think they have understood. (Book II, Chapter II, p. 118)

In that respect, logic does not tell us the whole story:

> Logic teaches us that on such and such a road we are sure of not meeting an obstacle; it does not tell us which is the road that leads to the desired end. (*Ibid.*, pp. 129–130)

Philosophers of science commonly distinguish between the 'logic of justification' and the 'logic of discovery'. Factors that guide the process of discovery also fall under the general category of 'understanding', and, indeed, understanding and discovery are often linked. For example, understanding a proof may involve, in part, seeing how the proof could have been discovered; or, at least, seeing how the train of inferences could have been anticipated.

> It seems to me, then, as I repeat an argument I have learned, that I could have discovered it. This is often only an illusion; but even then, even if I am not clever enough to create for myself, I rediscover it myself as I repeat it. (*Ibid.*, Part I, Chapter III, p. 50)

While knowing the relevant definitions may be enough to determine that a proof is correct, understanding is needed to find the definitions that make it possible to discover a proof. Poincaré characterized the process of discovery, in turn, as follows:

> Discovery consists precisely in not constructing useless combinations, but in constructing those that are useful, which are an infinitely small minority. Discovery is discernment, selection. (*Ibid.*, p. 51)

These musings provide us with some helpful metaphors. Mathematics presents us with a complex network of roads; understanding helps us navigate them, and find the way to our destination. Mathematics presents us with a combinatorial explosion of options; understanding helps us sift through them, and pick out the ones that are worth pursuing. Without understanding, we are lost in confusion, wandering blindly, unable to cope. When we do mathematics, we are like Melville's sailors, swimming in a vast expanse. Just as the sailors cling to sides of their ship, we rely on our understanding to guide us and support us.

12.2 Understanding and ability

Let us see if we can work these metaphors into something more definite. One thing to notice is that there seems to be some sort of reciprocal relationship between mathematics and understanding. That is, we speak of understanding theorems, proofs, problems, solutions, definitions, concepts, and methods; at the same time, we take all these things to contribute to our understanding. This duality is reflected in the two questions I posed at the outset.

The way the questions are posed seem to presuppose that understanding is a relationship between an agent, who understands, and mathematics, which is the object of that understanding. On the surface, talk of knowledge shares this feature; it is an agent that knows that a theorem is true. But the role of an agent is usually eliminated from the epistemological account; once knowledge of a theorem is analyzed in terms of possession of a proof, proofs become the central focus of the investigation. To adopt a similar strategy here would amount to saying that an agent understands X just in case the agent is in possession of Y. But what sort of thing should Y be?

Suppose I tell you that my friend Paolo understands group theory, and you ask me to explain what I mean. In response, I may note that Paolo can state the definition of a group and provide some examples; that he can recognize the additive group structure of the integers, and characterize all the subgroups; that he knows Lagrange's theorem, and can use it to show that the order of any element of a finite group divides the order of the group; that he knows what a normal subgroup is, and can form a quotient group and work with it appropriately; that he can list all the finite groups of order less than 12, up to isomorphism; that he can solve all the exercises in an elementary textbook; and so on.

What is salient in this example is that I am clarifying my initial ascription of understanding by specifying some of the abilities that I take such an understanding to encompass. On reflection, we see that this example is typical: when we talk informally about understanding, we are invariably talking about the *ability*, or a capacity, to do something. It may be the ability to solve a problem, or to choose an appropriate strategy; the ability to discover a proof; the ability to discern a fruitful definition from alternatives; the ability to apply a concept efficaciously; and so on. When we say that someone understands we simply mean that they possess the relevant abilities.

Ordinary language is sloppy, and it would be foolish to seek sharp accounts of notions that are used in vague and imprecise ways. But notions of understanding also play a role in scientific claims and policy decisions that should be subject to critical evaluation. In more focused contexts like these, philosophical clarification can help further inquiry.

What I am proposing here is that in such situations it is often fruitful to analyze understanding in terms of the possession of abilities. This is a straightforward extension of the traditional epistemological view: the ability to determine whether a proof is correct is fundamental to mathematics, and the standard theory has a lot to say as to what that ability amounts to. But this way of framing things enables us to address a much wider range of epistemological issues; verifying correctness is only a small part of understanding a proof, and we commonly speak of understanding other sorts of things as well. The task of answering our two main questions is thereby reduced to the task of describing and analyzing the interrelated network of abilities which constitute the practice of mathematics, *vis-à-vis* fields of inquiry that rely, either implicitly or explicitly, on models of that practice.

It may be argued that this proposal runs counter to our intuitions. Understanding is clearly needed to carry out certain mathematical tasks, but although understanding can explain the ability to carry out a task successfully, isn't it a mistake to conflate the two? Suppose I am working through a difficult proof. The argument is confusing, and I struggle to make sense of it. All of a sudden, something clicks, and everything falls into place—now I *understand*. What has just happened, and what has it got to do with ability?

We have all shared such 'Aha!' moments and the deep sense of satisfaction that comes with them. But surely the philosophy of mathematics is not supposed to explain this sense of satisfaction, any more than economics is supposed to explain the feeling of elation that comes when we find a $20 bill lying on the sidewalk. Economic theories describe agents, preferences, utilities, and commodities in abstract terms; such theories, when combined with social, political, psychological, or biological considerations, perfectly well explain the subjective appeal of cold, hard cash. In a similar way, we should expect a philosophical theory to provide a characterization of mathematical understanding that is consistent with, but independent of, our subjective experience.

Returning to the example above, what lies behind my moment of insight? Perhaps, all of a sudden, I see how to fill in a gap in the argument that had me puzzled. I may realize that the third line of the proof appeals to a prior lemma, which simply has to be instantiated appropriately; or that the claim follows easily from a general fact about, say, Hilbert spaces. Or perhaps, with Poincaré, I feel as though I understand how the proof could have been discovered; that is, I see why, in this situation, it is natural to consider the objects that have been introduced, or to express a term in the form in which is has been presented. Perhaps I see why a certain hypothesis in the theorem is necessary, and what would go wrong if the hypothesis were omitted. Perhaps I have grasped a general method in the structure of the argument, one that can fruitfully be applied in other situations. Perhaps I have realized that the argument is just like one that I am fully familiar with, straightforwardly adapted to the case at hand. These insights are perfectly well explained in terms of the acquisition of abilities to supply missing inferences, draw appropriate analogies, discover other theorems, and so on. And, in turn, these abilities are just the sort of thing that should explain our pleasure at having understood: it is simply the pleasure of having acquired a new skill, or of finding ourselves capable of doing something we could not do before.

According to this analysis, when we say that someone understands something—a theorem, a problem, a method, or whatever—what we mean is that they possess some general ability. Such uses of a particular to stand for something more general are familiar. Suppose I tell you that Rebecca, a bright girl in Ms Schwartz's second grade class, can multiply 34 by 51. What do I mean by that? Surely I am not just attributing to her the ability to utter '1734' in response to the corresponding query; even the dullest student in the class can be trained to do that. We are often tempted to say that what we mean is that Rebecca is able to arrive at the answer '1734' by the right process, but a metaphysical commitment to 'processes' is easily avoidable. What we really mean is that Rebecca is capable of carrying out a certain type of arithmetical operation, say, multiplying numbers of moderate size. Phrasing that in terms of the ability to multiply 34 by 51 is just a convenient manner of speaking.

This way of thinking about understanding is not novel. In the next section, I will show that the strategy I have outlined here accords well with Wittgenstein's views on language, and I will situate the framework I am describing here with respect to the general viewpoint of the *Philosophical Investigations* (Wittgenstein, 1953). In the section after that, I will address some of the concerns that typically attend this sort of approach.

12.3 Mathematics as a practice

An important segment of the *Philosophical Investigations* explores what it means to follow a rule, as well as related notions, like obeying a command or using a formula correctly.[1] The analysis shows that, from a certain philosophical perspective, it is fruitless to hope for a certain type of explanation of the 'meaning' of such judgments. Nor is it necessary: there are philosophical gains

---

[1] I have in mind, roughly, sections 143 to 242, though the topic is foreshadowed in sections 81 to 88. The literature on these portions of the *Investigations* is vast, much of it devoted to critiquing Kripke's interpretation (1982); see, for example, Goldfarb (1985) and Tait (1986). The present chapter is essentially devoted to showing that a 'straightforward' reading of the *Investigations* has concrete consequences for the development of a theory of mathematical understanding. I owe much of my interpretation of Wittgenstein on rule following, and its use in making sense of a mathematical 'practice', to discussions with Ken Manders.

to be had by exploring the relationships between such fundamental judgments, and unraveling problems that arise from confused or misguided applications of the terms.

From the perspective of the *Investigations*, language is a community practice. It determines what can meaningfully be said, while the meaning of a word or sentence is, reciprocally, determined by the way that word or sentence functions in the practice. On that view, meaning is closely linked with the possibilities for use. Philosophical difficulties arise, however, when we try to explain the relationship between the two.

> When someone says the word 'cube' to me, for example, I know what it means. But can the whole *use* of the word come before my mind, when I *understand* it in this way?
>
> Well, but on the other hand isn't the meaning of the word also determined by this use? And can these ways of determining meaning conflict? Can what we grasp *in a flash* accord with a use, fit or fail to fit it? And how can what is present to us in an instant, what comes before our mind in an instant, fit a *use*? (Wittgenstein, 1953, §139)

The problem is that understanding is only made manifest in an infinite array of uses. The common philosophical tendency is therefore to distinguish the two, and take understanding to be 'possession' of a meaning that somehow 'determines' the appropriate usage.

> Perhaps you will say here: to have got the system (or again, to understand it) can't consist in continuing the series up to *this* or *that* number: *that* is only applying one's understanding. The understanding itself is a state which is the *source* of correct use. (§146)

But Wittgenstein urges us against this way of thinking.

> If one says that knowing the ABC is a state of the mind, one is thinking of a state of a mental apparatus (perhaps of the brain) by means of which we explain the *manifestations* of that knowledge. Such a state is called a disposition. But there are objections to speaking of a state of mind here, inasmuch as there ought to be two different criteria for such a state: a knowledge of the construction of the apparatus, quite apart from what it does... (§149)

The discussion in the *Investigations* aims to convince us that this last way of framing matters is problematic. For example, we may attribute someone's ability to continue a sequence of numbers correctly to the fact that he has grasped the right pattern. But substituting the phrase 'grasping the correct pattern' for 'understanding' is little more than word play, unless we say more about what has been 'grasped'. The appropriate pattern may, perhaps, be described by an algebraic formula, and so, at least in some cases, we may explain the ability to continue the sequence in terms of knowing the correct formula. But then we are left with the task of explaining how he is able to apply the algebraic formula correctly. It is not enough to say that the formula simply 'occurs to him' as he produces the desired behavior; perhaps he will continue to think of the formula and do something entirely unexpected at the next step. So we have simply replaced the problem of explaining what it means to understand how to continue the sequence with the problem of explaining what it means to understand how to apply a formula correctly. To make matters worse, there may be other ways in which we can account for the person's ability to continue the sequence according to the pattern we have in mind; or we may find that the person is simply able to do it, without being able to explain how.

> We are trying to get hold of the mental process of understanding which seems to be hidden behind those coarser and therefore more readily visible accompaniments. But we do not succeed; or, rather, it does not get as far as a real attempt. For even supposing I had found something that happened in all those cases of understanding,—why should *it* be the understanding? ... (§153)

The solution is, in a sense, just to give up. In other words, we simply need to resist the temptation to find a suitable 'source' for the behavior.

> If there has to be anything 'behind the utterance of the formula' it is *particular circumstances,* which justify me in saying that I can go on—when the formula occurs to me.
> Try not to think of understanding as a 'mental process' at all.—For *that* is the expression which confuses you. But ask yourself: in what sort of case, in what kind of circumstances, do we say, 'Now I know how to go on,' when, that is, the formula *has* occurred to me?—
> In the sense in which there are processes (including mental processes) which are characteristic of understanding, understanding is not a mental process. (§154)

If our goal is to explain what it means to say that someone has understood a particular word, formula, or command, we simply need to describe the circumstances under which we are willing to make this assertion. In doing so, we may find that there is a good deal that we can say that will clarify our meaning. Giving up the attempt to identify understanding as some sort of *thing* doesn't mean that we cannot be successful, by and large, in explaining what understanding amounts to.

Thus, from a Wittgensteinian perspective, the philosopher's task is not to explain the feeling of having understood, or any underlying mental or physical processes. The challenge, rather, is to clarify the circumstances under we which we make our ascriptions.

> ... when he suddenly knew how to go on, when he understood the principle, then possibly he had a special experience ... but for us it is *the circumstances* under which he had such an experience that justify him in saying in such a case that he understands, that he knows how to go on. (§155)

We should not be distressed by the fact that our ascriptions of understanding are fallible. I may reasonably come to believe that a student understands the fundamental theorem of calculus, but then subsequently change my mind after grading his or her exam. This does not in any way preclude the utility of trying to explain what it means to 'understand the fundamental theorem of calculus', in terms of what we take such an understanding to entail.

The aim of the *Investigations* is to shape the way we think about language and thought. Here, I have proposed that this world view is relevant to the way we think about mathematics. The nature of our more specific goals does, however, impose some important differences in emphasis. In the excerpts I have quoted, Wittgenstein is primarily concerned with exploring what it means to follow a rule, obey a command, or use a word *correctly*. When it comes to the philosophy of mathematics, I believe it is also fruitful to explore what we take to constitute *appropriate* behavior, even in situations where we take a goal or a standard of correctness to be fixed and unproblematic. For example, we may agree, for the moment, to take provability in some formal axiomatic theory to provide an appropriate standard of correctness, and then ask what types of abilities are appropriate to finding the proofs we seek. Or we may be in a situation where we have a clear notion as to what counts as the solution to a particular problem, and then wonder what type of understanding is needed to guide a student to a solution. Of course, some goals of the practice, like finding 'natural' definitions or 'fruitful' generalizations, are harder to characterize. And the distinction between goals and the methods we use to achieve them blur; the goals of finding natural definitions and fruitful generalizations can also be interpreted as means to further goals, like solving problems and proving theorems. The network of goals is complex, but we need not chart the entire territory at once; by focusing on particular phenomena of interest we can start by mapping out small regions. The claim I am making here is simply that the terrain we are describing is best viewed as a network of abilities, or mechanisms and capacities for thought.

Indeed, the *Investigations* is not only concerned with questions of correctness. The work is, more broadly, concerned with the effective use of language with respect to our various goals and ends.

> Language is an instrument. Its concepts are instruments. Now perhaps one thinks that it can make no *great* difference *which* concepts we employ. As, after all, it is

> possible to do physics in feet and inches as well as in metres and centimetres; the difference is merely one of convenience. But even this is not true if, for instance, calculations in some system of measurement demand more time and trouble than is possible for us to give them. (§569)
>
> Concepts lead us to make investigations; are the expressions of our interest, and direct our interest. (§570)

One finds similar views on the role of specifically mathematical concepts in Wittgenstein's other works. For example, we find the following in the *Remarks on the Foundations of Mathematics*:

> The mathematical Must is only another expression of the fact that mathematics forms concepts.
>
> And concepts help us to comprehend things. They correspond to a particular way of dealing with situations.
>
> Mathematics forms a network of norms. (Wittgenstein, 1956, VI, §67)

This stands in contrast to the traditional view of mathematics as a collection of definitions and theorems. For Wittgenstein, a proposition is not just an object of knowledge, but, rather, something that shapes our behavior:

> The mathematical proposition says to me: Proceed like this! (§72)

With respect to propositions in general, we find in *On Certainty*:

> 204. Giving grounds, however, justifying the evidence, comes to an end;—but the end is not certain propositions' striking us immediately as true, i.e. it is not a kind of *seeing* on our part, it is our *acting*, which lies at the bottom of the language game. (Wittgenstein, 1969)

This way of thinking challenges us to view mathematics in dynamic terms, not as a body of knowledge, but, rather, as a complex system that guides our thoughts and actions. We will see in Part II of this essay that this provides a powerful and fundamentally useful way of thinking about the subject.

12.4 A functionalist epistemology

I have proposed that a theory of mathematical understanding should be a theory of mathematical abilities. In ordinary circumstances, when we say, for example, that someone understands a particular proof, we may take them to possess any of the following:

- the ability to respond to challenges as to the correctness of the proof, and fill in details and justify inferences at a skeptic's request;

- the ability to give a high-level outline, or overview of the proof;
- the ability to cast the proof in different terms, say, eliminating or adding abstract terminology;
- the ability to indicate 'key' or novel points in the argument, and separate them from the steps that are 'straightforward';
- the ability to 'motivate' the proof, that is, to explain why certain steps are natural, or to be expected;
- the ability to give natural examples of the various phenomena described in the proof;
- the ability to indicate where in the proof certain of the theorem's hypotheses are needed, and, perhaps, to provide counterexamples that show what goes wrong when various hypotheses are omitted;
- the ability to view the proof in terms of a parallel development, for example, as a generalization or adaptation of a well-known proof of a simpler theorem;
- the ability to offer generalizations, or to suggest an interesting weakening of the conclusion that can be obtained with a corresponding weakening of the hypotheses;
- the ability to calculate a particular quantity, or to provide an explicit description of an object, whose existence is guaranteed by the theorem;
- the ability to provide a diagram representing some of the data in the proof, or to relate the proof to a particular diagram;

and so on. The philosophical challenge is to characterize these abilities with clarity and precision, and fit them into a structured and informative theory. Thanks to our Wittgensteinian therapy, we will not let the phrase 'possess an ability' fool us into thinking that there is anything mysterious or metaphysically dubious about this task. We have serious work to do, and worrying about what sort of thing is being 'possessed' is an unnecessary distraction.

And yet we may still be plagued by qualms. Our analysis entails that understanding only becomes manifest in an agent's behavior across a range of contexts, and we seem to have come dangerously close to identifying understanding with the class of relevant behaviors. Such a 'dispositional' or 'behavioral' account of understanding has famously been put forth by Gilbert Ryle (1949) as part of a more general philosophy of mind. Since Ryle's approach is commonly viewed as having failed, it is worth reviewing some of the usual criticisms, to see what bearing they have on the more specific issues addressed here.[2]

---

[2] These criticisms are enumerated, for example, in Carr (1979).

Ryle intended his dispositional theory to account for ascriptions of a variety of mental states, including things like belief, desire, intent, and so on. He begins his account, however, with a discussion of ascriptions of 'knowing how'. He has been criticized, in this respect, for failing to distinguish between knowing how to perform an action, and the ability to do so. For example, it seems to reasonable to say that an arthritic piano player knows how to play the *Moonlight Sonata,* and that an injured gymnast knows how to perform a back flip, even if they are temporarily or permanently unable to do so. Here, we may have similar concerns that there may be situations under which it makes sense to say that someone understands a proof, but is unable to exhibit the expected behaviors. But the examples that come to mind are contrived, and it does not seem unreasonable to declare these outside the scope of a suitably focused theory of mathematical understanding. If we view at least the outward signs of mathematical activity as essentially linguistic, it seems reasonable to take verbal and written communication as reliable correlates of understanding.

A further critique of Ryle's analysis is a pragmatic one. Ryle imagines, for example, characterizing mental states, like hunger, in terms of dispositions to behave in certain ways, like opening the refrigerator door when one is in the kitchen. But one would not expect an agent to open the refrigerator door if he or she held the belief that the door was wired to an explosive device, and so circumstances like that need to be excluded. To start with, it is unsettling that one may have to characterize such contexts in terms of other mental states, like the agent's beliefs, when that is what a dispositional account is designed to avoid. But even more significantly, it seems unlikely that one can characterize the rogue circumstances that need to be excluded from the class of contexts in which we expect to observe the characteristic behavior. To be sure, mitigating factors like the one I described above are patently irrelevant to hunger, and one would like to exclude them with a suitable *ceteris paribus* clause. But this is exactly the point: it is hard to see how one can gauge relevance prior to having some kind of understanding of what it means to be hungry.

With respect to the topic at hand, these concerns translate to doubts that one can adequately characterize the behaviors that warrant attributions of understanding. But, once again, the problem can be mitigated by limitations of scope. Our theory need not account for the full range of human behaviors, as a theory of mind ought to do. We would like our theory to help explain why certain proofs are preferred in contemporary mathematics, why certain historical developments are viewed as advances, or why certain expository practices yield desired results. Moving from a general theory of mind to a more specific theory of mathematical understanding gives us great latitude in bracketing issues that we take to fall outside our scope. We should be able to screen off extraneous beliefs and desires, and assume nothing about an agent's intent beyond the intent to perform mathematically. I am certainly not claiming that it is obvious that one can provide adequate characterizations of the circumstances under which we tend to ascribe mathematical understanding; only that it is not obvious that attempts to do so are doomed to failure.

Perhaps the most compelling criticism of a dispositional account is that even if we could characterize the behaviors that are correlated with the mental states under consideration, identifying the mental states with the associated behaviors simply tells the wrong kind of story. We expect a philosophical theory to provide some sort of causal explanation that tells us how intelligent and intentional behavior is brought about; it seems unsatisfying to identify 'knowing how to play the piano' with successful performance, when what one really wants of a theory is an account of the mental activity that makes such performance possible. In the case at hand, we would like a theory that explains how a proper understanding enables one to function mathematically. This insistence has not only an intuitive appeal, but also a pragmatic one. For example, in so far as our theory is to be relevant to mathematical exposition and pedagogy, we would expect it not only to characterize the outward signs of mathematical understanding, but also provide some hints as to how they can be encouraged and taught. Similarly, I take it that a theory of mathematical understanding should be of service to computer scientists trying to write software that exhibits various types of competent mathematical behavior. Even if we set aside the question as to whether it is appropriate to attribute 'understanding' to a computer, we might expect a good philosophical theory not just to clarify and characterize the desired behaviors, but also to provide some guidance in bringing them about.

We are therefore tempted to renounce our therapy and try, again, to figure out just what understanding *really* is. What saves us, however, is the observation that our theory of mathematical abilities need not degenerate to a laundry list of behavioral cues. The abilities we describe will interact in complex ways, and will not always be cast in terms of behavioral manifestations. Consider our explanation of what it means for Paolo to understand group theory. Some of the relevant abilities may be cast in terms of behaviors, for example, the ability to state a theorem or answer a question appropriately. But others may be cast in more abstract terms, such as the ability to 'recognize' a group structure, 'determine' subgroups and cosets, 'apply' a lemma, or 'recall' a fundamental fact. In fact, we often take these abstract abilities to provide the 'mechanisms' that explain the observable behaviors. We may be relieved to learn that our Wittgensteinian training does not preclude talk of mechanisms, provided that we keep in mind that 'these mechanisms are only hypotheses, models designed to explain, to sum up, what you observe' (Wittgenstein, 1953, §156). What gives our theoretical terms meaning is the role they play in explaining the desired or observed behaviors; 'An "inner process" stands in need of outward criteria' (Wittgenstein, 1953, §580). While these outward criteria are necessary, they are also sufficient to give our terms meaning. So, we can once again set our metaphysical qualms aside.

What we are left with is essentially a functionalist approach to explaining various aspects of mathematical understanding. The fundamental philosophical challenge is to develop a language and conceptual framework that is appropriate to our goals. If you want an explanation of how a car works, a description of the subsystems and their components, situated against general understanding as to how these interact, may be just what you need to keep your car running smoothly, and to diagnose problems when they arise. A more fine-grained description is more appropriate if you are studying to be a mechanic or engineer. What we are seeking here are similar explanations of how mathematical understanding works. In this case, however, our intuitions as to how to talk about the relevant subsystems and components is significantly poorer. I have argued that a theory of mathematical abilities and their relationships should do the trick, but, at this point, the proposal is vague. The only way to make progress is to pay closer attention to the data that we are trying to explain, and to the particular aims that our explanations are to serve.

Part II. Formal verification

12.5 The nature of proof search

In Part I, I described a general way of thinking about mathematical understanding. My goal in Part II is to show that this way of thinking is fruitful in at least one scientific context where informal notions of understanding are used. In doing so, I will consider only one small aspect of mathematical understanding, with respect to one particular scientific practice. While I expect that the general perspective will be useful in other domains as well, and that the problems that arise share enough common structure that they can be supported by a unified conceptual framework, I cannot make this broader case here. So I ask you to keep in mind that, in what follows, we are considering only one restricted example.

We have seen that understanding an ordinary textbook proof involves, in part, being able to spell out details that are left implicit in the presentation. I have argued elsewhere (Avigad, 2006) that it is hard to make sense of this aspect of understanding in terms of traditional logical analyses. Formal axiomatic deduction provides a model of proof in which *every* detail is spelled out precisely, in such a way that correctness boils down to pattern matching against a manageable list of precisely specified rules. In contrast, ordinary textbook proofs proceed at a higher level, relying on the reader's ability to 'see' that each successive step is warranted. In order to analyze this capacity, we need a model of proof on which such 'seeing' is a nontrivial affair.

Coming to terms with the ability to understand higher-level proofs is a central task in the field of formal verification and automated reasoning. Since early in the 20th century, it has been understood that, at least in principle, mathematical proof can be modeled by formal axiomatic deduction. In recent years, a number of computational 'proof assistants' have been developed to make such formalization feasible in practice (see Wiedijk, 2006). In addition to verifying mathematical assertions, such systems are sometimes developed with the goal of verifying that (mathematical descriptions of) hardware and software systems meet their specifications, or are free from dangerous bugs. Since, however, such systems and specifications are modeled in mathematical terms, the two efforts overlap, to a large extent.

A user's interaction with such a system can be seen as an attempt to provide the computer with enough information to see that there is a formal axiomatic proof of the purported theorem. Alternatively, the formal 'proof scripts' that are given to the computer can be viewed (like informal proofs) as providing instructions as to how to find a formal axiomatic derivation. The task of the computational proof assistant is to use these scripts to construct such a derivation. When it has done so, the system indicates that it has 'understood' the user's proof by certifying the theorem as correct. In Avigad (2006), I consider a range of informal epistemological judgments that are not easily explicated on the standard logical models, and argue that 'higher-level' notions of proof, akin to the scripts just described, are better equipped to support the relevant judgments.

Proof assistants like Coq, Isabelle, and HOL-light provide a style of proof development that allows the user to view the task of theorem proving in a goal-driven manner. Stating a theorem can be seen as a way of announcing the goal of proving it. Each step in a proof then serves to reduce the currently open goals to ones that are (hopefully) simpler. These goals are often represented in terms of *sequents*. For example, if A, B, C, and D are formulas, the sequent $A, B, C \Rightarrow D$ represents the goal of showing that D follows from A, B, and C. The most basic steps correspond to logical rules. For example, the 'and introduction' rule reduces the goal $A, B, C \Rightarrow D \wedge E$ to the pair of goals $A, B, C \Rightarrow D$ and $A, B, C \Rightarrow E$. This corresponds to the situation where, in an ordinary proof, we have assumed or established A, B, and C, and need to prove an assertion of the form 'D and E'; we can do this by noting that 'it suffices to establish D, and then E, in turn' and then accomplishing each of these two tasks. We can also work forwards from hypotheses: for example, from $A \wedge B, C \Rightarrow D$ we can conclude $A, B, C \Rightarrow D$. In ordinary terms, if we have established or assumed 'A and B', we may use both A and B to derive our conclusion. A branch of the tree is closed off when the conclusion of a sequent matches one of the hypotheses, in which case the goal is clearly satisfied.

Things become more interesting when we try to take larger inferential steps, where the validity of the inference is not as transparent. Suppose, in an ordinary proof, we are trying to prove D, having established A, B, and C. In sequent form, this corresponds to the goal of verifying $A, B, C \Rightarrow D$. We write 'Clearly, from A and B we have E', thus reducing our task of verifying $A, B, C, E \Rightarrow D$. But what is clear to us may not be clear to the computer; the assertion that E follows from A and B corresponds to the sequent $A, B \Rightarrow E$, and we would like the computer to fill in the details automatically. 'Understanding' this step of the proof, in this context, means being able to justify the corresponding inference.

In a sense, verifying such an inference is no different from proving a theorem; a sequent of the form $A, B \Rightarrow E$ can express anything from a trivial logical implication to a major conjecture like the Riemann hypothesis. Sometimes, brute-force calculation can be used to verify inferences that require a good deal of human effort. But what is more striking is that there is a large class of inferences that require very little effort on our part, but are beyond the means of current verification technology. In fact, most textbook inferences have this character: it can take hours of painstaking work to get a proof assistant to verify a short proof that is routinely read and understood by any competent mathematician.

The flip explanation as to why competent mathematicians succeed where computers fail is simply that mathematicians understand, while computers don't. But we have set ourselves precisely the task of explaining how this understanding works. Computers can search exhaustively for an axiomatic derivation of the desired inference, but a blind search does not get very far. The problem is that even when the inferences we are interested in can be justified in a few steps, the space of possibilities grows exponentially.

There are a number of ways this can happen. First, excessive case distinctions can cause problems. Proving a sequent of the form $A \vee B, C, D \Rightarrow E$ reduces to showing that the conclusion follows from each disjunct; this results in two subgoals, $A, C, D \Rightarrow E$ and $B, C, D \Rightarrow E$. Each successive case distinction again doubles the number of subgoals, so that, for example, 10 successive case splits yield 1024 subgoals.

Second, proving an existential conclusion, or using a universal hypothesis, can require finding appropriate instances. For example, one can prove a sequent of the form $A, B, C \Rightarrow \exists x\, D(x)$ by proving $A, B, C \Rightarrow D(t)$ for some term t; or possibly $A, B, C \Rightarrow D(t_1) \vee \ldots \vee D(t_k)$ for a sequence of terms $t_1, \ldots, t_k$. Dually, proving a sequent of the form $\forall x\, A(x), B, C \Rightarrow D$ can involve finding appropriate instances $t_1, \ldots, t_k$ for the universal quantifier, and then proving $A(t_1), \ldots, A(t_k), B, C \Rightarrow D$. The problem is that there may be infinitely many terms to consider, and no *a priori* bound on k. In situations like this, search strategies tend to defer the need to choose terms as long as possible, using Skolem functions and a procedure known as 'unification' to choose useful instantiations. But even with such methods, there are choices to be made, again resulting in combinatorial explosion.

Finally, common mathematical inferences typically require background facts and theorems that are left implicit. In other words, justifying an inference usually requires a background theory in addition to purely logical manipulations. In that case, the challenge is to determine which external facts and theorems need to be imported to the 'local' list of hypotheses. Once again, the range of options can render the problem intractable.

All three factors—case disjunctions, term instantiation, and the use of a background theory—are constantly in play. For example, suppose we are reasoning about the real numbers. The trichotomy law states that for any x and y, one of the expressions $x < y$, or $x = y$, or $y < x$ must hold. At any point, a particular instantiation of this law may be just the thing needed to verify the inference at hand. But most of the time, choosing terms blindly and splitting across the disjunction will be nothing more than an infernal waste of time.

Despite all this, when we read a proof and try to fill in the details, we are somehow able to go on. We don't guess terms blindly, or pull facts at random from our shelves. Instead, we rely on our understanding to guide us. At a suitable level of abstraction, an account of how this understanding works should not only explain the efficacy of our own mathematical practices, but also help us fashion computational systems that share in the success. This last claim, however, is often a sticking point. One may object that the way that we, humans, understand proofs is different from the ways in which computers should search for proofs; determining the former is the task of cognitive psychologists, determining the latter is the task of computer scientists, and philosophy has nothing to do with it. Ordinary arithmetic calculation provides an analogy that seems to support this distinction. Cognitive psychologists can determine how many digits we can hold in our short term memory, and use that to explain why certain procedures for long multiplication are effective. Computer scientists have developed a clever method called 'carry save addition' that makes it possible for a microprocessor to perform operations on binary representations in parallel, and therefore multiply numbers more efficiently. Each theory is fruitfully applied in its domain of application; no overarching philosophy is needed.

But this objection misses the point. Both the experimental psychologist and the computer scientist presuppose a normative account of what it means to multiply correctly, and our foundational accounts of arithmetic apply equally well to humans and machines. In a similar way, a theory of mathematical understanding should clarify the structural aspects of mathematics that characterize successful performance for agents of both sorts. To be sure, when it comes to verifying an inference, different sorts of agents will have different strengths and weaknesses. We humans use diagrams because we are good at recognizing symmetries and relationships in information so represented. Machines, in contrast, can keep track of gigabytes of information and carry out exhaustive computation where we are forced to rely on little tricks. If we are interested in differentiating good teaching practice from good programming practice, we have no choice but to relativize our theories of understanding to the particularities of the relevant class of agents. But this does not preclude the possibility that there is a substantial theory of mathematical understanding that can address issues that are common to both types of agent. Lightning fast calculation provides relatively little help when it comes to verifying common mathematical inferences; blind search does not work much better for computers than for humans. Nothing, *a priori*, rules out that a general philosophical theory can help explain what makes it possible for either type of agent to understand a mathematical proof.

In the next four sections, I will consider four types of inference that are commonly carried out in mathematical proofs, and which, on closer inspection, are not as straightforward as they seem. In each case, I will indicate some of the methods that have been developed to verify such inferences automatically, and explore what these methods tell us about the mechanisms by which such proofs are understood.

12.6 Understanding inequalities

I will start by considering inferences that are used to establish inequalities in ordered number domains, like the integers or the real numbers. For centuries, mathematics was viewed as the science of magnitude, and judgments as to the relative magnitude of various types of quantities still play a key role in the subject. In a sense, methods of deriving *equalities* are better developed, and computer algebra systems carry out symbolic calculations quite effectively. This is not to say that issues regarding equality are trivial; the task of determining whether two terms are equal, in an axiomatic theory or in an intended interpretation, is often difficult or even algorithmically undecidable. One strategy for determining whether an equality holds is to find ways of putting terms into canonical 'normal forms', so that an assertion $s = t$ is then valid if and only if s and t have the same normal form. There is an elaborate theory of 'rewrite systems' for simplifying terms and verifying equalities in this way, but we will not consider this here.

Let Γ be a set of equalities and inequalities between, say, real or integer-valued expressions, involving variables $x_1, \ldots, x_n$. Asking whether an inequality of the form $s < t$ is a consequence of Γ is the same as asking whether Γ together with the hypothesis $t \leq s$ is *un*satisfiable. Here, too, there is a well-developed body of research, which provides methods of determining whether such a system of equations is satisfiable, and finding a solution if it is. This research falls under the heading 'constraint programming', or, more specifically, 'linear programming', 'nonlinear programming', 'integer programming', and so on. In automated reasoning, such tasks typically arise in connection to scheduling and planning problems, and heuristic methods have been developed to deal with complex systems involving hundreds of constraints.

But the task of verifying the entailments that arise in ordinary mathematical reasoning has a different character. To start with, the emphasis is on finding a proof that an entailment is valid, rather than finding a counterexample. Often the inequality is tight, which means that conservative methods of approximation will not work; and the structure of the terms is often more elaborate than those that arise in industrial applications. On the other hand, the inference may involve only a few hypotheses, in the presence of suitable background knowledge. So the problems are generally smaller, if structurally more complex.

Let us consider two examples of proofs involving inequalities. The first comes from a branch of combinatorics known as *Ramsey theory*. An (undirected) *graph* consists of a set of vertices and a set of edges between them, barring 'loops', i.e. edges from a vertex to itself. The *complete graph on n vertices* is the graph in which between any two vertices there is an edge. Imagine coloring every edge of a complete graph either red or blue. A collection of k points is said to be *homogeneous* for the coloring if either all the edges between the points are red, or all of them are blue. A remarkable result due to F. P. Ramsey is that for any value of k, there is an n large enough, so that no matter how one colors the the complete graph of n vertices edges red and blue, there is a homogeneous subset of size k.

This raises the difficult problem of determining, for a given value of k, how large n has to be. To show that, for a given k, a value of n is not large enough means showing that there is a graph of size n with no homogeneous subset of size k. Paul Erdös pioneered a method, called *the probabilistic method*, for providing such lower bounds: one imagines that a coloring is chosen at random from among all colorings of the complete graph on n vertices, and then one shows that with nonzero probability, the graph will have no homogeneous subset of size k.

Theorem 1. *For all $k \geq 2$, if $n < 2^{k/2}$, there is a coloring of the complete graph on n vertices with no homogeneous subset of size k.*

Proof. For $k = 2$ this is trivial, and for $k = 3$ this is easily verified by hand. So we can assume $k \geq 4$.

Suppose $n < 2^{k/2}$, and consider all red–blue colorings, where we color each edge independently red or blue with probability $1/2$. Thus all colorings are equally likely with probability $2^{-\binom{n}{2}}$. Let A be a set of vertices of size k. The probability of the event A_R that the edges in A are all colored red is then $2^{-\binom{k}{2}}$. Hence it follows that the probability p_R for some k-set to be colored all red is bounded by

$$p_R = \text{Prob} \bigcup_{|A|=k} A_R \leq \sum_{|A|=k} \text{Prob}(A_R) = \binom{n}{k} 2^{-\binom{k}{2}}.$$

Now for $k \geq 2$, we have $\binom{n}{k} = \frac{n(n-1)(n-2)\cdots(n-k+1)}{k(k-1)\cdots 1} \leq \frac{n^k}{2^{k-1}}$. So for $n < 2^{k/2}$ and $k \geq 4$, we have

$$\binom{n}{k} 2^{-\binom{k}{2}} \leq \frac{n^k}{2^{k-1}} 2^{-\binom{k}{2}} < 2^{\frac{k^2}{2} - \binom{k}{2} - k + 1} = 2^{-\frac{k}{2}+1} \leq 1/2.$$

Since $p_R < 1/2$, and by symmetry $p_B < 1/2$ for the probability of some k vertices with all edges between them colored blue, we conclude that $p_R + p_B < 1$ for $n < 2^{\frac{k}{2}}$, so there *must* be a coloring with no red or blue homogeneous subset of size k. □

The text of this proof has been reproduced with only minor modifications from Aigner and Ziegler's *Proofs from the Book* (2001). (For sharper bounds and more information see Graham et al., 1994.) To make sense of the proof, remember that $\binom{n}{k} = \frac{n!}{k!(n-k)!}$ is the number of ways of choosing a subset of k elements from a set of n objects. The details of the argument are not so important; I am specifically interested in the chain of inequalities. You may wish to pause here to reflect on what it would take to verify these inferences axiomatically.

Before discussing that issue, let us consider a second example. The following fact comes up in a discussion of the Γ function in Whittaker and Watson (1996).

Lemma 1. *For every complex number z, the series $\sum_{n=1}^{\infty} |\log(1 + \frac{z}{n}) - \frac{z}{n}|$ converges.*[3]

Proof. It suffices to show that for some N, the sum $\sum_{n=N+1}^{\infty} |\log(1 + \frac{z}{n}) - \frac{z}{n}|$ is bounded by a convergent series. But when N is an integer such that $|z| \leq \frac{1}{2}N$, we have, if $n > N$,

$$\left|\log\left(1 + \frac{z}{n}\right) - \frac{z}{n}\right| = \left|-\frac{1}{2}\frac{z^2}{n^2} + \frac{1}{3}\frac{z^3}{n^3} - \cdots\right|$$

$$\leq \frac{|z|^2}{n^2}\left\{1 + |\frac{z}{n}| + |\frac{z^2}{n^2}| + \cdots\right\}$$

$$\leq \frac{1}{4}\frac{N^2}{n^2}\left\{1 + \frac{1}{2} + \frac{1}{2^2} + \cdots\right\}$$

$$\leq \frac{1}{2}\frac{N^2}{n^2}.$$

Since the series $\sum_{n=N+1}^{\infty}\{N^2/(2n^2)\}$ converges, we have the desired conclusion. □

Once again, the text is only slightly modified from that of Whittaker and Watson (1996), and the chain of inequalities is exactly the same. The equality involves a Taylor series expansion of the logarithm. The first inequality follows from properties of the absolute value, such as $|xy| = |x| \cdot |y|$ and $|x + y| \leq |x| + |y|$, while the second inequality makes use of the assumption $|z| < \frac{1}{2}N$ and the fact that $n > N$.

Inequalities like these arise in all branches of mathematics, and each discipline has its own bag of tricks for bounding the expressions that arise (see, for example Hardy et al., 1988; Steele, 2004). I have chosen the two examples above because they invoke only basic arithmetic reasoning, against some general background knowledge. At present, providing enough detail for a computer to verify even straightforward inferences like these is a burdensome chore (see Avigad et al., 2007).

In sufficiently restricted contexts, in fact, decision procedures are often available. For example, in 1929 Presburger showed that the theory of the

---

[3] Here we are taking the principal value of $\log(1 + \frac{z}{n})$.

integers in a language with 0, 1, +, and < is decidable, and in the early 1930s, Tarski showed that the theory of the real numbers in a language with 0, 1, +, ×, and < is decidable (the result was not published, though, until 1948, and was soon reprinted as Tarski, 1951). These decision procedures work for the full first-order language, not just the quantifier-free fragments, and have been implemented in a number of systems. But decision procedures do not tell the whole story. Thanks to Gödel, we know that decidability is the exception rather than the rule; for example, the theory of the integers becomes undecidable in the presence of multiplication, and the theory of the reals becomes undecidable in the presence of, say, the sine function. Even in restricted settings where problems are decidable, full decision procedures tend to be inefficient or even infeasible. In fact, for the types of inferences that arise in practice, short justifications are often possible where general decision procedures follow a more circuitous route.

There has therefore been a fair amount of interest in 'heuristic procedures', which search for proofs by applying a battery of natural inferences in a systematic way (for some examples in the case of real arithmetic, see Beeson, 1998; Hunt et al., 2003; Tiwari, 2003). One strategy is simply to work backwards from the goal inequality. For example, one can prove an inequality of the form $1/(1+s) < 1/(1+t)$ by proving $1+s > 1+t > 0$, and one can do that, in turn, by proving $s > t > -1$. This amounts to using basic rules like

$$x > y, y > 0 \Rightarrow 1/x < 1/y$$

to work backwards from the goal, a technique known as 'backchaining'.

Backchaining has its problems, however. For example, one can also prove $1/x < 1/y$ by proving that x and y are negative and $x > y$, or that x is negative and y is positive. Trying all the possibilities can result in dreaded case splits. Search procedures can be nondeterministic in more dramatic ways; for example, one can prove $q + r < s + t + u$, say, by proving $q < t$ and $r \leq s + u$, or by proving $q < s + t + u$ and $r \leq 0$. Similarly, one can prove $s + t < 5 + u$ by proving, say, $s < 3$ and $t \leq 2 + u$.

So, simply working backwards is insufficient on its own. In the proof of Theorem 1, we used bounds on $\binom{N}{k}$ and N to bound composite expressions using these terms, and in the proof of Lemma 1, we used the bounds $|z| \leq N/2$ and $N < n$ to bound the terms $|z^i/n^i|$. Working forwards in this way to amass a store of potentially useful inequalities is also a good idea, especially when the facts may be used more than once. For example, when all the terms in sight are positive, noting this once and for all can cut down the possibilities for backwards search dramatically. But, of course, deriving inequalities blindly won't help in general. The choice of facts we derive must somehow be driven by the context and the goal.

A final observation is that often inferences involving inequalities are verified by combining methods from more restricted domains in which it is clear how one should proceed. For example, the additive theory of the reals is well understood, and the multiplicative fragment is not much more complicated; more complex inferences are often obtained simply by combining these modes of reasoning. There is a body of methods stemming from a seminal paper by Nelson and Oppen (1979) that work by combining such 'local' decision procedures in a principled way.

These three strategies—backward-driven search, forward-driven search, and combining methods for handling more tractable problems—are fundamental to automated reasoning. When it comes to real-valued inequalities, an analysis of the problem in these terms can be found in Avigad and Friedman (2006). But what we really need is a general theory of how these strategies can be combined successfully to capture common mathematical inferences. To some extent, it will be impossible to avoid the objection that the methods we devise are merely '*ad hoc* and heuristic'; in real life, we use a considerable amount of mucking around to get by. But in so far as structural features of the mathematics that we care about make it possible to proceed in principled and effective ways, we should identify these structural features. In particular, they are an important part of how we understand proofs involving inequalities.

12.7 Understanding algebraic reasoning

The second type of inference I would like to consider is that which makes use of algebraic concepts. The use of such concepts is a hallmark of modern mathematics, and the following pattern of development is typical. Initially, systems of objects arising in various mathematical domains of interest are seen to share a common structure. Abstracting from the particular examples, one then focuses on this common structure, and determines the properties that hold of all systems that instantiate it. This infrastructure is then applied in new situations. First, one 'recognizes' an algebraic structure in a domain of interest, and then one instantiates facts, procedures, and methods that have been developed in the general setting to the case at hand. Thus algebraic reasoning involves complementary processes of abstraction and instantiation. We will consider an early and important example of such reasoning, namely the use of the concept of a group to establish a proposition in elementary number theory.

A group consists of a set, G, an associative binary operation, $\cdot$, on G, and an identity element, 1, satisfying $1 \cdot a = a \cdot 1 = a$ for every a in G. In a group, every element a is assumed to have an *inverse*, a^{-1}, satisfying $a \cdot a^{-1} = a^{-1} \cdot a = 1$. It is common to use the letter G to refer to both the group and the underlying set of elements, even though this notation is ambiguous. We will also write ab instead of $a \cdot b$. Exponentiation a^n is defined to be the the product $a \cdot a \cdots a$ of n copies of a, and if S is any finite set, $|S|$ denotes the number of elements in S.

Proposition 1. *Let G be a finite group and let a be any element of G. Then $a^{|G|} = 1$.*

This proposition is an easy consequence of a theorem known as *Lagrange's theorem*, but there is an even shorter and more direct proof when the group is *abelian*, that is, when the operation satisfies $ab = ba$ for every a and b.

Proof of Proposition 1 when G is abelian. Given G and a, consider the operation $f_a(b) = ab$, that is, multiplication by a on the left. This operation is injective, since $ab = ab'$ implies $b = a^{-1}ab = a^{-1}ab' = b'$, and surjective, since for any b we can write $b = a(a^{-1}b)$. So f_a is a bijection from G to G.

Suppose the elements of G are $b_1, \ldots, b_n$. By the preceding paragraph, multiplication by a simply permutes these elements, so $ab_1, \ldots, ab_n$ also enumerates the elements of G. The product of these elements is therefore equal to both

$$(ab_1)(ab_2) \cdots (ab_n) = a^n(b_1 b_2 \cdots b_n),$$

and $b_1 b_2 \cdots b_n$. This implies $a^n = 1$, as required. □

To apply Proposition 1 to number theory, we need only find a suitable group. Write $a \equiv b(m)$, and say that a and b are *congruent modulo m*, if a and b leave the same remainder on division by m, that is, if m divides $a - b$. The relationship of being congruent modulo m is symmetric, reflexive, and transitive. It also respects addition and multiplication: if $a \equiv a'(m)$ and $b \equiv b'(m)$, then $(a + a') \equiv (b + b')(m)$ and $aa' \equiv bb'(m)$. Any integer, a, is congruent to a number between 0 and $m - 1$ modulo m, namely, its remainder, or 'residue', modulo m. Thus congruence modulo m divides all the integers into m 'equivalence classes', each represented by the corresponding residue. We can think of addition and multiplication as operations on these equivalence classes, or, alternatively, as operations on residues, where after each operation we take the remainder modulo m. (For example, under clock arithmetic, we only care about the values $0, \ldots, 11$, and we are adding modulo 12 when we say that '5 hours after 9 o'clock, it will be 2 o'clock'.)

Two integers a and m are said to be *relatively prime* if they have no factors in common other than ± 1. If two numbers a and b are relatively prime to m, then so is their product, ab. Also, if a and m are relatively prime, the Euclidean algorithm tells us that there are integers x and y such that $ax + my = 1$. In the language of congruences, this says that if a and m are relatively prime, there is an x such that $ax \equiv 1(m)$.

What this means is that the residues that are relatively prime to m form a group under the operation of multiplication modulo m. For $m \geq 2$, *Euler's φ function* is defined by setting $\varphi(m)$ equal to the number of these equivalence classes, that is, the number of integers between 0 and $m-1$ that are relatively prime to m. The following theorem, known as *Euler's theorem*, is then an immediate consequence of Proposition 1.

Theorem 2. *For any $m \geq 2$ and any number, a, relatively prime to m, $a^{\varphi(m)} \equiv 1(m)$.*

In particular, if p is prime, then $1, \ldots, p-1$ are all relatively prime to p, and $\varphi(p) = p - 1$. This special case of Euler's theorem is known as Fermat's little theorem:

Corollary 1. *If p is prime and does not divide a, then $a^{p-1} \equiv 1(p)$.*

Neither Euler's theorem nor Fermat's little theorem has anything to do with groups *per se*. Nonetheless, these theorems are usually understood as reflections of the underlying algebraic structure on residues. In fact, Euler published the first proof of Fermat's little theorem in 1736, and a proof of the more general Theorem 2 in 1760, well before the first axiomatic definition of a group was given by Cayley in 1854. Chapter III of Dickson's three-volume *History of the Theory of Numbers* (1966) enumerates dozens of proofs of Fermat's and Euler's theorems; see also Wussing (1984) for a discussion of early algebraic proofs.

A comparison of some of the various proofs can be used to illustrate the advantages of the algebraic method. The most commonly cited advantage is, of course, generality: a single abstract theorem about groups has consequences everywhere a group can be identified, and it is more efficient to carry out a general argument once than to have to repeat it in each specific setting. But this does not tell the whole story; for example, algebraic methods are often deemed useful in their *initial* application, before they are applied in other settings. Sometimes the benefits are terminological and notational: group theory gives us convenient methods of calculating and manipulating expressions where other authors have to resort to cumbersome locutions and manners of expression. But algebraic abstraction also has a way of focusing our efforts by suppressing distracting information that is irrelevant to the solution of a problem. For example, in the last step of the argument above, it suffices to know that $b_1 b_2 \cdots b_n$ has a multiplicative inverse modulo m; in other words, that there is number c such that $cb_1 b_2 \cdots b_n$ is equal to 1 modulo m. Recognizing that fact eliminates the need to clutter the proof with calculations that produce the particular c. Finally, an important feature of algebraic methodology is that it enables us to discover notions that are likely to be fruitful elsewhere. It provides a uniform way of 'seeing' analogies in otherwise disparate settings. Echoing Wittgenstein, algebraic concepts 'lead us to make investigations; are the expressions of our interest, and direct our interest'.

From a traditional logical perspective, algebraic reasoning is easily explained. Proposition 1 makes a universal assertion about groups, giving it the logical form $\forall G\ (Group(G) \to \ldots)$. Later, when we have defined a particular object G and shown that it is a group, applying the proposition requires nothing more than the logical rules of universal instantiation and modus ponens.

But somehow, when we read a proof, we are not conscious of this continual specialization. Once we recognize that we are dealing with a group, facts about groups are suddenly ready to hand. We know how to simplify terms, and what properties are potentially relevant. We are suddenly able to think about the objects in terms of subgroups, orbits, and cosets; our group-theoretic understanding enables us to 'see' particular consequences of the abstract theory. The logical story does not have much to say about how this works. Nor does it have much to say about how we are able to reason in the abstract setting, and how this reasoning differs from that of the domain of application.

In contrast, developers of mechanized proof assistants have invested a good deal of effort in understanding how these inferences work. In formal verification, the notion of a collection of facts and procedures that are 'ready to hand' is sometimes called a 'context'. Proof assistants provide various methods of reasoning within such a context; Isabelle implements the notion of a 'locale' (Ballarin, 2006), while Coq supports a system of 'modules' (Bertot and Castéran, 2004). Here is a localic proof of Proposition 1, which is understood by Isabelle:

lemma (in *comm-group*) *power-order-eq-one:*
 assumes *finite (carrier G)* **and** *a:carrier G*
 shows *a* ($^\wedge$) *card(carrier G) = one G*
proof−
 have ($\bigotimes$ *x:carrier G. x*) = ($\bigotimes$ *x:carrier G. a ⊗ x*)
  by (*subst (2) finprod-reindex* [*symmetric*],
   auto simp add: Pi-def inj-on-const-mult surj-const-mult prems)
 also have ... =($\bigotimes$ *x:carrier G. a*) ⊗ ($\bigotimes$ *x: carrier G. x*)
  by (*auto simp add: finprod-multf Pi-def prems*)
 also have ($\bigotimes$ *x:carrier G. a*) = *a* ($^\wedge$) *card(carrier G)*

>   **by** (*auto simp add: finprod-const prems*)
>   **finally show** *?thesis*
>   **by** (*auto simp add: prems*)
> **qed**

The notation (**in** *comm-group*) indicates that this is a lemma in the commutative group locale. The notation ($\wedge$) denotes exponentiation, the expression *carrier G* denotes the set underlying the group, and the notation $\bigotimes$ *x:carrier G* denotes the product over the elements of that set. With this lemma in place, here is a proof of Euler's theorem:

> **Theorem** *euler-theorem:*
>   **assumes** $m > 1$ **and** $zgcd(a, m) = 1$
>   **shows** $[a \wedge phi\ m = 1]\ (mod\ m)$
> **proof**–
>   **interpret** *comm-group* [*residue-mult-group m*]
>   **by** (*rule residue-group-comm-group*)
>   **have** (*a mod m*) {$\wedge$}*m* (*phi m*) = {1}*m*
>   **by** (*auto simp add: phi-def power-order-eq-one prems*)
>   **also have** (*a mod m*) {$\wedge$}*m* (*phi m*) = ($a \wedge phi\ m$) *mod m*
>   **by** (*rule res-pow-cong*)
>   **finally show** *?thesis*
>   **by** (*subst zcong-zmod-eq, insert prems auto simp add: res-one2-eq*)
> **qed**

The proof begins by noting that the residues modulo m from an abelian group. The notation {$\wedge$}*m* then denotes the operation of exponentiation modulo m on residues, and {1}*m* denotes the residue 1, viewed as the identity in the group. The proof then simply appeals to the preceding lemma, noting that exponentiating in the group is equivalent to applying ordinary integer exponentiation and then taking the remainder modulo m.

The relative simplicity of these proof scripts belies the work that has gone into making the locales work effectively. The system has to provide mechanisms for identifying certain theorems as theorems about groups; for specializing facts about groups, automatically, to the specific group at hand; and for keeping track of the calculational rules and basic facts that enable the automated tools to recognize straightforward inferences and calculations in both the general and particular settings. Furthermore, one needs mechanisms to manage locales and combine them effectively. For example, every abelian group is a group; the multiplicative units of any ring form a group; the collection of subgroups of a group has a lattice structure; the real numbers as an ordered field provide instances of an additive group, a multiplicative group, a field, a linear ordering, and so on. Thus complex bookkeeping is needed to keep track of the facts and procedures that are immediately available to us, in ordinary mathematics, when we recognize one algebraic structure as present in another.

The work that has been done is a start, but proof assistants are still a long way from being able to support algebraic reasoning as it is carried out in introductory textbooks. Thus, a good deal more effort is needed to determine what lies behind even the most straightforward algebraic inferences, and how we understand simple algebraic proofs.

12.8 Understanding Euclidean geometry

Our third example will take us from modern mathematics to the mathematics of the ancients. Over the centuries, the style of diagram-based argumentation of Euclid's *Elements* was held to be the paradigm of rigor, and presentations much like Euclid's are still used today to introduce students to the notion of proof. In the 19th century, however, Euclidean reasoning fell from grace. Diagrammatic reasoning came to be viewed as imperfect, lacking sufficient mathematical rigor, and relying on a faculty of intuition that has no place in mathematics. The role of diagrams in licensing inference was gradually reduced, and then, finally, eliminated from most mathematical texts. Axiomatizations by Pasch (1882) and Hilbert (1899), for example, were viewed as 'corrections' or improvements to the flawed methods of Euclid, filling in gaps that Euclid overlooked.

As Manders points out in his contributions to this volume, this view of Euclidean geometry belies the fact that Euclidean reasoning was a remarkably stable and robust practice for more than two thousand years. Sustained reflection on the *Elements* shows that there are implicit rules in play, and norms governing the use of diagrams that are just as determinate as the norms governing modern proof. Given the importance of diagrammatic reasoning not just to the history of mathematics but to geometric thought today, it is important to understand how such reasoning works.

In order to clarify the types of inferences that are licensed by a diagram, Manders distinguishes between 'exact' conditions, such as the claim that two segments are congruent, and 'coexact' conditions, such as the claim that two lines intersect, or that a point lies in a given region. Coexact conditions depend only on general topological features of the diagram, and are stable under perturbations of a diagram, whereas exact conditions are not. Manders observes that only coexact claims are ever inferred from a diagram in a Euclidean proof; text assertions are required to support an inference that results in an exact claim. Thus, diagrams serve as useful representations of certain types of coexact data, and these representations are used in regimented ways.

Observations like these have led Nathaniel Miller (2001) and John Mumma (2006) to develop formal systems for diagram-based reasoning that are more faithful reflections of Euclidean reasoning than Hilbert's and Pasch's axiomatizations. In both systems, diagrams and formulas together bear the burden of representing assertions, and precise rules govern the construction of diagrams and the inferences that can be drawn. In Miller's system, a diagram is a more abstract object, namely, a graph representing the topological data that is relevant to a Euclidean argument. In Mumma's system, a diagram consists of labeled geometric elements with coordinates on a grid of points.

Both systems face the problem of explaining how a diagrammatic proof can ensure generality. Suppose that to prove a general theorem about triangles, I begin by drawing a particular triangle, *ABC*. Aside from being imperfect, this triangle will have specific properties that play no role in the proof: it may, for example, be acute or obtuse, or (more or less) a right triangle. The problem to be addressed is how reasoning about this *particular* triangle can warrant conclusions about *all* triangles, even ones with different coexact properties.

Let us consider, for example, the second proposition in Book I of the *Elements*, which shows that if *BC* is any segment and *A* is any point, it is possible to construct a segment congruent to *BC* with an endpoint at *A*. Aside from the definitions, common notions, and postulates, the proof relies only

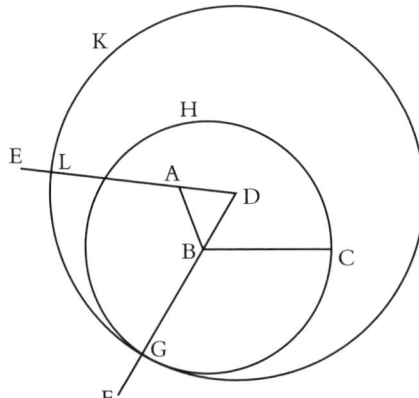

Fig. 12.1.

on the first proposition of Book I, which shows that it is possible to construct an equilateral triangle on a given side. The text below is taken from Heath's translation of Euclid. The example, and the gist of the subsequent discussion, are taken from Mumma (2006).

Proposition 2. *To place at a given point (as an extremity) a straight line equal to a given straight line.*

Proof. Let A be the given point, and BC the given straight line (Fig. 12.1). Thus it is required to place at the point A (as an extremity) a straight line equal to the given straight line BC.

From the point A to the point B, let the straight line AB be joined; and on it let the equilateral triangle DAB be constructed. Let the straight lines AE, BF be produced in a straight line with DA, DB; with centre B and distance BC let the circle CGH be described; and again, with centre D and distance DG let the circle GKL be described.

Then, since the point B is the centre of the circle CGH, BC is equal to BG. Again, since the point D is the centre of the circle GKL, DL is equal to DG. And in these DA is equal to DB; therefore the remainder AL is equal to the remainder BG. But BC was also proved equal to BG; therefore each of the straight lines AL, BC is equal to BG. And things that are equal to the same thing are also equal to one another; therefore AL is also equal to BC.

Therefore at the given point A the straight line AL is placed equal to the given straight line BC. (Being) what it was required to do. □

The conclusion of the proof is valid. But how does it work? As Mumma points out, the position of the point A with respect to BC is indeterminate, and different ways of placing that point would result in diagrams with different coexact properties. For example, if A were chosen so that AB is longer than BC, then the point A would lie *outside* the circle CGH. The diagram is used to license the conclusion that circle CGH intersects the line DF, and, in fact, this does hold in general. It is similarly used to license the conclusion that GKL will intersect the line GE. Moreover, to warrant the conclusion that AL is equal to BG, one needs to know that the point A lies between D and L, and that the point B lies between D and G. The question is, how can a particular diagram warrant these general conclusions?

In Miller's system, whenever a representational choice can affect the diagram in a way that may influence the inferences that can be drawn, one simply carries out a case split. Thus, in Miller's system, one needs to check the result against each topological configuration that can arise. But, as Mumma points out, this can yield a combinatorial explosion of possibilities, and such case splits are notably absent from the *Elements*. Euclid is, somehow, able to draw the right general consequences from a specific representation, without having to distinguish irrelevant cases.

Mumma (2006) provides a more subtle explanation of how this works, by providing rules by which certain features of a diagram can be justified as general consequences of the construction, irrespective of the choices made. Note that Mumma is not claiming to describe the procedure by which Euclid and his peers justified their inferences. All we have, in the *Elements*, is a body of text, with no record of the cognitive mechanisms that produced them. But these texts provide us with a collection of inferences (which, today, we recognize as being sound for a certain semantics). Understanding Euclidean proof means, in part, being able to determine which inferences are allowed. Mumma has explained what this understanding amounts to, both by characterizing the allowable inferences, and providing a mechanism by which they can be validated.

What does all this have to do with automated reasoning and formal verification? Geometric theorem proving has, for the most part, followed modern mathematics in utilizing algebraic treatments of the subject. In other words, most modern theorem provers express geometric assertions in algebraic terms and use algebraic methods to justify them. In fact, via the usual embedding of Euclidean geometry in the Cartesian plane, a decision procedures for real closed fields is a decision procedure for geometry as well. Moreover, specialized algebraic procedures (Chou *et al.*, 1994; Wu, 1994) have proved to be remarkably effective in verifying ordinary geometric theorems.

But analyses like Miller's and Mumma's may point the way to developing computational methods for verifying Euclidean inferences on their own terms. Not only would this be interesting in its own right, but it would provide important insights into the inner workings of modern methods as well. For example, in the 16th century, the geometric method of 'analysis and synthesis' foreshadowed Descartes' algebraic treatment of geometry (Bos, 2000). In fact, this method of naming an object sought and then analyzing its properties can be seen as a geometric version of the method of Skolemization and unification described in Section 12.5. For another example, note that we often prove theorems in analysis—say, when dealing with Hilbert spaces, which are infinite-dimensional generalizations of Euclidean geometry—by drawing a diagram. So, even in light of the attenuated role of diagrams in modern mathematical proofs, understanding the nature of geometric inference is still an important part of understanding how such modern proofs work.

12.9 Understanding numeric substructures

As a fourth and final example, I would like to consider a class of inferences that involve reasoning about the domains of natural numbers, integers, rational numbers, real numbers, and complex numbers, with respect to one another. From a foundational perspective, it is common to take the natural numbers as basic, or to construct them from even more basic objects, like sets. Integers can then be viewed as equivalence classes of pairs of natural numbers; rational numbers can be viewed as equivalence classes of pairs of integers; real numbers can be viewed, say, as sequences of rational numbers; and complex numbers can be viewed as pairs of reals. As McLarty points out in the first of his essays in this collection, however, this puts us in a funny situation. After all, we think of the natural numbers as a *subset* of the integers, the integers as a subset of the rationals, and so on. On the foundational story, this is not quite true; really, each domain is *embedded* in the larger ones. Of course, once we have the complex numbers, we may choose to baptize the complex image of our original copy of the natural numbers as our new working version. Even if we do this, however, we still have to keep track of where individual elements 'live'. For example, we can apply the principle of induction to the complex copy of the natural numbers, but not to the complex numbers as a whole; and we can divide one natural number by another, but the result may not be a natural number. If we think of the natural numbers as embedded in the complex numbers, we have to use an embedding function to make our statements literally true; if we think of them as being a subset of the complex numbers, all of our statements have to be carefully qualified so that we have specified what types of objects we are dealing with.

The funny thing is that in ordinary mathematical texts, all this happens under the surface. Textbooks almost never tell us which of these two foundational options are being followed, because the choice has no effect on the subsequent arguments. And when we read a theorem that combines the different domains, we somehow manage to interpret the statements in such a way that everything makes sense. For example, you probably did not think twice about the fact that Lemma 1 above involved three sorts of number domain. The fact that the series is indexed by n means that we have to think of n as a natural number (or a positive integer). Similarly, there is an implicit indexing of terms by natural numbers in the use of '...' in the expression for the logarithm. The proof, in fact, made use of properties of such sums that are typically proved by induction. In the statement of the theorem, the variable z is explicitly designated to denote a complex number, so when we divide z by n in the expression z/n, we are thinking of n as a complex number as well. But the absolute value function converts a complex value to a real value, so expressions like $|z/n|$ denote real numbers, and many of the products and sums in that proof denote real multiplication and addition. In the proof of the lemma, it is crucial that we keep track of this last fact: the $\leq$ ordering only makes sense on the real numbers, so to invoke properties of the ordering we need to know that the relevant expressions are real-valued.

The elaborate structure of implicit inferences comes to the fore when we try to formalize such reasoning. With a formal verification system, these inferences need to be spelled out in detail, and the result can be painstaking and tedious (see Avigad *et al.*, 2007). In Isabelle, for example, one has to use a function *real(n)* to cast a natural number, n, as a real. Coq has mechanisms that apply such 'coercions' automatically, but, in both cases, the appropriate simplifications and verifications are rarely carried out as automatically as one would like.

The fact that methods of reasoning that we are barely conscious of when we read a mathematical proof requires so much effort to formalize is one of the scandals of formal verification, and a clear sign that more thought is needed as to how we understand such inferences. I suspect that the structural perspective described in McLarty's article, combined with the locale mechanisms described in Section 12.7, holds the germ of a solution. When we prove theorems about the natural numbers, that is the structure of interest; but when we identify the natural numbers as a substructure of the reals, we are working in an expanded locale, where both structures are present and interact. To start with, everything we know in the context of the natural numbers and the real numbers individually is imported to the combined locale, and is therefore available to us. But there are also new facts and procedures that govern the combination of the two domains. Figuring out how to model such an understanding so that proof assistants can verify the proof of Lemma 1, as it stands, will go a long way in explaining how we understand proofs that make use of mixed domains.

12.10 Conclusions

I have described four types of inference that are found in ordinary mathematical proofs, and considered some of the logical and computational methods that have been developed to verify them. I have argued that these efforts are not just pragmatic solutions to problems of engineering; rather, they address core issues in the epistemology of mathematics, and should be supported by broader philosophical reflection.

Mathematics guides our thought in deep and powerful ways, and deserves a philosophy that recognizes that fact. When we focus on particular features of mathematical practice, metaphysical concerns often seem petty and irrelevant, and we find, instead, a rich assortment of issues that have concrete bearing upon what we do and say about the subject. Our task is to develop a conceptual framework in which we can fruitfully begin to address these issues, and to narrow our focus to the point where discernible progress can be made. I hope the present chapter serves as encouragement.

Added in Proof. After the article had been sent to the publisher, I came across work by Jody Azzouni which bears directly on many of the issues raised here. Although, I cannot explore the points of agreement and disagreement now, the reader may wish to compare my views to those of Azzouni (2005) and Azzoumi (2006).

Bibliography

AIGNER, Martin and ZIEGLER, Günter M. (2001), *Proofs from The Book*, 2nd edn (Berlin: Springer-Verlag).

AVIGAD, Jeremy (2006), 'Mathematical method and proof', *Synthese*, 153, 105–159.

AVIGAD, Jeremy, DONNELLY, Kevin, GRAY, David, and RAFF, Paul (2007) 'A formally verified proof of the prime number theorem', *ACM Transactions on Computational Logic*, 9(1:2).

AVIGAD, Jeremy and FRIEDMAN, Harvey (2006), 'Combining decision procedures for the reals', *Logical Methods in Computer Science*, 2(4:4).

AZZOUMI, Jody (2005), 'Is there a sense in which mathematics can have foundations?' In G. Sica (ed.), Essays in the Foundations of Mathematics and Logic, 9–47 (Monza: Polimetrica).

—— (2006), Tracking Reason: Proof, Consequence, and Truth (Oxford: Oxford University Press).

BALLARIN, Clemens (2006), 'Interpretation of locales in Isabelle: theories and proof contexts', in J. M. Borwein and W. M. Farmer (eds.), *Mathematical Knowledge Management: Proceedings of the Fifth International Conference, MKM 2006*, 31–43 (Berlin: Springer-Verlag).

BEESON, Michael (1998), 'Design principles of Mathpert: software to support education in algebra and calculus', in N. Kajler (ed.), *Computer—Human Interaction in Symbolic Computation*, 89–115 (Berlin: Springer-Verlag).

BERTOT, Yves and CASTÉRAN, Pierre (2004), *Interactive Theorem Proving and Program Development: Coq'art: the Calculus of Inductive Constructions* (Berlin: Springer-Verlag).

BOS, Henk J. M. (2000), *Redefining Geometrical Exactness: Descartes' Transformation of the Early Modern Concept of Construction* (New York: Springer-Verlag).

CARR, David (1979), 'The logic of knowing how and ability', *Mind*, 88, 394–409.
CHOU, C. C., GAO, X. S., and ZHANG, J. Z. (1994), *Machine Proofs in Geometry* (Singapore: World Scientific).
DICKSON, Leonard Eugene (1966), *History of the Theory of Numbers*, vols. I–III (New York: Chelsea Publishing Co.).
EUCLID, *The Thirteen Books of Euclid's Elements Translated from the Text of Heiberg*, trans. with introduction and commentary by Thomas L. Heath, 2nd edn, vols. I–III (New York: Dover Publications).
GOLDFARB, Warren (1985), 'Kripke on Wittgenstein on rules', *Journal of Philosophy*, 82, 471–488.
GRAHAM, Ronald L., KNUTH, Donald E., and PATASHNIK, Oren (1994), *Concrete Mathematics: a Foundation for Computer Science*, 2nd edn (Reading, MA: Addison-Wesley).
HARDY, G. H., LITTLEWOOD, J. E., and PÓLYA, G. (1988), *Inequalities* (Cambridge: Cambridge University Press), repr. of 1952 edn.
HILBERT, David (1899), 'Grundlagen der Geometrie', in *Festschrift zur Feier der Enthüllung des Gauss-Weber Denkmals in Göttingen* (Leipzig: Teubner). Trans. by Leo Unger (1971) as *Foundations of geometry* (La Salle: Open Court).
HUNT, Warren A., KRUG, Robert Bellarmine, and MOORE, J. (2003), 'Linear and nonlinear arithmetic in ACL2', in Daniel Geist and Enrico Tronci (eds.), *Correct Hardware Design and Verification Methods, Proceedings of CHARME 2003*, 319–333 (Berlin: Springer-Verlag).
KRIPKE, Saul (1982), *Wittgenstein on Rules and Private Language* (Cambridge, Mass.: Harvard University Press).
MILLER, Nathaniel (2001), *A Diagrammatic Formal System for Euclidean Geometry*, Ph.D. thesis, Cornell University.
MUMMA, John (2006), *Intuition Formalized: Ancient and Modern Methods of Proof in Elementary Geometry*, Ph.D. thesis, Carnegie Mellon University.
NELSON, Greg and OPPEN, Derek C. (1979), 'Simplification by cooperating decision procedures', *ACM Transactions of Programming Languages and Systems*, 1, 245–257.
PASCH, Moritz (1882), *Vorlesungen über neueren Geometrie* (Leipzig: Teubner).
POINCARÉ, Henri (1908), *Science et Méthode* (Paris: Flammarion). Trans. by Francis Maitland in 1914 as *Science and Method* (London: T. Nelson and Sons, 1914), republished in 2003 (Mineola, NY: Dover Publications).
RYLE, Gilbert (1949), *The Concept of Mind* (Chicago, IL: University of Chicago Press).
STEELE, J. Michael (2004), *The Cauchy–Schwarz Master Class: an Introduction to the Art of Mathematical Inequalities* (Washington, DC: Mathematical Association of America).
TAIT, William W. (1986), 'Wittgenstein and the "skeptical paradoxes"', *Journal of Philosophy*, 83, 475–488.
TARSKI, Alfred (1951), *A Decision Procedure for Elementary Algebra and Geometry*, 2nd edn (University of California Press). Reprinted in B. F. Caviness and J. R. Johnson (eds.) *Quantifier Elimination and Cylindrical Algebraic Decomposition*, 24–84 (Vienna: Springer-Verlag, 1988).

TIWARI, A. (2003), 'Abstraction based theorem proving: an example from the theory of reals', in C. Tinelli and S. Ranise (eds.), *Proceedings of the CADE-19 Workshop on Pragmatics of Decision Procedures in Automated Deduction, PDPAR 2003*, 40–52 (Berlin: Springer-Verlag).

WHITTAKER, E. T. and WATSON, G. N. (1996), *A Course of Modern Analysis*, (Cambridge: Cambridge University Press), reprinting of the 4th edn, 1927.

WIEDIJK, Freek (2006), *The Seventeen Provers of the World* (Berlin: Springer-Verlag).

WITTGENSTEIN, Ludwig (1953), *Philosophical Investigations* (New York: Macmillan), ed. and trans. G. E. M. Anscombe, 3rd edn (2001) (Oxford: Blackwell Publishers).

—— (1956), *Remarks on the Foundations of Mathematics*, ed. G. H. von Wright, R. Rhees and G. E. M. Anscombe, trans. G. E. M. Anscombe (Oxford: Blackwell). Revised edition (1978) (Cambridge, MA: MIT Press).

—— (1969), *On Certainty*, ed. G. E. M. Anscombe and G. H. von Wright, trans. G. E. M. Anscombe and D. Paul (Oxford: Blackwell).

WU, Wen Tsün (1994), *Mechanical Theorem Proving in Geometries*, trans. from the 1984 Chinese original by Xiao Fan Jin and Dong Ming Wang (Vienna: Springer-Verlag).

WUSSING, Hans (1984), *The Genesis of the Abstract Group Concept: A Contribution to the History of the Origin of Abstract Group Theory*, trans. Abe Shenitzer and Hardy Grant (Cambridge, MA: MIT Press).

13

What Structuralism Achieves

COLIN MCLARTY

> For some time mathematicians have emphasized that mathematics is concerned with structures involving mathematical objects and not with the 'internal' nature of the objects themselves. They have recognized that we are not given mathematical objects in isolation but rather in structures.
>
> (Resnik 1981, p. 529).

Resnik does not say mathematicians *could* deal with structures rather than the 'internal' nature of objects. He says they *do* and he is right. The basic idea goes back to the first extant mathematics textbook, Euclid's *Elements*. The lines and vertices in a Euclidean diagram have specified relations to each other—and no meaning at all outside that diagram. His 'definitions' of points and lines are famously irrelevant to his proofs, and it is senseless to ask where the vertex A of a triangle ABC in one diagram is located in space or to ask whether it is equal to or distinct from vertex A of a square $ABCD$ in another diagram.[1] Modern mathematics deals much more extensively, and more explicitly, with structures where 'the "elements" of the structure have no properties other than those relating them to other "elements" of the same structure' (Benacerraf, 1965, p. 70). While someone might try to stipulate that the number 2, for example, *is* the Zermelo–Fraenkel set $\{\{\phi\}\}$ or the set $\{\phi, \{\phi\}\}$, such an identification is irrelevant for the current practice of mathematics. So Benacerraf asked for a philosophical account that would define the natural numbers 0, 1, 2, ... *structurally*, by their arithmetic relations to one another, without saying what any one of them *is*.[2]

---

[1] If the points were distinct, then by postulate a unique line would join them; but a line between two diagrams is senseless in Euclid's practice.

[2] He did not consider the answers by mathematicians Dedekind (1888) and Lawvere (1964).

Philosophers have pursued their own versions of structural mathematics without always specifying which of their ideas aim at purely philosophical concerns and which are meant to explicate existing practice.[3] Yet even the purely philosophical projects can benefit from seeing how mathematicians make structure work. The philosophical efforts face problems that were solved long ago by mathematicians who rely daily on the answers: How can separate structures relate to one another? Is there a coherent way to define structures as themselves places in larger structures? Since Emmy Noether, in fact, mathematicians routinely describe structures entirely by the structural relations among them. This indeed gives structural definitions such as Benacerraf sought but that is incidental to the goals of practice. In practice, structural mathematics achieves the discovery and proof of great theorems.

We might worry that mathematicians are pragmatists, not concerned with our issues, and so mathematical practice might be a poor guide for philosophers in general and structuralists in particular:

- Mathematics might lack the conceptual rigor of philosophy.
- Practice might violate some favored ontology or epistemology.
- Mathematics might be structural in principle yet not in practice.

But these concerns underestimate the practical need for rigorous principles. For example Andrew Wiles proved Fermat's Last Theorem as a step in the Langlands Program, the largest project ever posed in mathematics. He cites scores of theorems so advanced that he cannot assume an audience of experts knows them or knows offhand where to find them. In mathematics on this scale it is completely out of the question for one author or even a small number of authors to give a major proof in full starting from just concepts and theorems found in graduate textbooks.[4]

So there are feasibility conditions:

- Each theorem must be concise and fully explicit, so as to be easily and rigorously detachable from its original context. Proofs are often very long, and for just this reason the statement of each theorem must be brief and rigorously self-contained.
- Theorems must be stated structurally. If, *per impossibile*, authors could be made to choose set theoretic definitions of each structure they use,

---

[3] Chihara seeks ways mathematics *could* be 'nominalist' although it is not in practice (2004, p. 65). Hellman (1989) and Shapiro (1997) say little relating their ideas of modality, coherence, and structure to practice. Resnik (1997), Awodey (2004), and Carter (2005) offer explications.

[4] A book aimed at readers already specialized in number theory takes twenty-two authors and 570 pages not to prove but merely to introduce the results that Wiles uses (Cornell *et al.*, 1997).

then they could not use each other's results without also verifying the equivalence of their chosen definitions. Any major paper uses scores of structures, each of which has many different set theoretic reductions (compare *homology* in Section 13.3.1). Checking so many equivalences is infeasible. Leaving them implicit would court disaster. A set theoretic specification may be used in some step of a proof but those details must be rigorously irrelevant to the statement of the theorem (see discussion of Theorem 4).

- Methods must handle structures at every level from natural numbers to functors by comparable means that readily relate any two levels.

These methods are used in practice, by necessity, with deductive rigor, even if they violate some philosophical theories.[5] Practice also dispels the idea that structuralism abstracts away from intuitive content. Structuralist tools give the most direct known path from 'pure' content to rigor.

Section 13.1 extends Resnik's structuralism by the standard practice of identifying some structures as parts of others. Certain injections $\mathbb{S}_1 \rightarrowtail \mathbb{S}_2$ of one structure $\mathbb{S}_1$ into another $\mathbb{S}_2$ are taken as *identity preserving* and thus as making $\mathbb{S}_1$ a part of $\mathbb{S}_2$. We use textbook treatments of the real and complex numbers to argue that such identity is not defined by any logical principles but by stipulation or tradition. This opens up a philosophical topic of explaining how particular cases come to be accepted as identity preserving. Section 13.1.2 argues that Shapiro's distinction of systems and structures does not help to understand current practice.

Section 13.2 pursues the original point of structuralist methods—defining structures as themselves places in patterns of structures rather the way that Resnik describes in his later chapters. It takes polynomials as an example and discusses foundations. The ontology so far as it goes is far from philosophically classical: The Leibniz law of identity fails as many individuals lack fully individuating properties. Nothing here prevents philosophers from going beyond practice and attributing a more classical ontology to mathematical objects. We know that mathematics *can* be re-interpreted in various ways to give it a classical ontology. But a further task faces anyone who claims that only classical ontology is conceptually sound: Either show that contrary to the appearances practice concretely does recognize such ontology; or explain how conceptually unsound practice can succeed so well. Whatever position one takes in debates over foundations and 'structuralism,' it remains

---

[5] See related arguments from practice against Quine's ontology in Maddy (2005, p. 450). Compare Krömer (2007, p. xv) on the way category theory became standard mathematics despite some clash with official ZF foundations.

that the tools used in textbooks and research are more specific than most philosophical proposals and are daily tested for rigor in principle and feasibility in practice.

Section 13.3 gives sample achievements and looks at intuition, purity, and unification. While some philosophical structuralisms may make any unity of mathematics unintelligible, structuralist practice makes that unity ever more productive.

13.1 Beginning structuralism

13.1.1 Separate structures

Following the imagery of Euclid's *Elements*, Resnik calls mathematical objects *positions* in *patterns* and says they have no properties except relations to others in the same pattern. But this 'threatens to conflict with mathematical practice' (MacBride, 2005 p. 568). If, for example, real numbers have only relations to each other, how do they gain relations to complex numbers? The answer is given in practice.

Innumerable textbooks say something like:

> We suppose you understand the real numbers! The complex numbers are formal expressions $x_0 + x_1 i$ with x_0, x_1 real, combined by
>
> $$(x_0 + x_1 i) + (y_0 + y_1 i) = (x_0 + y_0) + (x_1 + y_1)i$$
>
> $$(x_0 + x_1 i)(y_0 + y_1 i) = (x_0 y_0 - x_1 y_1) + (x_0 y_1 + x_1 y_0)i$$
>
> (Conway and Smith, 2003, p. 1)

If Conway and Smith were Zermelo–Fraenkel set theorists this would mean they do not embed the real numbers $\mathbb{R}$ in the complex numbers $\mathbb{C}$. A real number defined in ZF is not a formal expression made from two real numbers and a letter i. But Conway and Smith are not set theorists. They freely equate each real number x with the complex number $x + 0i$. Virtually all mathematicians do. This extends to all uses of the numbers, so real polynomials are considered a kind of complex polynomial as mentioned in (Carter, 2005, p. 298). Elementary textbooks typically define complex numbers in some way which would imply in ZF set theory that none are real numbers, then add explicitly: 'we identify $\mathbb{R}$ with its image in $\mathbb{C}$' (Lang, 2005, p. 347). For some philosophers 'the notion of "identity by fiat" makes dubious sense' (MacBride, 2005, p. 578). But mathematics thrives on violating this dictum.

To put it generally, positions in one pattern or structure $\mathbb{S}_1$ can be compared to those in another $\mathbb{S}_2$ by way of a function $f : \mathbb{S}_1 \to \mathbb{S}_2$.[6] If f is one-to-one it is an *injection* and we say it *identifies* each position $x \in \mathbb{S}_1$ with the value $f(x) \in \mathbb{S}_2$. This is a three place relation: x is identified with y via f, or as an equation $f(x) = y$. The equation lies squarely in $\mathbb{S}_2$ as $f(x)$ and y are in $\mathbb{S}_2$. Some important injections, though, are taken as *identity preserving*. When an element $x \in \mathbb{S}_1$ is identified with an element $y \in \mathbb{S}_2$ via an identity-preserving injection $\mathbb{S}_1 \rightarrowtail \mathbb{S}_2$ then mathematicians will equate x with y. They state as a two place relation $x = y$ and in this way they *embed* $\mathbb{S}_1$ in $\mathbb{S}_2$. There is no definition of *identity preserving*. Mathematicians stipulate which injections are identity preserving only taking care never to stipulate two distinct identity-preserving injections between the same patterns. Nearly everyone takes $\mathbb{R} \rightarrowtail \mathbb{C}$ as identity preserving.

As a contrary example, mathematicians generally do not identify the complex number $x_0 + x_1 i$ with the quaternion

$$x_0 + x_1 \mathrm{i} + 0\mathrm{j} + 0\mathrm{k}$$

(Conway and Smith, 2003)

One reason is that we could just as well identify it with

$$x_0 + 0\mathrm{i} + x_1\mathrm{j} + 0\mathrm{k} \quad \text{or} \quad x_0 + 0\mathrm{i} + 0\mathrm{j} + x_1\mathrm{k}$$

or infinitely many other quaternions. All these injections preserve the usual structure and are very useful and so no one is taken as identity preserving. In some contexts the software package *Mathematica*® does identify $x_0 + x_1 i$ with $x_0 + x_1 \mathrm{i} + 0\mathrm{j} + 0\mathrm{k}$. So does the subject of *crossed modules* in group representation theory (Mac Lane, 2005, pp. 95f.). It is a matter of convention. Someday mathematicians may generally identify the complex numbers with some quaternions. It will be decided by the demands of practice and not by principles of set theory.

Mathematicians do not use the phrase *identity preserving*, Rather, they say for example that 'it is customary to identify' certain elements of many different algebras with 'the integers' or the rational numbers, real numbers, etc., quite apart from set theoretic definitions (Lang, 1993, p. 90). Definitions are rarely given in Zermelo–Fraenkel set theoretic terms and when such terms are given they usually conflict with this custom. Yet the custom takes priority as with the complex numbers. This obviates an historical problem MacBride sees for some structuralists (2005, p. 580). When two mathematicians posit 'the' natural

---

[6] See discussion on (MacBride, 2005, p. 573), but we do not only use isomorphisms.

numbers we need no criterion to tell if they posit the same ones. We can take that as stipulated.

13.1.2 Comparison with Resnik and Shapiro

Resnik tends to reject embeddings like $\mathbb{R} \hookrightarrow \mathbb{C}$, but he says we might justify this one by 'likening the historical development of the complex number system to the step by step construction of a complicated pattern through adding positions' to the real number system (Resnik, 1997, p. 215). He rightly offers this as simile and not fact and so it is close to our appeal to stipulation. Over the centuries some mathematicians took the complex numbers as including the reals, some took them as a different kind of thing, and some considered them not genuine things.[7] The history as a whole explains why today's textbooks identify $\mathbb{R}$ with a part of $\mathbb{C}$. But the history is ambivalent and not widely known. The textbook stipulations are clear, overt, and authoritative.

The textbook treatment of complex numbers might be analyzed into two steps based on Shapiro's distinction of *system* and *structure*: First a ZF definition produces a *system* $\mathbb{C}$ of complex numbers, where each complex number is some specified ZF set. Then abstraction forms a *structure* $\mathbb{C}$ of complex numbers 'ignoring any features of them' except their algebraic interrelations (Shapiro, 1997, p. 73).[8] But then which are Lang's complex numbers, or Conway and Smith's?

Lang defines a complex number as an ordered pair $\langle x_0, x_1 \rangle$ where Conway and Smith write $x_0 + x_1 i$ (Lang, 2005, p. 346). On the 'system' approach using ZF sets, Lang's equation $\langle x, 0 \rangle = x$ is literally false for any real number x. When Lang says to identify real numbers with the corresponding complex numbers, is he revising his definition of complex numbers in some unspecified way? Or does he mean to treat some false equations as true? There are ways to do both but Lang never gives one. Conway and Smith are even more ambiguous as there is no standard definition of 'formal expressions' in ZF. At any rate, if complex numbers are sets in a 'system' then 'structures' do not appear and at least for this quite typical case we would have to say mathematics is not structuralist in practice.

We can better understand complex numbers as places in a structure. Then '$x_0 + x_1 i$' and '$\langle x_0, x_1 \rangle$' are not names of different sets. They are merely different notations for the same position in the complex number structure. This explains why textbook authors routinely feel they can and must specify that they identify each real number x with the complex $x + 0i$. Structural

---

[7] See the expert essay (Mazur, 2003) and the detailed exposition (Flament, 2003).

[8] ZF is the relevant set theory here. In categorical set theory each set is itself both 'system' and 'structure' as its elements have no individuating features in the first place (McLarty, 1993).

definitions of real and complex numbers neither imply nor preclude any identity relation between them until we stipulate one.[9] But on this view there are no 'systems' in Lang, or in Conway and Smith, or nearly anywhere outside of textbooks on ZF set theory. It remains to understand structuralist practice in terms of 'structures' themselves.

13.2 Working structuralism

> Patterns themselves are *positionalized* by being identified with positions of another pattern, which allows us to obtain results about patterns which were not even previously statable. It is [this] sort of reduction which has significantly changed the practice of mathematics. (Resnik, 1997, p. 218).

One good example is the structure, specifically a *ring* $\mathbb{R}[X]$, of all real polynomials. Lang says there are 'several devices' for reducing polynomials to sets and suggests one for undergraduates and another for graduate students.[10] What does not change are the rules for adding, subtracting, and multiplying polynomials. To collect the real polynomials into one structure Lang turns to rings and ring morphisms. A *commutative ring* is a structure with 0,1, addition, subtraction, and multiplication following the familiar formal rules.[11] A *ring morphism* $f : A \to B$ is a function which preserves 0,1, and the ring operations:

$$f(0) = 0 \qquad f(x-y) = f(x) - f(y) \qquad \text{etc.}$$

We will say 'ring' to mean commutative ring. Lang characterizes $\mathbb{R}[X]$:

Fact 1. $\mathbb{R}[X]$ *is a ring with element* $X \in \mathbb{R}[X]$ *and a morphism* $c : \mathbb{R} \to \mathbb{R}[X]$ *called the insertion of constants.*[12] *For each ring morphism* $f : \mathbb{R} \to A$ *and element* $a \in A$, *there is a unique morphism* $u_a : \mathbb{R}[X] \to A$

with $u_a(X) = a$ and agreeing with f on constants—that is $u_a c = f$.

---

[9] Shapiro's notion of *offices* does not bear on embeddings like this since it only connects isomorphic structures (Shapiro, 1997, p. 82).

[10] The quote is in both Lang (2005, p. 106) and Lang (1993, p. 97).

[11] See Mac Lane (1986, p. 98) or any modern algebra textbook.

[12] Is a real number the same thing as a constant real polynomial? Lang identifies them, although a ZF reading of either of his definitions of polynomials says no.

This says nothing to specify what polynomials are beyond the paradigmatically uninformative assertion that one of them is called X. It does say $\mathbb{R}[X]$ is a ring, so there are other elements, like $X^2 = X \cdot X$, and $3X^2 + X$, described by their algebraic relation to X. Intuitively u_a evaluates these at $X = a$. That is, since u_a is a ring morphism:

$$u_a(X^2) = a^2 \quad \text{and} \quad u_a(3X^2 + X) = 3a^2 + a \quad \text{and so on.}$$

But nothing in Fact 1 specifies the elements of $\mathbb{R}[X]$ as sets. It specifies where $\mathbb{R}[X]$ fits in the pattern of rings and morphisms.

Lang states Fact 1 in italics without calling it a definition or theorem or anything else (Lang, 1993, p. 99). If pressed, he would probably call it a theorem. Indeed either one of his official definitions of polynomials can be made rigorous and used to prove the Fact. But Lang does not like using them in proofs. Fact 1 itself gives a rigorous definition of $\mathbb{R}[X]$ up to isomorphism which he does like to use.

Philosophers usually define an *isomorphism* of structured sets as a one-to-one onto function preserving structure in each direction, as in (Bourbaki, 1939). Mathematicians today use a simpler and more general definition, which however agrees with Bourbaki's in this case (Lang, 1993, p. 54). For the sake of generality it speaks of *morphisms* instead of *structure preserving functions*: Every structure $\mathbb{S}_1$ has an *identity morphism* $1_{\mathbb{S}_1} : \mathbb{S}_1 \to \mathbb{S}_1$. A morphism $f : \mathbb{S}_1 \to \mathbb{S}_2$ is an *isomorphism* if it has an *inverse*, that is a morphism $g : \mathbb{S}_2 \to \mathbb{S}_1$ which composes with f to give identities:

The need for generality will appear in Section 13.3.1 and the accompanying case study. For now, note how directly this concept of isomorphism suits Fact 1.

Theorem 3. *Suppose a ring $\mathbb{R}[X]$ and function c satisfy Fact 1, and so do another $\mathbb{R}[X']$ and c'. Then there is a unique ring isomorphism $u : \mathbb{R}[X] \to \mathbb{R}[X']$ such that $u(X) = X'$ and $uc = c'$.*

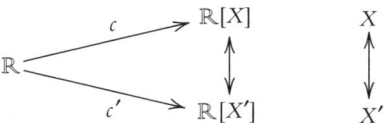

Proof. By assumption there are unique morphisms u with $u(X) = X'$ and $uc = c'$ and $u : \mathbb{R}[X'] \to \mathbb{R}[X]$ with $v(X') = X$ and $vc' = c$. The composite vu is a ring morphism with $vu(X) = X$. But the identity $1_{\mathbb{R}[X]}$ also takes X to X, so uniqueness implies $vu = 1_{\mathbb{R}[X]}$. Similarly $uv = 1_{\mathbb{R}[X']}$. □

Lang deduces everything he needs about $\mathbb{R}[X]$ from Fact 1 plus properties of the real numbers. A ZF set theoretic definition of 'polynomial' is rather beside the point—except that we still need to prove:

Theorem 4. *There are rings with the properties given in Fact 1.*

Here the foundations matter. This theorem can be proved by specifying a set $\mathbb{R}[X]$ with these properties, or from more general theorems. Either way, on ZF foundations, it will come down to specifying a set by specifying which sets are its elements. On ZF foundations to specify $\mathbb{R}[X]$ is to specify which sets are polynomials. Proofs in ZF cannot all be structural because the axioms are not structural. Proofs in categorical set theory are entirely structural. Categorical set theory specifies any set, including any candidate for $\mathbb{R}[X]$, only up to isomorphism entirely by the pattern of functions between it and other sets (McLarty, 2004).

Lang (1993) does not specify a set theory. In particular he nowhere says what functions *are*. He only says they take values. He calls them 'mappings' and his only account of them reads:

If $f : A \to B$ is a mapping of one set into another, we write

$$x \mapsto f(x)$$

to denote the effect of f on an element x of A. (Lang, 1993, p. ix)

ZF set theory formalizes this by taking elementhood as primitive and defining a function f as a suitable set of ordered pairs, and defining $f(x) = y$ to mean the ordered pair $\langle x, y \rangle$ is an element of the set f. Categorical set theory formalizes it by taking function composition as primitive and defining elements as suitable functions, and defining $f(x) = y$ to mean y is the composite of f and x. In Resnik's words this set theory treats functions 'as positions in the pattern generated by the composition relation' (1997, p. 218). Either approach will work for Lang.

The point is that structural practice isolates foundations in a few proofs, namely existence proofs as for Theorem 4. Definitions, theorems, and the great majority of proofs proceed rigorously without ever specifying the elements of any set at all. So it does not matter to them if elements are:

- other sets, as in ZF set theory.
- global elements $x : 1 \to S$ as in categorical set theory.
- entities lower in type by 1 than the entity S as in Russell's type theory.
- etc.

Even existence proofs usually specify elements only 'in principle.' Lang never actually spells out in ZF detail what a polynomial is.

So far the pattern or *category* of rings and morphisms is defined by a prior set theoretic definition of rings and morphisms. But standard tools can also describe the pattern simply as a pattern ignoring, as Resnik says, the 'internal' nature of rings and morphisms.[13] Then a ring A is just a place in the pattern so that, depending on the choice of foundations, A either has no elements in a set theoretic sense or it has them but they are unspecified and irrelevant. Either way morphisms are not functions defined on set theoretic elements. They are just *morphisms* or more pictorially *arrows* in the pattern. The pattern includes a place or places identifiable up to isomorphism as the ring $\mathbb{Z}$ of integers, and a place or places identifiable up to isomorphism as the ring $\mathbb{Z}[X]$ of integer polynomials. Fact 1 has an analogue with $\mathbb{Z}$ in place of the real numbers $\mathbb{R}$ which says we can *define* 'elements' of any ring A in a pattern theoretic sense to be the morphisms $\mathbb{Z}[X] \to A$. These correspond exactly to the classical set theoretic elements so that all the classical axioms on elements are verified, and thus all the classical theorems.

The pattern of rings and ring morphisms, i.e. the category **Ring**, can be defined in turn as a position in the pattern of categories and functors i.e. in the category of categories. Each ring A appears as a functor $A : \mathbf{1} \to \mathbf{Ring}$ where **1** is the *terminal* category. Ring morphisms f appear as functors $f : \mathbf{2} \to \mathbf{Ring}$ where **2** is the *arrow category* (Lawvere, 1966). Ring elements can still be defined by morphisms as in the previous paragraph. All the classical theorems on rings, elements, and morphisms reappear at this level. There is no obstacle to taking any one of these levels as basic with the others derived from it.

13.3 The unifying spirit

13.3.1 *A few uses of patterns*

These structural methods have deep roots in practice. As a notable example, topology by the 1920s relied on a booming but chaotic notion of *homology*. There were many versions.[14] Topologists assumed that results proved for one version would normally hold for all—but they could not say just what was 'normal' here. Some versions of homology were later proved exactly equivalent

---

[13] Universal algebra defines the category up to isomorphism (Lawvere, 1963). Monads define it not even up to isomorphism but up to *equivalence* (Mac Lane, 1998, Ch. 6).

[14] e.g. the Čech, Vietoris, and singular theories of Chapter 8 in Hocking and Young (1961) all existed before the unifying category theoretic language of Chapter 7.

to each other. Some were already known in the 1920s to be inequivalent to each other when applied to some more arcane spaces. Topologists had all too many nuts and bolts definitions of homology but no structural characterization. They had no homological 'Fact 1'.

The Fact, when it was found by Eilenberg and Steenrod (1945), turned out to characterize homology (up to isomorphism) as a pattern of *functors* from a certain category of topological spaces and continuous maps to the category of Abelian groups. This characterization enabled massive progress in homology. It also suggested useful variants such as *K-theory* which began by solving the classical problem of *independent vector fields* on an n-dimensional sphere S^n. It showed how many *everywhere different* ways there are to 'comb the hair flat' on S^n. There is only one way on the circle S^1: combing all clockwise is not different in this sense from combing all counterclockwise. There is no way to comb it all flat on the sphere S^2 or any even-dimensional S^{2n}. The matter is complicated in higher odd dimensions. Carter (2004) looks at K-theory and philosophical structuralism.

Noether's number theory was simplified and extended by similar functors applied to groups in place of topological spaces. This *group cohomology* lies behind the Langlands Program, and so behind Wiles on Fermat's Last Theorem, and much more. One earlier classical result was Gerd Faltings's proof of the Mordell conjecture, a problem that had drawn top number theorists for over sixty years: Any algebraic curve above a certain low level of complexity has at most finitely many rational points (Monastyrsky, 1998, p. 71). For philosophic discussion of these ideas see Krieger (2003), especially on Robert Langlands and André Weil.

All these proofs ignore nuts and bolts as far as possible—which is very far indeed by the conventional, functorial methods. That made the proofs feasible. Even when the morphisms of a category are structure preserving functions, these methods abstract away from the 'internal' nature of functions in favor of the patterns they form. The categorical definition of isomorphism given in Section 13.2 is a textbook example. In other important categories the morphisms are not functions but the useful pattern-theoretic properties still hold for them. The accompanying case study gives examples, and see Corfield (2003, esp. ch. 10).

13.3.2. Intuition and purity

Olga Taussky-Todd joked that she had trouble with Noether's courses on number theory because 'my own training was in ... the kind of number theory that involves numbers' (Taussky-Todd, 1981, p. 79). Noether created radical new methods on the board as she lectured and everyone needed time to get the intuition. There is no serious question of whether the methods are intuitive. They are to people who grasp them. Keep in mind how intuition functions in mathematics. Italian algebraic geometry into the 1920s was famously intuitive. Focusing on complicated four-dimensional configurations in higher-dimensional spaces, it relied on long years of expert experience to replace explicit proof or calculation. Outsiders were unable to follow it. It rose to a level which 'even the *Maestri* were unable to sustain' (Reid, 1990, p. 114). One person's intuition can be another's nightmare, and *expert* intuition by definition does not come easily.

The more interesting question is: are the new methods *number theory* and *topology* or a foreign import? Coordinate geometry is a paradigm of 'impure' mathematics where questions about points and lines are answered using real numbers and differential calculus. It is an evident change of topic while no one denies its value. Homology and other categorical methods do not change the topic.

Compare computer spreadsheets. Some people find them ungainly monsters. But spreadsheets are merely a format for what has to be done anyway: Data must be categorized, and some relations between categories must be routinized. The general theory of homological methods today is thousands of pages long and growing—and deep new facts are revealed by it. Yet experts who wrote those pages and use cohomology throughout their work say it should be used easily as an organizing device. Functorial homology *is* pure topology, or group theory, or number theory, or other mathematics, depending on which functors it uses.

The main stake in discussions of purity is that pure proofs are meant to be desirable for their own sake, as being more intrinsic to their subjects. One century of great efforts proved the *prime number theorem* in arithmetic by using deep complex analysis. Many people still wanted a purely arithmetic proof and Chapter 7 of this volume shows how another fifty years of great efforts found one pp. 189–191.[15] All purely arithmetic proofs of it to date are longer and harder than the analytic proof, but they are valued for their purity and genuinely new insights. The situation in homology is just the reverse. Routine methods can eliminate homology from specific number theoretic or topological theorems. Mathematicians eliminate it in practice only when they have an easier alternative.

For example, a 'very utilitarian description of the Galois cohomology needed in Wiles's proof' defines certain cohomology groups by elementary constructions known well before group cohomology—indeed since Noether in the

---

[15] In Detlefsen's terms I claim homology introduces no *remote* ideas but merely organizes ideas *already directly* involved in number theory and topology.

1930s (Washington, 1997, p. 101). This could eliminate Galois cohomology from the proof but Washington takes it the other way around as introducing that cohomology to the reader. The elimination would be routine and would replace conceptual arguments by opaque calculations. Of course, to short-cut Wiles's use of cohomology by genuinely simpler new insights would be wonderful—like shortening any proof by any simpler new insight.

13.3.3 Languages

Some philosophical theories of structuralism allow 'no *intelligible* attempt—of the kind that Zermelo and Fraenkel undertook—to unify mathematical practice' (MacBride, 2005, p. 572). Structural methods in practice, though, stand out for making the unity of mathematics work more deeply and powerfully than Zermelo and Fraenkel ever did. Barry Mazur well expressed this in accepting the 2000 Steele Prize. He names the founders of category theory Eilenberg, Mac Lane, and Grothendieck, not for category theory but also not by coincidence. He names them for what they did with the theory:

> I came to number theory through the route of algebraic geometry and before that, topology. The unifying spirit at work in those subjects gave all the new ideas a resonance and buoyancy which allowed them to instantly echo elsewhere, inspiring analogies in other branches and inspiring more ideas. One has only to think of how the work of Eilenberg, Mac Lane, Steenrod, and Thom in topology and the early work in class field theory as further developed in the hands of Emil Artin, Tate, and Iwasawa was unified and amplified in the point of view of algebraic geometry adopted by Grothendieck, a point of view inspired by the Weil conjectures, which presaged the inextricable bond between topology and arithmetic. One has only to think of the work of Serre or of Tate.[16] But mathematics is one subject, and surely every part of mathematics has been enriched by ideas from other parts. (NAMS, 2000, p. 479)

These methods cultivate each branch of mathematics in its own terms, and highlight how each one precisely in its own terms has bonds to others.

Poincaré defended geometry and specifically his beloved *analysis situs* now called *topology* against those who would reduce all mathematics to differential analysis. We must not 'fail to recognize the importance of well constructed language':

> The problems of analysis situs might not have suggested themselves if the analytic language alone had been spoken; or, rather, I am mistaken, they would have occurred surely, since their solution is essential to a crowd of questions in analysis.

---

[16] See references to Artin, Iwasawa, Tate, and Thom in Monastyrsky (1998). Serre and the Weil conjectures are central to the accompanying case study.

But they would have come singly, one after another, and without our being able to perceive their common bond. (Poincaré, 1908, p. 380)

Each language has its own deep results and must be spoken for its own sake, but each can also take specific results and general analogies from the other. Structuralism has aided and not eliminated such languages. Categorical concepts make an easy format for each one, and give uniform tools for relating formats, so they facilitate ever new interactions in substance. Before she went to university Emmy Noether was certified as a language teacher (Tollmien, 1990, p. 155).

Acknowledgments. Thanks to John Mayberry for many conversations on axiomatics and structure, and to John Cobb, Fraser MacBride, Michael Resnik, and Jamie Tappenden and students in his seminar especially Michael Docherty, for comments that have improved this paper.

Bibliography

AWODEY, Steven (2004), 'An answer to G. Hellman's question: "Does category theory provide a framework for mathematical structuralism?"', *Philosophia Mathematica*, 12, 54–64.

BENACERRAF, Paul (1965), 'What numbers could not be', *Philosophical Review*, 74, 47–73.

BOURBAKI, N. (1939), *Théorie des Ensembles, Fascicules de résultats* (Paris: Hermann).

CARTER, Jessica (2004), 'Ontology and mathematical practice', *Philosophia Mathematica*, 12, 244–267.

———(2005), 'Individuation of objects: a problem for structuralism?', *Synthese*, 143, 291–307.

CHIHARA, Charles (2004), *A Structural Account of Mathematics* (Oxford: Oxford University Press).

CONWAY, John and SMITH, Derek (2003), *On Quaternions and Octonians* (Natick, MA: A. K. Peters).

CORFIELD, David (2003), *Towards a Philosophy of Real Mathematics* (Cambridge: Cambridge University Press).

CORNELL, Gary, SILVERMAN, Joseph, and STEVENS, Glenn (eds.) (1997), *Modular Forms and Fermat's Last Theorem* (New-York: Springer-Verlag).

DEDEKIND, Richard (1888), *Was sind und was sollen die Zahlen?* (Braunschweig: Vieweg). Reprinted in *Essays on the Theory of Numbers* (New York: Dover Reprints, 1963).

EILENBERG, Samuel and STEENROD, Norman (1945), 'Axiomatic approach to homology theory', *Proceedings of the National Academy of Science*, 31, 117–120.

FLAMENT, Dominique (2003), *Histoire des Nombres Complexes: Entre Algèbre et Géométrie* (CNRS éditions, Histoire des sciences).

HELLMAN, Geoffrey (1989), *Mathematics Without Numbers* (Oxford: Oxford University Press).

HOCKING, John and YOUNG, Gail (1961), *Topology* (Reading: Addison-Wesley).

KRIEGER, Martin (2003), *Doing Mathematics* (London: World Scientific).

KRÖMER, Ralf (2007), *Tool and Object: A History and Philosophy of Category Theory* (New York: Birkhäuser).

LANG, Serge (1993), *Algebra* (Reading: Addison-Wesley).

—— (2005), *Undergraduate Algebra* (New York: Springer-Verlag).

LAWVERE, F. William (1963), *Functorial Semantics of Algebraic Theories*, Ph.D. thesis, Columbia University. Republished with commentary by the author in: *Reprints in Theory and Applications of Categories*, No. 5 (2004) pp. 1–121. <http://138.73.27.39/tac/reprints/articles/5/tr5abs.html>.

—— (1964), 'An elementary theory of the category of sets', *Proceedings of the National Academy of Science of the USA*, 52, 1506–1511.

—— (1966), 'The category of categories as a foundation for mathematics', in Samuel Eilenberg et al. (eds.), *Proceedings of the Conference on Categorical Algebra, La Jolla, 1965*, 1–21 (New York: Springer-Verlag).

—— (1998), *Categories for the Working Mathematician*, 2nd edn (New York: Springer-Verlag).

—— (2005), *A Mathematical Autobiography* (Natich, MA: A. K. Peters).

MACBRIDE, Fraser (2005), 'Structuralism reconsidered', in S. Shapiro (ed.), *The Oxford Handbook of Philosophy of Mathematics and Logic* (Oxford: Oxford University Press) pp. 563–589.

MAC LANE, Saunders (1986), *Mathematics: Form and Function* (New York: Springer-Verlag).

MCLARTY, Colin (1993), 'Numbers can be just what they have to', *Noûs*, 27, 487–498.

—— (2004), 'Exploring categorical structuralism', *Philosophia Mathematica*, 37–53.

MADDY, Penelope (2005), 'Three forms of naturalism', in S. Shapiro (ed.), *Oxford Handbook of Philosophy of Logic and Mathematics* (Oxford: Oxford University Press) pp. 437–459.

MAZUR, Barry (2003), *Imagining Numbers (Particularly the Square Root of Minus Fifteen)* (New York: Farrar, Straus and Giroux).

MONASTYRSKY, Michael (1998), *Modern Mathematics in the Light of the Fields Medals* (Wellesley, MA: A. K. Peters).

NAMS (2000), 'Steele Prize for a Seminal Contribution to Research: Barry Mazur', *Notices of the Amer. Math. Soc.*, 47, 478–479.

POINCARÉ, Henri (1908), 'L'Avenir des mathématiques', in *Atti del IV Congresso Internazionale dei Matematici*, 167–182 (Accademia dei Lincei). Translated in heavily edited form as 'The future of mathematics' in G. Halstead (ed.) *The Foundations of Science.* (New York: The Science Press, 1921), pp. 369–382.

REID, Miles (1990), *Undergraduate Algebraic Geometry* (Cambridge: Cambridge University Press).

RESNIK, Michael (1981), 'Mathematics as a science of patterns: ontology and reference', *Noûs*, 15, 529–550.

—— (1997), *Mathematics as a Science of Patterns* (Oxford: Oxford University Press).

SHAPIRO, Stewart (1997), *Philosophy of Mathematics: Structure and Ontology* (Oxford: Oxford University Press).

TAUSSKY-TODD, Olga (1981), 'My personal recollections of Emmy Noether', in James Brewer and Martha Smith (eds.), *Emmy Noether: A Tribute to Her Life and Work* (New York: Marcel Dekker) pp. 79–92.

TOLLMIEN, Cordula (1990), 'Sind wir doch der Meinung dass ein weiblicher Kopf nur ganz ausnahmweise in der Mathematik schöpferische tätig sein kann', *Göttinger Jahrbuch*, 153–219.

WASHINGTON, L. (1997), 'Galois cohomology', in G. Cornell, J. Silverman, and S. Stevens (eds.), *Modular Forms and Fermat's Last Theorem* (New York: Springer-Verlag) pp. 101–120.

14

'There is No Ontology Here': Visual and Structural Geometry in Arithmetic

COLIN MCLARTY

> In Diophantine geometry one looks for solutions of polynomial equations which lie either in the integers, or in the rationals, or in their analogues for number fields. Such polynomial equations $\{F_i(T_1, \ldots T_n)\}$ define a subscheme of affine space $\mathbb{A}^n$ over the integers which can have points in an arbitrary commutative ring R. (Faltings, 2001, p. 449)

Structuralists in philosophy of mathematics can learn from the current heritage of the ancient arithmetician Diophantus. A list of polynomial equations defines a kind of geometric space called a *scheme*. By one definition these schemes are countable sets built from integers in very much the way that Leopold Kronecker approached pure arithmetic. In another version every scheme is a functor as big as the universe of sets. The two versions are often mixed together because they give precisely the same structural relations between schemes. The practice was vividly put by André Joyal in conversation: 'There is no ontology here'. Mathematicians work rigorously with relations among schemes without choosing between the definitions. The tools which enable this in principle and require it in practice grew from topology.

The three great projects for 20th century mathematics were to absorb Richard Dedekind's and David Hilbert's algebra, to absorb Henri Poincaré's and Luitzen Brouwer's topology, and to create functional analysis.[1] Algebra and topology made explosive advances when Emmy Noether initiated a series of ever-deeper structural unifications. Her group theory became the method of

---

[1] Functional analysis also joined the structural unification (Dieudonné, 1981). Leading workers in all three projects contributed to mathematical logic.

homology of spaces. When abstract algebra spread from advanced number theory into basic topology it became a contender for organizing all of mathematics, as Noether intended. Her protégé Bartel van der Waerden advanced the new hegemony in his *Moderne Algebra* (1930). Bourbaki based their encyclopedic work on van der Waerden's text (Corry, 1996, pp. 309ff.). This 'structural' mathematics was the research norm by the 1950s and the textbook norm by the 1960s. Homology itself continues expanding and linking Dedekind and Poincaré to the latest Fields Medals.[2]

This case study looks at *schemes*, which arose largely in pursuit of a single problem, namely the *Weil conjectures*, a series of elegant conjectures on counting the solutions to certain arithmetic equations (Weil, 1949).[3] Weil sketched a fascinating strategy for a proof if only the *Lefschetz fixed point theorem* from topology would apply in arithmetic. It was a brilliant idea with repercussions all across mathematics, but only a handful of leading mathematicians thought it could possibly work.

Philosophical ideas of 'structuralism' in and out of mathematics could go deeper than they have before by absorbing some general features of scheme theory, the proof of the Weil conjectures, and much other 20th century mathematics:

- Structuralist tools are the feasible method of highlighting intuitive and relevant information on each structure as against technical nuts and bolts.
- Single structures matter less than the *maps* or *morphisms* between them.
- Maps and structures are often best understood by placing them in higher level structures: 'Patterns themselves are *positionalized* by being identified with positions of another pattern' (Resnik, 1997, p. 218).
- Maps are often richer and more flexible than functions. In set theoretic terms they are often more complicated than functions while in structural terms they often form simpler more comprehensible patterns.

Intuition develops as knowledge does. The *Chinese remainder theorem* is an accessible example from number theory. Logicians use it in arithmetizing syntax (Gödel, 1967, p. 611). Georg Kreisel saw its 'mathematical core in the combinatorial or constructive aspect of its proof' which suits its role in proving Gödel's incompleteness theorem but he added:

> I realize that there are other points of view. E.g. a purely abstract point of view: Jean-Pierre Serre once told me that he saw the mathematical core of the Chinese

---

[2] See notably Numford and Tate (1978); Lafforgue (2003); Soulé (2003).
[3] Chapters 7–11 of Ireland and Rosen (1992) introduce the arithmetic aspect of the conjectures.

remainder theorem in a certain result of cohomology theory. (Kreisel, 1958, p. 158)

We will see, though, that cohomology is not 'abstract.' It is geometrical.

Section 14.1 sketches Kronecker's and Noether's arithmetic. Section 14.2 shows how Noether's algebra organized Poincaré's topology and illustrates the centrality of maps. Poincaré emphasized single spaces. Noether emphasized the pattern of spaces and maps which was soon captured as the category **Top** of topological spaces which in turn became one object in the pattern of categories and functors, i.e. the category **CAT** of categories. Homology today uses patterns where each single position is a functor from one category to another and the maps are *natural transformations* (Mac Lane, 1986, p. 390). The rising levels are sketched in Sections 14.2.2 and 14.2.4 while Section 14.2.3 illustrates the Lefschetz fixed point theorem. Sections 14.3 and 14.4 introduce schemes and give the Chinese remainder theorem in classical and cohomological forms.

The final section focuses on three points that philosophy of mathematics ought to learn from the past century's practice: first, the rising levels of structure directly aid intuition; second, practice relies on the generality of categorical morphisms as against the set theoretic functions favored by most philosophers of mathematics today; and third, the interplay of levels raises conceptual questions on identity. In particular, categorists currently debate the importance of equality versus isomorphism, as discussed in Sections 14.5.3–5.5.

14.1 Diophantine equations

14.1.1 Kronecker's treatment of irrationals

Diophantus sought positive integer and rational solutions to arithmetic equations. Sometimes he would explore a problem far enough to show in our terms that the solution (or its square) is negative. Then he says there is no solution. When a problem leads towards a positive irrational solution 'he retraces his steps and shows how by altering the equation he can get a new one that has rational roots' (Kline, 1972, p. 143). Perhaps he rejected irrationals although they had been studied for centuries before him (Fowler, 1999). Perhaps he just enjoyed rational arithmetic. Perhaps he was like modern mathematicians who appreciate the irrational solutions to $X^n + Y^n = Z^n$ but also worked for centuries to see if it has nonzero rational ones. His motives are as hard to tell today as the date of his life, which is only known to lie between 150BC and 350AD (Knorr, 1993).

His book *Arithmetica* was cutting edge mathematics in 1650, though, as Fermat worked from it and sparked a rebirth of Diophantine arithmetic. Today this uses irrational and complex numbers in two different ways. The subject called *analytic number theory* uses complex function theory. Even now it is mysterious why deep theorems of calculus should reveal so much arithmetic, although the formal techniques are well understood.[4] The other use of irrational and complex numbers, *algebraic number theory*, was unmysterious by the mid-19th century. Or, better, it was no more mysterious than arithmetic itself. It only added algebra.

For an infamously seductive example let ω be a complex cube root of 1 so that $1 + \omega + \omega^2 = 0$.[5] The degree 3 Fermat equation $X^3 + Y^3 = Z^3$ factors as:

$$(X + Y)(X + Y\omega)(X + Y\omega^2) = Z^3$$

Just multiply out the left-hand side and use the equation on ω. In fact prime factorization holds for numbers of the form $a_0 + a_1\omega$ with ordinary integers a_0, a_1 and this helps to prove that the degree 3 Fermat equation has no non-trivial solutions. Prime factorization fails for the numbers formed using some primes p in place of 3 so this reasoning cannot prove Fermat's Last Theorem. But the method made great advances on Fermat and many other problems (Kline, 1972, p. 819f.).

Kronecker would replace ω with the arithmetic of polynomials

$$P(X) = a_0 + a_1 X + \cdots + a_{n-1} X^{n-1} + a_n X^n$$

where $a_0 \ldots a_n$ are ordinary integers; and X is merely a variable. We say polynomials $P(X)$ and $Q(X)$ are congruent *modulo* $1 + X + X^2$, and we write

$$P(X) \equiv Q(X) \pmod{1 + X + X^2}$$

if and only if the difference $Q(X) - P(X)$ is divisible by $1 + X + X^2$. In particular

$$X \not\equiv 1 \pmod{1 + X + X^2}$$

since clearly $1 - X$ is not divisible by $1 + X + X^2$. And yet

$$X^3 \equiv 1 \pmod{1 + X + X^2} \quad \text{because} \quad 1 - X^3 = (1 - X)(1 + X + X^2).$$

---

[4] Mazur conveys the depth of both mystery and knowledge in one famous case (1991).
[5] Cube roots of 1 are the roots of $1 - X^3$ which factors as $(1 - X)(1 + X + X^2)$. So 1 is a cube root of itself, and the two complex cube roots satisfy $1 + \omega + \omega^2 = 0$. The quadratic formula shows the complex roots are $\omega = (-1 \pm \sqrt{-3})/2$.

This arithmetic reproduces the algebra of ω only writing X for ω and congruence modulo $1 + X + X^2$ for $=$. Of course the complex number ω also fits into an analytic theory of the whole complex number plane. This analytic context is lost, as Kronecker intended, when we restrict attention to integer polynomials.

As another example, the *Gaussian integers* are complex numbers $a_0 + a_1 i$ where a_0, a_1 are ordinary integers. Replace them with polynomials modulo $1 + X^2$, writing

$$P(X) \equiv Q(X) \quad (\text{mod } 1 + X^2)$$

if and only if $Q(X) - P(X)$ is divisible by $1 + X^2$. In particular

$$X^2 \equiv -1 \quad (\text{mod } 1 + X^2)$$

Other congruence relations give all the *number fields* mentioned in Faltings' quote above. Kronecker would even banish negative numbers by replacing -1 with a variable Y modulo the positive polynomial $1 + Y$ (Kronecker, 1887a, b).

14.1.2 *The theology of numbers*

It is worth a moment to put Kronecker's famous saying in context:

> Many will recall his saying, in an address to the 1886 Berliner Naturforscher-Versammlung, that 'the whole numbers were made by dear God (*der liebe Gott*), the rest is the work of man.' (Weber, 1893, p. 15)

Kronecker elsewhere reversed this to say we make the whole numbers and not the rest. He endorsed Gauss in print:

> The principal difference between geometry and mechanics on one hand, and the other mathematical disciplines we comprehend under the name of 'arithmetic,' consists according to Gauss in this: the object of the latter, number, is a *pure* product of our mind, while space as well as time has reality also *outside* of our mind which we cannot fully prescribe a priori. (Kronecker, 1887b, p. 339)

The late Walter Felscher has pointed out that:

> 'lieber Gott' is a colloquial phrase usually used only when speaking to children or illiterati.[6] Addressing grown-ups with it contains a taste of being unserious, if not condescending...; no priest, pastor, theologian or philosopher would use it when expressing himself seriously. There is the well known joke of Helmut Hasse

---

[6] The phrase is famous in classical music but searches of WorldCat confirm that 'lieber Gott' in 19th century prose was generally folkloric or aimed at children.

who, having quoted Kronecker's dictum on page 1 of his yellow *Vorlesungen über Zahlentheorie* (1950), added to the index of names at the book's end under the letter L the entry 'Lieber Gott p. 1.' (26 May 1999 post to the list *Historia Matematica* archived at <http://www.mathforum.org.>)

Kronecker was not serious about the theology of numbers. He was serious about replacing irrational numbers with the 'pure' arithmetic of integer polynomials.

14.1.3 One Diophantine equation

Consider this Diophantine equation:

$$Y^2 = 3X + 2 \qquad (14.1)$$

Calculation modulo 3 will show it has no integer solutions. Say that integers a, b are congruent modulo 3, and write

$$a \equiv b \pmod{3}$$

if and only if the difference $a - b$ is divisible by 3. The key here is that congruent numbers have congruent sums and products.

Theorem 5. *Suppose* $a \equiv b$ *and* $c \equiv d$ (mod 3). *Then*

$$(a + c) \equiv (b + d) \quad \text{and} \quad (a \cdot c) \equiv (b \cdot d) \pmod{3}$$

Proof. Suppose 3 divides both $a - b$ and $c - d$. The claim follows because

$$(a + c) - (b + d) = (a - b) + (c - d)$$
$$(a \cdot c) - (b \cdot d) = (a - b) \cdot c + b \cdot (c - d) \qquad \square$$

If Equation (14.1) had an integer solution $X = a$, $Y = b$ then the sides would also be congruent modulo 3

$$b^2 \equiv 3a + 2 \equiv 2 \pmod{3}$$

But b is congruent to one of $\{0, 1, 2\}$ modulo 3. By Theorem 5 the square b^2 is congruent to the square of one of these. And none of these squares is 2 modulo 3:

$$0^2 \equiv 0 \quad 1^2 \equiv 1 \quad 2^2 \equiv 4 \equiv 1 \pmod{3}$$

So a, b cannot be an integer solution to Equation (14.1).

For future reference notice that Equation (14.1) does have solutions modulo other integers. For example $X = 2$, $Y = 1$ is a solution modulo 7 since

$$1^2 \equiv 1 \equiv 8 \equiv 3 \cdot 2 + 2 \pmod{7}$$

And the equation has solutions of other forms. For example, in numbers of the form $a+b\sqrt{2}$ with ordinary integers a, b there is a solution $X=0$, $Y=\sqrt{2}$ since

$$(\sqrt{2})^2 = 3 \cdot 0 + 2$$

14.1.4 Arithmetic via morphisms

This reasoning can be organized in the ring $\mathbb{Z}/(3)$, the *quotient ring* of $\mathbb{Z}$ by 3, which is the set $\{0, 1, 2\}$ with addition and multiplication by casting out 3s:

+	0	1	2
0	0	1	2
1	1	2	0
2	2	0	1

·	0	1	2
0	0	0	0
1	0	1	2
2	0	2	1

Construing Equation (14.1) in $\mathbb{Z}/(3)$ makes $3 = 0$ and so gives $Y^2 = 2$. The multiplication table shows 0,1 are the only squares in $\mathbb{Z}/(3)$. So Equation (14.1) has no solutions in $\mathbb{Z}/(3)$. Next, consider the function $r : \mathbb{Z} \to \mathbb{Z}/(3)$ taking each integer to its remainder on division by 3. So

$$r(0) = r(3) = 0 \quad r(2) = r(8) = 2 \quad \text{and so on}$$

By Theorem 5, this preserves addition and multiplication. So it would take any integer solution to any polynomial equation over to a solution for that same equation in $\mathbb{Z}(3)$. Since Equation (14.1) has no solutions in $\mathbb{Z}(3)$, it cannot have any in $\mathbb{Z}$ either.

The watershed in 'modern algebra' came when Noether reversed the order of argument. Instead of beginning with arithmetic she would deduce the arithmetic from purely structural descriptions of structures like $\mathbb{Z}/(3)$ (Noether, 1926). In the case of $\mathbb{Z}/(3)$ this meant looking at arbitrary rings and ring morphisms: A ring R is any set with selected elements 0,1 and operations of addition, subtraction, and multiplication satisfying the familiar associative, distributive, and commutative laws.[7] A *ring morphism* $f : R \to R'$ is a function preserving 0,1 addition and multiplication. Noether would rely on the following:

Fact on $\mathbb{Z}/(3)$ *The ring morphism* $r : \mathbb{Z} \to \mathbb{Z}$ *has* $r(3) = 0$ *and: For any ring R and ring morphism* $f : \mathbb{Z} \to R$ *with* $f(3) = 0$ *there is a unique morphism* $u : \mathbb{Z}/(3) \to R$ *with* $ur = f$.

---

[7] See Mac Lane (1986, pp. 39, 98) or any algebra text. All rings in this paper are commutative.

'THERE IS NO ONTOLOGY HERE' 377

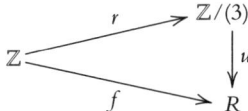

This Fact in no way *identifies* the elements of $\mathbb{Z}/(3)$. It says how $\mathbb{Z}/(3)$ fits into the pattern of rings and ring morphisms. It implies there are exactly three elements and they have specific algebraic relations to each other. It does not say what the elements are. Noether's school specified them in various ways: as the integers 0,1,2, or else as congruence classes of integers, or else as integers taken with congruence modulo 3 as a new equality relation. Saunders Mac Lane contrasted those last two approaches.[8] Noether could have defined $\mathbb{Z}/(3)$ in some such way and proved the Fact from the definition; but in practice the Fact was her working definition. She knew it characterized $\mathbb{Z}/(3)$ up to isomorphism:

Theorem 6. *Suppose a ring $\mathbb{Z}/(3)$ and morphism r satisfy the Fact on $\mathbb{Z}/(3)$, as do another $\mathbb{Z}/(3)'$ and r'. Then there is a unique ring isomorphism $u : \mathbb{Z}/(3) \to \mathbb{Z}(3)'$ such that $ur = r'$.*

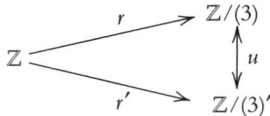

Proof. Since $r'(3) = 0$, the assumption on $\mathbb{Z}/(3)$ and r says there is a unique morphism u with $ur = r'$. Since $r(3) = 0$, the assumption on $\mathbb{Z}/(3)'$ and r' says there is a unique $v : \mathbb{Z}/(3)' \to \mathbb{Z}/(3)$ with $vr' = r$. The composite vu is a ring morphism with $vur = vr' = r$. In other words, $vur = 1_{\mathbb{Z}/(3)}r$, so uniqueness implies $vu = 1_{\mathbb{Z}/(3)}$. Similarly $uv = 1_{\mathbb{Z}/(3)'}$.

She took the Fact as a case of the *homomorphism theorem*: a theorem or family of theorems on quotient structures which she was rapidly expanding through the ten years up to her death.[9] From the homomorphism theorem she deduced *isomorphism theorems* very much the way we deduced Theorem 6. She recast problems of arithmetic as problems about morphisms, as for example problems of arithmetic modulo 3 became problems about morphisms between $\mathbb{Z}/(3)$ and related structures.

Equation 14.1 has solutions in other rings such as the quotient $\mathbb{Z}/(7)$ or the ring $\mathbb{Z}[\sqrt{2}]$ of numbers of the form $a + b\sqrt{2}$ with a, b ordinary integers.

---

[8] See (Noether, 1927, n. 6) and discussion in McLarty (2007b, Section 5).
[9] She knew she had not yet measured the scope of this theorem (McLarty, 2006, Section 4). We could call the present paper a case study of Noether's homomorphism theorem.

Kronecker's way of eliminating the irrational $\sqrt{2}$ amounts to treating $\mathbb{Z}[\sqrt{2}]$ as a quotient of the ring $\mathbb{Z}[X]$ of integer polynomials in one variable X.

$$r_{2-X^2} : \mathbb{Z}[X] \to \mathbb{Z}[X]/(2 - X^2) \cong \mathbb{Z}[\sqrt{2}]$$

Noether's homomorphism and isomorphism theorems describe this and all quotients up to isomorphism by their places in the pattern of ring morphisms. That means arbitrary commutative rings as in the epigraph from Faltings. Restricting attention to 'the rings that occur in practice' would be pointless and unworkable. Just stating the restriction would mean focusing on details irrelevant to most proofs. Plus, too many rings of too many kinds are already in use and more are constantly brought in. Important results often refer to specific rings, notably the integers $\mathbb{Z}$ and rationals $\mathbb{Q}$, but no such focus fits into the basic theorems or definitions.

14.2 The homology of topological spaces

14.2.1 *The sphere and the torus*

Homology theory began pictorially enough. Compare the sphere S^2, i.e. the surface of a ball, to the torus T, i.e. the surface of a doughnut. A small circle on either one can bound a patch of it:

Fig. 14.1.

The difference is that every circle on the sphere bounds a region but not all circles on the torus do. Draw a vertical circle around the small circumference of the torus, another larger horizontal circle around the top of the torus, and a diagonal circle that spirals around in both ways:

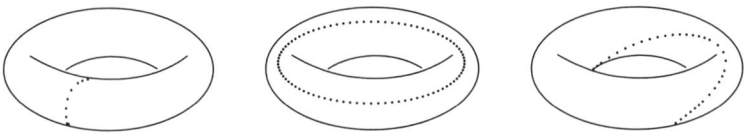

Fig. 14.2.

'THERE IS NO ONTOLOGY HERE' 379

Intuitively the vertical circle on the left is not wrapped around any region on the surface because it is wrapped around the 'hole' that runs through the interior of the torus. The larger horizontal circle wraps around the hole through the center of the torus. The spiraling circle wraps around both holes.

A hole is called 1-dimensional if a circle can wrap around it. The hole inside the sphere is not 1-dimensional since any circle on the sphere can slip off to one side and (for example) shrink down to the small bounding circle in Figure 14.1. That hole is 2-dimensional as the sphere surface wraps around it. The holes in each dimension say a great deal about a topological space (Atiyah, 1976).

To organize this information, Henri Poincaré called a circle C on a surface *homologous to* 0 if it bounds a region, and then wrote $C \sim 0$ (Sarkaria, 1999). The two small circles in Figure 14.1 are both homologous to 0, while none of the three on the torus in Figure 14.2 are. Name those three C_1, C_2, C_3 respectively. Together they do bound a region.[10] Poincaré says their sum is homologous to 0:

$$C_1 + C_2 + C_3 \sim 0$$

He would use the usual rules of arithmetic to rewrite this as

$$C_3 \sim -C_1 - C_2$$

In fact he would consider all the circles C on the torus.[11] He called any formal sum

$$a_j C_j + a_k C_k + \cdots + a_n C_n$$

of circles C_i on the torus with integer coefficients a_i a *1-cycle*. If those circles with those multiplicities form the boundary of some sum of regions then the 1-cycle is homologous to 0:

$$a_j C_j + a_k C_k + \cdots + a_n C_n \sim 0$$

A pair of 1-cycles are homologous to each other

$$a_i C_i + \cdots + a_k C_k \sim a_n C_n + \cdots + a_p C_p$$

if and only if their formal difference is homologous to 0:

$$a_i C_i + \cdots + a_k C_k - a_n C_n - \cdots - a_p C_p \sim 0$$

---

[10] Cut the torus along the small circle C_1 to form a cylinder, and then along the horizontal circle C_2 to get a flat rectangle with the spiral circle C_3 as diagonal. Either triangular half of the rectangle is bounded by one vertical side, one horizontal, and the diagonal.

[11] Here a 'circle' in any space M is any continuous map $C : S^1 \to M$ from the isolated circle or *1-sphere* S^1 to M. It may be quite twisted and run many times around M in many ways.

The key to the homology of the torus is that every 1-cycle on the torus is homologous to a unique sum of C_1 and C_2:

$$a_j C_j + a_k C_k + \cdots + a_n C_n \sim a_1 C_1 + a_2 C_2$$

for some unique $a_1, a_2 \in \mathbb{Z}$. The pair $\{C_2, C_3\}$ serves as well

$$a_j C_j + a_k C_k + \cdots + a_n C_n \sim a_2 C_2 + a_3 C_3$$

for some unique $a_2, a_3 \in \mathbb{Z}$. The pair $\{C_1, C_3\}$ also serves as do infinitely many others. No single cycle will do. So the *first Betti number* of the torus is 2. The first Betti number of the sphere S^2 is 0, since all 1-cycles on it are homologous to 0.

Higher Betti numbers are defined using higher dimensional figures in place of circles. A good n-dimensional space M has an ith Betti number, counting the i-dimensional holes in M, for every i from 0 to n. The Betti numbers of a space say much about its topology but, in fact, Poincaré and all topologists of the time knew that the numbers alone omit important relations between cycles.

14.2.2 *Brouwer to Noether to functors*

By 1910 the standard method in topology was algebraic calculation with cycles (Herreman, 2000). Everyone from Poincaré on knew that the cycles formed groups. The 1-cycles on the torus add and subtract and they satisfy all the axioms for an Abelian group, when the relation $\sim$ is taken as equality. The same holds for the i-dimensional cycles on any n-dimensional space M, for each $0 \leq i \leq n$. Only no one wanted to use group theoretic language in topology. Brouwer would not even calculate with cycles. But he was a friend of Noether's and they shared some students.

As Noether emphasized morphisms in algebra, so Brouwer organized his topology around *maps* or continuous functions $f : M \to N$ between topological spaces (van Dalen, 1999). His most famous theorem, the *fixed point theorem*, says: Let D^n be the n-dimensional solid disk, that is all points on or inside the unit sphere in n-dimensional space R^n, then every map $f : D^n \to D^n$ has a fixed point, a point x such that $f(x) = x$. Many of his theorems are explicitly about maps, and essentially all of his proofs are based on finding suitable maps.

Noether's homomorphism and isomorphism theorems unified her viewpoint with Brouwer's. Given any topological space M, she would explicitly form the group $Z_i(M)$ of i-cycles on M, as the formal sums of circles shown above form the group $Z_1(T)$ of 1-cycles on the torus. Then she would form the subgroup $B_i(M) \subseteq Z_i(M)$ of *n-boundaries* on M, in other words the i-cycles

which bound $(i+1)$-dimensional regions in M. She formed the *ith homology group* as the quotient:
$$H_i(M) \cong Z_i(M)/B_i(M)$$

Intuitively $H_i(M)$ counts the i-cycles on M, but counting a cycle as 0 if it bounds a region. In effect it counts the cycles that surround holes and thus counts the i-dimensional holes in M. Her whole approach to algebra led her to focus on:

Fact on $H_n(M)$. *There is a group morphism $q : Z_i(M) \to H_i(M)$ which kills boundaries in the sense that $q(\beta) = 0$ for every $\beta \in B_i(M)$ and: For any Abelian group A and $f : Z_i(M) \to A$ which kills boundaries there is a unique $u : H_i(M) \to A$ with $uq = f$.*

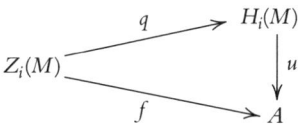

This defines $H_i(M)$ up to Abelian group isomorphism just as the Fact on $\mathbb{Z}/(3)$ defined $\mathbb{Z}/(3)$ up to ring isomorphism in Section 14.1.4.

Crucially, the Fact specifies group morphisms from $H_i(M)$. With some trivial facts of topology[12] it implies that each map $f : M \to N$ induces morphisms

$$H_i(f) : H_i(M) \to H_i(N)$$

It articulates what had been *ad hoc* and implicit before: While a map f carries points of M to points of N, the morphism $H_i(f)$ carries i-dimensional holes in M to i-dimensional holes in N and preserves intricate relations between them. This works in every dimension i relevant to the spaces M and N. Paul Alexandroff, Noether's student and also Brouwer's, seized on this.

When Hilbert published his lectures on intuitive geometry he asked Alexandroff for a section on topology. The section did not appear in Hilbert and Cohn-Vossen (1932) but became the brilliant fifty page Alexandroff (1932) which has never gone out of print. It does topology with thoroughly group theoretic tools and may be the only mathematics text ever dedicated jointly to Hilbert and Brouwer. Alexandroff and Hopf (1935) became the standard topology textbook for the next twenty years and has also never been out of print.

The explicit correlation of maps to group morphisms immediately proved new topological theorems. But it also simplified the creation of new homology theories using new technical definitions of *cycle* and *boundary*. So it brought new

---

[12] Each map $f : M \to N$ takes M-cycles to N-cycles, and M-boundaries to N-boundaries.

complexities and some disorder over the next twenty years until topologists found a way to organize the subject by bypassing all the nuts and bolts:

> In order that these algebraic techniques not remain a special craft, the private reserve of a few virtuosos, it was necessary to put them in a broad, coherent, and supple conceptual setting. This was accomplished in the 1940s and 1950s through the efforts of many mathematicians, notably Samuel Eilenberg at Columbia University, Saunders Mac Lane of the University of Chicago, the late Norman Steenrod, and Henri Cartan. (Bass, 1978, p. 505)

They axiomatized homology as a correlation between patterns of continuous maps and patterns of group morphisms. For each dimension i, the i-dimensional homology group became a functor H_i. This means:

- Homology preserves domain and codomain.

$$f : M \to N \quad \text{gives} \quad H_i(f) : H_i(M) \to H_i(N)$$

- Each identity map $1_M : M \to M$ (which, intuitively, does not affect the holes of M) has identity homology.

$$1_M : M \to M \quad \text{gives} \quad 1_{H_i(M)} : H_i(M) \to H_i(M)$$

- The homology of a composite gf is the composite of the homologies.

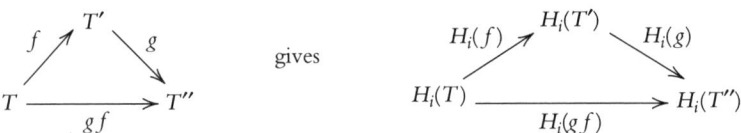

The axioms require more which we will not go into.[13]

The structuralist point is that all the groups and morphisms are defined only up to isomorphism. Topologists still use nuts-and-bolts descriptions of cycles and boundaries but textbooks use the axioms to define homology. The axioms make it easier to focus on geometry and they show how different nuts and bolts all yield the same calculations.

14.2.3 The Lefschetz fixed point theorem

The *Lefschetz fixed point theorem* applies to especially nice spaces M, the *orientable topological manifolds*, where each small enough region of M looks like a continuous piece of some Euclidean coordinate space $\mathbb{R}^n$ as for example small

---

[13] See Eilenberg and Steenrod (1945), Hocking and Young (1961, Chapter 7).

regions of a sphere or torus look like pieces of the plane $\mathbb{R}^2$. A sphere can turn on an axis the way the earth does. A torus can turn around a central axis the way a bicycle tire turns around its axle:

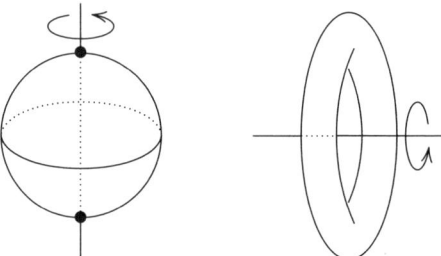

Fig. 14.3.

A rotating sphere has two fixed points, call them the North and South poles. The rotating torus has none—obviously because the axis passes through a hole. Of course the matter is more complex with general continuous functions rather than just rigid rotations. It is more complex yet for manifolds with more holes or in higher dimensions. In general the theorem relates fixed points of a map $f : M \to M$ to the way f acts on holes in all dimensions, that is to the morphisms $H_i(f)$.

On its face the fixed point theorem counts fixed points, which are solutions to equations of the form $f(x) = x$. Weil saw that if he could apply it to suitable arithmetic spaces then he could use this plus Galois theory to count solutions to his polynomials. There was one crying problem: it was nearly inconceivable that arithmetic spaces could be defined so as to support such a topological theorem.

14.2.4 Cohomology

The route to scheme theory ran through a variant of homology called *cohomology* and the key to schemes is that they admit *coverings* analogous to the topological case. For example, the torus can be covered by overlapping cylindrical sleeves, U_1, U_2, U_3, drawn here in solid outline:

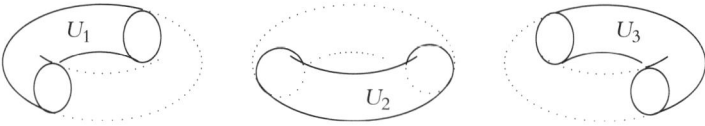

Fig. 14.4.

The route from U_1 to U_2 to U_3 travels around the hole in the center of the torus. The hole is revealed, very roughly, by the fact that every two components of this cover have non-empty intersection

$$U_{12} = U_1 \cap U_2 \quad U_{23} = U_2 \cap U_3 \quad U_{31} = U_3 \cap U_1$$

while the triple intersection is empty:

$$U_{123} = U_1 \cap U_2 \cap U_3 = \emptyset$$

There is no point in the center where the components would all overlap. In the center is a hole. The precise relation between holes and covers is complex. For example the cover $\{U_1, U_2, U_3\}$ does not reveal the hole inside the torus, which is hidden inside each sleeve. Other covers reveal that one. Cohomology summarizes covers of a space M to produce cohomology groups $H^n(M)$ which are close kin to the homology groups $H_n(M)$. The standard tool for it is *sheaves*.

On one definition, a sheaf on a space M is another space with a map $S \to M$ where S is a union of partially overlapping partial covers of M, which may (or may not) patch together in many different ways to form many covers.[14] Another definition says a sheaf $\mathcal{F}$ on M assigns a set $\mathcal{F}(U)$ to each open subset $U \subseteq M$, thought of heuristically as a set of 'patches' covering U. There are compatibility conditions among the sets over different open subsets of M which we will not detail. If the sets $\mathcal{F}(U)$ are Abelian groups in a compatible way then $\mathcal{F}$ is an *Abelian sheaf*.

The cohomology of a space M became an infinite series $H^1, H^2, \ldots$ of functors from the category $\mathbf{Ab}_M$ of Abelian sheaves on M to the category $\mathbf{Ab}$ of ordinary Abelian groups. Then the entire series $H^1, H^2, \ldots$ was conceived as a single object called a δ-*functor*, which can be defined (up to isomorphism) by its place within a category of series of functors $\mathbf{Ab}_M \to \mathbf{Ab}$. Today's tools were introduced in one of the most cited papers in mathematics (Grothendieck, 1957). In topology or complex analysis the functorial patterns are always invoked although lower-level constructions are often taken as defining cohomology. Textbooks on cohomological number theory take the functorial pattern as definition. Of course the theory of derived functors was developed to get at geometry, arithmetic, etc. and not to call attention to itself. So for example Washington on Galois cohomology never mentions derived functors. But each time he writes 'for proof see…', the reference uses them.[15]

---

[14] See the light introduction in Mac Lane (1986, pp. 252–256, 351–355).
[15] See Huybrechts (2005) on complex geometry, Hartshorne (1977) and Milne (1980) on number theory. Compare Washington (1997, pp. 102, 104, 105, 107, 108, 118, 120).

14.3 Arithmetic schemes

14.3.1 Geometry over the integers

Classical algebraic geometry looked at real and complex number solutions to lists of polynomials. Around 1940, though, André Weil took up Kronecker's vision of arithmetic and called for 'algebraic geometry over the integers' which Grothendieck achieved by the theory of *schemes* (Weil, 1979, vol. 1, p. 576). So now, for example, Faltings in the epigraph writes of 'affine space over the integers', written $\mathbb{A}^n_{\mathbb{Z}}$ to make the integers explicit. What is it?

Classical real affine n-space $\mathbb{R}^n$, or $\mathbb{A}^n_{\mathbb{R}}$, is the space of n-tuples $\langle x_1, \ldots x_n \rangle$ of real numbers.[16] But the set $\mathbb{Z}^n$ of n-tuples of integers captures little arithmetic. Recalling Section 14.1.3, the equation

$$Y^2 = 3X + 2 \tag{14.2}$$

defines the empty subset of $\mathbb{Z}^2$ the same as, say,

$$1 = 0$$

But Equation (14.2) is not arithmetically trivial while $1 = 0$ is. Equation (14.2) has solutions in many rings including paradigmatically arithmetic examples like $\mathbb{Z}/(7)$ and $\mathbb{Z}[\sqrt{2}]$ as in Sections 14.1.3–14.1.4. The affine space $\mathbb{A}^2_{\mathbb{Z}}$ is more subtle than $\mathbb{Z}^2$, so that Equation (14.2) can define a subscheme of it containing these solutions, while $1 = 0$ defines the empty subscheme. There are two basic approaches.

14.3.2 Schemes as variable sets of solutions

Consider the equation

$$X^2 + Y^2 = 1 \tag{14.3}$$

It has four integer solutions

$$X = \pm 1 \text{ while } Y = 0; \text{ or else } X = 0 \text{ while } Y = \pm 1$$

Clearly the equation points towards a circle, but these four solutions do little to show it. So the idea is to look not only at integer solutions. The equation has infinitely many rational solutions. Each rational number q gives a rational solution

---

[16] This usage contrasts affine spaces to projective spaces, not to vector spaces.

$$X = \frac{q^2-1}{q^2+1} \quad Y = \frac{2q}{q^2+1}$$

as straightforward calculation shows. Every real number θ gives a real solution

$$X = \cos(\theta) \quad Y = \sin(\theta)$$

The ring $\mathbb{Z}/(7)$ of integers modulo 7 has eight solutions, including $X = 2$, $Y = 5$:

$$2^2 + 5^2 = 29 \equiv 1 \pmod{7}$$

So the equation has not one set of solutions, but for each ring R it has a set of solutions in R which we call $V_{S^1}(R)$. Here 'V' recalls the classical geometer's term, the *variety* of points defined by an equation, while the subscript is to suggest that this is the equation of the circle S^1. Ring morphisms preserve solutions. Given any ring morphism $f : R \to R'$ and any solution $a, b \in R$ to Equation (14.3) then $f(a), f(b) \in R'$ is also a solution since by definition of morphism

$$f(a)^2 + f(b)^2 = f(a^2 + b^2) = f(1) = 1$$

So V_{S^1} is a *variable set*, specifically a functor $V_{S^1} : \mathbf{Ring} \to \mathbf{Set}$ from the category of rings to the category of sets. It takes each ring R to the set $V_{S^1}(R)$, and each ring morphism $f : R \to R'$ to the corresponding function. It preserves identities:

$$1_R : R \to R \quad \text{gives} \quad 1_{V_{S^1}(R)} : V_{S^1}(R) \to V_{S^1}(R)$$

And it preserves composition:

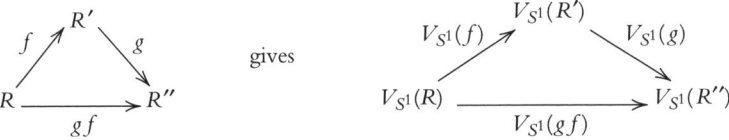

Call V_{S^1} the *affine scheme* of the Equation (14.3). From this viewpoint the affine n-space $\mathbb{A}_{\mathbb{Z}}^n$ is the functor corresponding to n variables and the trivial equation $1 = 1$. Any list of equations on n variables defines a similar functor, a *subscheme* of $\mathbb{A}_{\mathbb{Z}}^n$. An equation may have no variables, as for example the equation $7 = 0$ gives subschemes expressing arithmetic modulo 7. An *arithmetic scheme* is any functor $\mathbf{Ring} \to \mathbf{Set}$ which contains suitably overlapping parts, where each part is isomorphic to the affine scheme of some finite list of integer polynomials. In general there are many ways to choose these parts and no one covering is intrinsic to the scheme.

A *scheme map*, on this approach, is simply any natural transformation between the functors. This is the start of many pay-offs as geometrically important ideas find quite direct functorial expression. It may not even seem plausible that such an abstract definition of maps could have geometric content until you have seen it work; as we do not have space to show here. In fact, though, it gives exactly the same patterns of maps as the more evidently geometrical definition in the following section. The functorial apparatus looked heavy to some algebraic geometers but it merely made a central fact of algebra explicit: Every integer polynomial has a (possibly empty) set of solutions in every commutative ring, and every ring morphism preserves solutions. The techniques spread as geometers learned the practical value of using arbitrary commutative rings to prove theorems on the ring $\mathbb{Z}$ of integers.

Grothendieck and Dieudonné (1971) used this version of schemes in the introduction. Such a scheme is 'basically a structured set' (Mumford, 1988, p. 113) but strictly it is a structured proper class. There are technical means to avoid proper classes—by specific tricks in specific situations, or generally by adding *Grothendieck universes* to set theory (Artin et al., 1972, pp. 185ff.). But standard textbook presentations use functors on all commutative rings and these are proper classes in either ZF or categorical set theory.

14.3.3 Schemes as Kroneckerian spaces

The other approach to schemes constructs a space from the data. If the equation

$$X^2 + Y^2 = 1 \tag{14.3}$$

is to define a space then intuitively a *coordinate function* on that space should be given by a polynomial in X and Y, while two polynomials $P(X,Y), Q(X,Y)$ give the same function if they agree all over the space. To a first approximation we would say $P(X,Y)$ and $Q(X,Y)$ give the same function on this space if

$$P(a,b) = Q(a,b) \quad \text{for all } \langle a,b \rangle \in \mathbb{Z}^2 \text{ with } a^2 + b^2 - 1 = 0$$

But this puts too much stress on integer solutions. In the case of equations with no integer solutions this poor definition would make every two polynomials give the same function trivially. So we actually use a stronger condition taken from the defining polynomial: Polynomials $P(X,Y)$ and $Q(X,Y)$ are called congruent modulo $X^2 + Y^2 - 1$, written

$$P(X, Y) \equiv Q(X, Y) \pmod{X^2 + Y^2 - 1}$$

if and only if

$$Q(X, Y) - P(X, Y) \text{ is divisible by } X^2 + Y^2 - 1$$

A coordinate function on the space is a congruence class of polynomials. For example X^2Y defines a coordinate function on this space, the same function as $Y - Y^3$, since light calculation shows

$$(X^2Y) - (Y - Y^3) = Y.(X^2 + Y^2 - 1)$$

Altogether the ring of coordinate functions is the ring of integer polynomials in two variables, $\mathbb{Z}[X, Y]$, modulo $X^2 + Y^2 - 1$, i.e. it is the quotient ring

$$\mathbb{Z}[X, Y]/(X^2 + Y^2 - 1)$$

These 'functions' are constructed from integer polynomials just as Kronecker would construct irrationals as in Section 14.1.1.

The next step is to use the functions to define points of this space. The motivation is that each point p should have a set of functions

$$p \subseteq \mathbb{Z}[X, Y]/(X^2 + Y^2 - 1)$$

which take value 0 at p. This set should be an *ideal*, closed under addition and under multiplication by arbitrary functions. That is, if the functions $P_1(X, Y)$ and $P_2(X, Y)$ are both construed as taking value 0 at p then $P_1(X, Y) + P_2(X, Y)$ must also be; and so must the product $R(X, Y) \cdot P_1(X, Y)$ for any polynomial $R(X, Y)$. Further, the ideal $\mathfrak{p}$ should be *prime*, in the sense that whenever a product lies in $\mathfrak{p}$ then at least one factor already lies in it.[17] More formally:

If $P_1(X, Y) \cdot P_2(X, Y) \in \mathfrak{p}$ then either $P_1(X, Y) \in \mathfrak{p}$ or $P_2(X, Y) \in \mathfrak{p}$.

Scheme theory demands no more than that for a point. It says there is a point for each prime ideal. Textbooks say the points *are* the prime ideals. So our space has a point for each integer solution $a, b \in \mathbb{Z}$ to Equation (14.3), but also a point for the mod 7 solution $2, 5 \in \mathbb{Z}/(7)$. It has one point combining

---

[17] This expresses the idea that if a product is 0 then at least one factor must be—which *does not* hold in every ring—but this definition makes it hold for function values at points of a scheme.

the two real algebraic solutions

$$\sqrt{2}/2, \sqrt{2}/2 \in \mathbb{R} \quad \text{and} \quad -\sqrt{2}/2, -\sqrt{2}/2 \in \mathbb{R}$$

and another point combining these two[18]

$$\sqrt{2}/2, -\sqrt{2}/2 \in \mathbb{R} \quad \text{and} \quad -\sqrt{2}/2, \sqrt{2}/2 \in \mathbb{R}$$

It has one point for each pair of conjugate complex algebraic solutions such as

$$2, \pm\sqrt{-3} \in \mathbb{C}$$

It has no points for real or complex transcendental solutions. This elegant algebraic definition gives our space points for precisely those solutions to Equation (14.3) given by roots of integer polynomials (possibly modulo some prime number) and it distinguishes only those points given by roots of distinct polynomials.

Intuitively a *closed set* should be the set of all points where some function is 0, or where some list of functions are all 0. Formally, an affine scheme has a *closed set* for each ideal of coordinate functions on it, and in the case of arithmetic schemes each ideal is defined by a finite list of polynomial equations:

$$P_1(X, Y) = 0 \quad \ldots \quad P_k(X, Y) = 0$$

Then we name the space after its coordinate ring, calling it

$$\text{Spec}(\mathbb{Z}[X, Y]/(X^2 + Y^2 - 1))$$

or the *spectrum* of the ring $\mathbb{Z}[X, Y]/(X^2 + Y^2 - 1)$.

Every commutative ring R has a spectrum $\text{Spec}(R)$. The coordinate function ring on $\text{Spec}(R)$ is just the ring R. So again the 'functions' are generally not functions in the set-theoretic sense. They are any elements of any ring. The points are the prime ideals of R. The spectrum has a topology where closed sets correspond to ideals. The spectra of rings are the *affine* schemes. Notably, the affine n-space $\mathbb{A}^n_\mathbb{Z}$ is the spectrum of the integer polynomial ring in n variables $\mathbb{Z}[X_1, \ldots X_n]$ subject to no equation or, if you prefer, the trivially true equation $1 = 1$.

$$\mathbb{A}^n_\mathbb{Z} = \text{Spec}(\mathbb{Z}[X_1, \ldots X_n])$$

A scheme is patched together from affine schemes. More fully, a *scheme* is a topological space X together with a sheaf of rings $\mathcal{O}_X$ called the *structure sheaf*

---

[18] Notice the first two solutions also satisfy $2XY = 1$. The second two satisfy $2XY = -1$.

of the scheme. This sheaf assigns a ring $\mathcal{O}_X(U)$ of 'coordinate functions' to each open subset $U \subseteq X$, and to each inclusion of open subsets $U \subseteq V$ of X a ring morphism

$$r_{U,V} : \mathcal{O}_X(V) \to \mathcal{O}_X(U)$$

The whole must be made of parts isomorphic to the spectra $\mathrm{Spec}(R)$ of rings R with their coordinate functions. A scheme map

$$f : (X, \mathcal{O}_X) \to (Y, \mathcal{O}_Y)$$

consists of a continuous function $f : X \to Y$ in the ordinary sense plus a great many ring morphisms in the opposite direction: By continuity of f, each open subset $V \subset Y$ has inverse image $f^{-1}(V)$ open in X, and the scheme map includes a suitable ring morphism

$$\mathcal{O}_Y(V) \to \mathcal{O}_X(f^{-1}(V))$$

for each open $V \subset Y$, showing how $f^{-1}(V)$ maps algebraically into V. This version of schemes dominates Grothendieck and Dieudonné (1971) and Hartshorne (1977), though Grothendieck favored the functorial version in his work.

An arithmetic scheme is a scheme pasted from finitely many parts defined by finite lists of integer polynomials. Each integer polynomial ring $\mathbb{Z}[X_1, \ldots X_n]$ is countable. Since each of its ideals is generated by a finite list of polynomials there are only countably many, thus countably many points and closed sets and functions on them. Altogether the Kroneckerian version of any arithmetic scheme is countable.

14.3.4 Scheme cohomology

Schemes were born for cohomology. In fact they were born and re-born for it. Jean-Pierre Serre introduced structure sheaves into algebraic geometry so as to produce the cohomology theory today called *coherent cohomology*. These structure sheaves were 'the principle of the right definition' of schemes (Grothendieck, 1958, p. 106). Then Serre took the first step towards the sought-after 'Weil cohomology'. Using ideas from differential geometry he defined covers and he proved they gave good 1-dimensional Weil cohomology groups $H^1(M)$ for algebraic spaces M. Notably, Serre proved his groups gave the first non-trivial step in the infinite series of a δ-functor as in our Section 14.2.4.[19]

For Grothendieck the functorial pattern was decisive. An idea that gave the *first* step had to give *every* step. He made it work by producing the general

[19] Serre (1958, esp. §1.2 and §3.6).

theory of schemes and lightly altering Serre's covers into a frankly astonishing theory of *étale* maps. The purely algebraic definition of a *finite étale* map $X \to S$ between schemes does a brilliant job of saying the space X lies smoothly stacked over S even when there is no very natural geometric picture of X or S alone. Working with Serre, Pierre Deligne, and others over several years Grothendieck proved that these étale covers yield a cohomology theory, called étale cohomology, satisfying enough classical topological theorems for the Weil conjectures and much more.

14.4 An example

14.4.1 Integers as coordinate functions

The arithmetic scheme Spec($\mathbb{Z}$), the affine scheme of the ring of integers $\mathbb{Z}$, is given by the trivially true equation $1 = 1$ in no variables. There is exactly one 'solution' to $1 = 1$—and that is to say it is *true*—and indeed the equation remains true in any ring R. As a variable set this scheme is the functor Spec($\mathbb{Z}$) = $V_{1=1}$ where for each ring R the set of solutions $V_{1=1}(R)$ is a singleton which we may think of as:

$$V_{1=1}(R) = \{true\} \quad \text{for every ring } R$$

This is perfectly simple and even too simple. It does nothing to reveal the arithmetic of the integers. But that is because we have looked at this scheme in isolation. Looking at its maps to and from other schemes we find it is *terminal* in the category of schemes: Every scheme has exactly one scheme map to Spec($\mathbb{Z}$). This is a very simple specification of the place of Spec($\mathbb{Z}$) in the pattern or category of all schemes but it does invoke the pattern or category of all schemes, and that very large pattern reveals all of the arithmetic of the integers! The arithmetic of any ring R does not lie *inside* Spec(R) but in the pattern of all scheme maps *to* Spec(R).

On the other hand, look at Spec($\mathbb{Z}$) as a space. The points are the prime ideals and these are of two kinds: The singleton 0 is a prime ideal $\{0\} \subseteq \mathbb{Z}$ and for each prime number $p \in \mathbb{Z}$ the set of all integer multiples of p is a prime ideal usually written as $(p) \subseteq \mathbb{Z}$. We may also write $\{0\} = (0)$ since 0 is indeed the only multiple of 0. Algebraic geometers often draw these points as a kind of line.

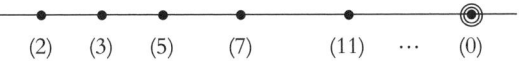

Then the idea is to think of each integer $m \in \mathbb{Z}$ as a function defined on this line. For very good reasons, the values of the function over the point (0) are rational numbers while the values over any point (p) are integers modulo p. The integer $m \in \mathbb{Z}$ is a function whose value at the point (0) is m and value at each point (p) is m modulo p. For example the integer 9 has

$$9 \equiv 1 \quad (\text{mod } 2) \qquad 9 \equiv 0 \quad (\text{mod } 3)$$

$$9 \equiv 4 \quad (\text{mod } 5) \qquad 9 \equiv 2 \quad (\text{mod } 7)$$

and we can graph 9 as a function this way:

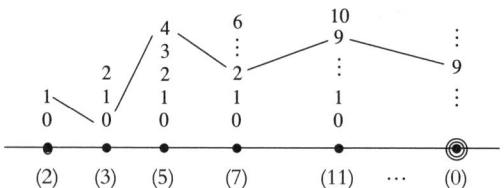

The spatial structure of Spec($\mathbb{Z}$) is determined by the 'coordinate functions' on it—namely by the integers! But the picture is only suggestive. Rigorous proofs show schemes have structural relations parallel to those in geometry and these relations return major new arithmetic theorems. We sketch one simple example.

14.4.2 The Chinese remainder theorem

Consider congruences modulo 4 and 6:

$$a \equiv 2 \quad (\text{mod } 4) \quad \text{and} \quad a \equiv 1 \quad (\text{mod } 6)$$

These two have no solution since the first implies a is even and the second implies it is odd. On the other hand consider

$$a \equiv 1 \quad (\text{mod } 4) \quad \text{and} \quad a \equiv 3 \quad (\text{mod } 6)$$

These agree modulo 2, as both say a is odd, and clearly $a = 9$ is a solution. Then too $a = 21$ is a solution. Adding any multiple of 12 will give another solution since 12 is the smallest common multiple of 4 and 6. One classical statement of the Chinese remainder theorem is:

Theorem. *Take any integer moduli m_1, m_2. Then for any integer remainders r_1, r_2 consider the congruences on an unknown integer a*

$$a \equiv r_1 \quad (\text{mod } m_1) \quad \text{and} \quad a \equiv r_2 \quad (\text{mod } m_2)$$

There are solutions a if and only if

$$r_1 \equiv r_2 \pmod{\operatorname{GCD}(m_1, m_2)}$$

where $\operatorname{GCD}(m_1, m_2)$ *is the greatest common divisor of* m_1 *and* m_2. *And in that case the solution is unique modulo* $\operatorname{LCM}(m_1, m_2)$, *the least common multiple of* m_1 *and* m_2. *In a formula, if a and b are both solutions then*

$$a \equiv b \pmod{\operatorname{LCM}(m_1, m_2)}$$

Note that $2 = \operatorname{GCD}(4, 6)$ and $12 = \operatorname{LCM}(4, 6)$ in the examples above.

This theorem supports thinking of integers as functions on the scheme $\operatorname{Spec}(\mathbb{Z})$. Compare this familiar fact about the real line $\mathbb{R}$:

Theorem. *Take any intervals on the line, say* $[0, 2] \subseteq \mathbb{R}$ *and* $[1, 3] \subseteq \mathbb{R}$. *Then for any continuous functions* f_1 *and* f_2 *consider these conditions on an unknown continuous function a:*

$$a(x) = f_1(x) \text{ for all } x \in [0, 2] \quad \text{and} \quad a(x) = f_2(x) \text{ for all } x \in [1, 3]$$

There are solutions a if and only if the functions f_1 *and* f_2 *agree on the intersection of the intervals:*

$$f_1(x) = f_2(x) \quad \text{for all } x \in [1, 2]$$

And in that case the solution is unique over the union of the intervals: if a and b are both solutions then

$$a(x) = b(x) \quad \text{for all } x \in [0, 3]$$

In the Chinese remainder theorem the given numbers r_1 and r_2 and the sought number a are taken as functions on the line $\operatorname{Spec}(\mathbb{Z})$. The equation

$$a \equiv r_1 \pmod{m_1}$$

says that a must agree with r_1 not necessarily over the whole line but at least at all points (p) of the line corresponding to prime factors p of m_1.[20] The other equation

$$a \equiv r_2 \pmod{m_2}$$

says a must agree with r_2 at all points (p) corresponding to prime factors p of m_2. The necessary condition is that r_1 and r_2 must agree at all points (p) where p divides both m_1 and m_2. But p divides both m_1 and m_2 if and only if it divides their greatest common divisor so this condition can be expressed in an equation:

$$r_1 \equiv r_2 \pmod{\operatorname{GCD}(m_1, m_2)}$$

---

[20] For each prime power p^n that divides m_1, the functions a and r_1 must agree on the nth order infinitesimal neighborhood of the point (p), or in other words $a \equiv r_1 \pmod{p^n}$.

394 COLIN MCLARTY

The proof shows this condition is also sufficient. And the function a is uniquely determined at all points (p) where p divides either one of m_1 and m_2. But those are the prime factors of LCM(m_1, m_2) so the unique determinacy says any two solutions a, b have

$$a \equiv b \pmod{\text{LCM}(m_1, m_2)}$$

The Chinese remainder theorem patches part of one function r_1 with part of another r_2 to get a single function a on Spec($\mathbb{Z}$) just as you can patch parts of continuous functions f_1, f_2 into a single continuous function a on the real line $\mathbb{R}$. But the Chinese remainder theorem only deals with parts given by finite numbers of primes. It follows from a more general result with a very similar proof: for every ring R you can patch compatible partial functions on the affine scheme $X = \text{Spec}(R)$. In technical terms: every structure sheaf $\mathcal{O}_X$ is actually a sheaf, so it agrees with its own 0-dimensional Čech cohomology.[21]

The point is simple. Many arithmetic problems are easily solved 'locally' in some sense, while we want 'global' solutions. In the Chinese remainder theorem the list of congruences is easily solved 'locally at each prime' and the trick is to patch together one solution on the whole line.[22] Cohomology is all about patching.

14.5 The philosophy of structure

14.5.1 Understanding sheaves and schemes

Intuitive ideas available in some way to some experts often become explicit and publicly available by passing to higher levels of structure, and they often grow further in the passage. Section 14.2.2 gave homology group morphisms and homology theories as two examples. For another, there was a serious problem in the early 1950s of how to understand sheaves and work with them. Several nuts-and-bolts definitions were known but none were as simple as the ideas seemed to be. And the specialists seeking a Weil cohomology to prove the Weil conjectures saw that none of these definitions was vaguely suitable for that. One thing was clear. People who worked with sheaves would draw vast commutative diagrams 'full of arrows covering the whole blackboard', dumbfounding the student Grothendieck (1985-87, p. 19). The vertices were Abelian sheaves and the arrows were sheaf morphisms between them. One

---

[21] Compare Hartshorne (1977, Proposition II.2.2) and Tamme (1994, p. 41).
[22] Eisenbud (1995, Exercise 2.6) gives a basically homological proof.

'THERE IS NO ONTOLOGY HERE' 395

key to getting the diagrams right was to think of them as diagrams of ordinary Abelian groups and group morphisms. Abelian sheaves were a lot like Abelian groups.

Grothendieck followed his constant belief that the simplest most unified account of the intuition would itself provide the needed generality.[23] Rather than focus on individual sheaves he would describe the 'vast arsenal' of sheaves on a given space:

> We consider this 'set' or 'arsenal' as equipped with its most evident structure, the way it appears so to speak 'right in front of your nose'; that is what we call the structure of a 'category'. (Grothendieck, 1985–87, p. 38)

The diagrams do not *stand for* some *other* information about sheaves. They *are* the *relevant, categorical* information. They are the information used in practice. So Grothendieck extended ideas from Mac Lane to produce a short list of axioms. A category satisfying these axioms is called an *Abelian category*:

1. The category **Ab** of Abelian groups is an Abelian category, as is the category $\mathbf{Ab}_M$ of Abelian sheaves on any space M. So are many categories used in other cohomology theories apparently quite unlike these. This was promising for a Weil cohomology and later did work for étale cohomology.
2. The Abelian category axioms themselves suffice to define cohomology in terms of derived functors and prove the general theorems on cohomology (Grothendieck, 1957).

The axioms explicated how Abelian sheaves are like Abelian groups. They quickly entered textbooks as the easiest way to work with sheaves. And they opened the way to an even more intuitive understanding as they led Grothendieck to analogous axioms for the category $\mathbf{Sh}_M$ of all sheaves on M.

He produced new axioms on a category, and called any category which satisfies them a *topos* (Artin et al., 1972, pp. 322ff.):

1. The category **Set** of all sets is a topos, as is the category $\mathbf{Sh}_M$ of all sheaves on any space M.
2. We can interpret mathematics inside any topos very much the way we do in sets. The 'Abelian groups' in the topos $\mathbf{Sh}_M$ of sheaves on any space M turn out to be just the Abelian sheaves on M. The 'Abelian groups' in other toposes play similar roles in other cohomology theories.

---
[23] For more detail see McLarty (2007a).

In short the objects of any topos are *continuously variable sets* as for example the topos $\mathbf{Sh}_M$ contains the sets varying continuously over the space M.[24] Mathematics in a topos is like classical mathematics with some differences reflecting the variation. Specifically mathematics in $\mathbf{Sh}_M$ differs from classical by reflecting the topology of M. The cohomology of M measures the difference between classical Abelian groups and Abelian groups varying continuously over M. This measure also works in other toposes to give other cohomology theories and it was the key to producing étale cohomology. For one expert view of how far toposes can be eliminated from étale cohomology, and yet how they help in understanding it, see Deligne (1977, 1998).

Categorical thinking is also central to understanding schemes. At the basic level of the spaces to be studied, on either definition of schemes, the geometry of a scheme is poorly revealed by looking at it as a set of points with geometric relations among them. It may even be 'bizarre' on that level:

> [In many schemes] the points... have no ready to hand geometric sense.... When one needs to construct a scheme one generally does not begin by constructing the set of points.... [While] the decision to let every commutative ring define a scheme gives standing to bizarre *schemes*, allowing it gives a *category of* schemes with nice properties. (Deligne, 1998, pp. 12–13)

One generally constructs a scheme from its relations to other schemes. The relations are geometrically meaningful. The category of schemes captures those relations and incorporates Noether's insight that the simplest arithmetic definitions often relate a desired ring to arbitrary commutative rings as in Section 14.1.4.

14.5.2 Several reasons morphisms are not functions

Functions are examples of morphisms, but overreliance on this example has been a major obstacle to understanding category theory (McLarty, 1990, §7). Indeed schemes typify the failure of Bourbaki's structure theory based on structure preserving functions and so they witness the inadequacy of current philosophical theories of 'structure'. It is not a problem with set theory. Set theory handles schemes easily enough. We handle schemes set-theoretically here. But even on set theoretic foundations scheme maps are not 'structure-preserving functions' in the set theoretic sense—that is in Bourbaki's sense or in the closely related sense of the model-theoretic structuralist philosophies of mathematics.[25]

---

[24] William Lawvere has made this much more explicit but Grothendieck already found each topos is 'like' the universe of sets (McLarty, 1990, p. 358).

[25] Bourbaki (1958); Hellman (1989); Shapiro (1997).

In the plainest technical sense neither of our definitions of schemes has set theoretic functions as maps. On the structure sheaf definition a map from $(X, \mathcal{O}_X)$ to $(Y, \mathcal{O}_Y)$ consists of a continuous function $f : X \to Y$ plus many ring morphisms in the opposite direction (in general, infinitely many). For the variable set definition each map is a natural transformation and thus a structure-preserving proper class of functions. Either way a map is more complex than a set theoretic function.

Topology gives another kind of morphism more complex than functions. Continuous functions $f, g : S \to S'$ between topological spaces are called *homotopic* if each one can be continuously deformed into the other (Mac Lane, 1986, p. 323). Write $\bar{f}$ for the homotopy class of f, that is the set of functions $S \to S'$ homotopic to f. The *homotopy category* **Toph** has topological spaces S as objects, and homotopy classes $\bar{f} : S \to S'$ as morphisms. The morphisms of **Toph** are not functions but equivalence classes of functions. By the global axiom of choice we can select one representative function from each class. But there is provably no way to select representatives so that the composite of every pair of selected representatives is also selected. The categorical structure of **Toph** cannot be defined by composition of functions but requires composition of equivalence classes of functions (Freyd, 1970).

A case study by Leng (2002) includes an example from functional analysis. C^*-algebras originated in quantum mechanics and the paradigms are algebras of operators on function spaces. The mathematician George Elliott sought a theorem characterizing certain ones. The theorem used *inductive limits*, where the inductive limit of an infinite sequence of C^*-algebras and morphisms

$$A_1 \longrightarrow A_2 \longrightarrow \cdots \longrightarrow A_n \longrightarrow \cdots$$

is a single C^*-algebra A_∞ combining all of the A_n in a way compatible with the morphisms. The inductive limit is an actual C^*-algebra. On the way to the theorem, though, Elliott first treated each infinite sequence of C^*-algebras as a single new object, and defined morphisms between sequences in just such a way as to make the sequences act like their own inductive limits.[26] As Leng says: 'Defining the morphisms between these sequences turned out to require some ingenuity' (2002, p. 21). But it worked. Elliott's formal inductive limits supported the proofs needed to imply the theorem for the actual C^*-algebra inductive limits. It would be triply useless to think of the ingeniously defined morphisms of sequences as 'structure-preserving functions' in the logician's

---

[26] Compare Grothendieck's *ind-objects* and *pro-objects* in (Artin et al., 1972, Exposé I).

or Bourbaki's sense. They are technically more complex than functions, the properties desired of them were categorical rather than set theoretic, and it was far from clear what structure they should preserve.

Carter (Forthcoming) gives a similar example in geometry, and urges a modal logic interpretation of it. Examples from analysis and geometry influenced the founders of category theory (McLarty, 2007b, §2).

Technicalities aside, it is a bad idea to think of scheme maps just one way. We have two fundamentally different definitions. This is quite different from cases familiar to philosophers such as the two definitions of real numbers from the rationals by Cauchy sequences and by Dedekind cuts. Neither one of those definitions is much used in practice.[27] Both definitions of scheme maps are used. Intuition draws on both and must hang on neither. Each is fitted to some proofs or calculations and clumsy for others. There is no ontology here.

Finally, categorical axioms patently ignore ontology. When the leader of a graduate student session on spectral sequences says 'Let **A** be your favorite Abelian category...,' then all we know of **A** is what kind of patterns its morphisms can form. In some useful Abelian categories the morphisms are structure-preserving functions, in others they are more complex, and in others the morphisms are finite arrays of numbers (Mac Lane, 1998, p. 11). It cannot matter which they are since we do not know which example anyone prefers. When Lang (1993) proves a theorem for all Abelian categories it does not matter what the morphisms *are*. And when Lawvere poses axioms for a category of categories as foundation (1963) or a category of sets as a foundation (1965) then the morphisms are what the axioms say they are—they are not 'functions' defined in any prior set theory.

14.5.3 'The' category of schemes

Section 14.3 gave two definitions of schemes, and so two definitions of the category of schemes. A structuralist might hope these would be isomorphic categories but the truth is more interesting. They are *equivalent*. Roughly speaking, equivalent categories contain the same structures in the sense that they contain all the same patterns of objects and arrows—but they may have different numbers of copies of each pattern (Mac Lane, 1998, § IV.4).

Let **Scheme**$_O$ be the category of schemes $X, \mathcal{O}_X$ as point sets with sheaves of rings and call these *ringed space schemes*. Let **Scheme**$_V$ be the category of

---

[27] Practice usually defines the real numbers up to isomorphism by their algebra and the least upper bound property (Rudin, 1953, p. 8).

schemes as functors, and call these *functor schemes*. There is a functor

$$h : \mathbf{Scheme}_{\mathcal{O}} \to \mathbf{Scheme}_V$$

with two nice properties (Mumford, 1988, §. II.6).[28]

1. The scheme maps from any $X, \mathcal{O}_X$ to any $Y, \mathcal{O}_Y$ in $\mathbf{Scheme}_{\mathcal{O}}$ correspond exactly to the scheme maps $h(X) \to h(Y)$ in $\mathbf{Scheme}_V$.
2. Every functor scheme is isomorphic to the image $h(X)$ of at least one ringed space scheme $X, \mathcal{O}_X$.

We say that a functor scheme V *corresponds* to the ringed space scheme X if V is isomorphic to $h(X)$. But then V also corresponds to every ringed space scheme isomorphic to X. The key to working this way is that all the ringed space schemes corresponding to X are already isomorphic (as ringed space schemes). For many purposes you need not decide which category you are in. You can go back and forth by the functor h. As one germane application this passage preserves cohomology. Cohomology of schemes can be defined purely in terms of categorical relations between schemes. Exactly the same relations exist in the category of ringed space schemes and the category of functor schemes. Many calculations of cohomology, and applications of cohomology, work identically verbatim in the two categories.

14.5.4 *Isomorphism versus equivalence*

Like the two definitions of the category of schemes, so the two definitions in Section 14.2.4 of the category of sheaves on a space M agree only up to equivalence. Equivalence is isomorphism *of* categories up to isomorphism *in* the categories.

Isomorphism is defined the same way for categories as for any kind of structure: A functor $\mathbf{F} : \mathbf{C} \to \mathbf{D}$ is an isomorphism if it has an inverse, namely a functor $\mathbf{G} : \mathbf{D} \to \mathbf{C}$ composing with $\mathbf{F}$ to give the identity functors $\mathbf{GF} = 1_{\mathbf{C}}$ and $\mathbf{FG} = 1_{\mathbf{D}}$. In other words, for all objects C of $\mathbf{C}$, and D of $\mathbf{D}$:

$$\mathbf{GF}(C) = C \qquad \mathbf{FG}(D) = D$$

and the same for morphisms. Categories are isomorphic if there is an isomorphism between them.

A functor $\mathbf{F} : \mathbf{C} \to \mathbf{D}$ is an *equivalence* if it has a quasi-inverse, a functor $\mathbf{G}$ with composites isomorphic to identity functors: $\mathbf{GF} \cong 1_{\mathbf{C}}$ and $\mathbf{FG} \cong 1_{\mathbf{D}}$. In

---

[28] These imply equivalence as defined in the next section, given a suitable axiom of choice.

other words for all objects C of **C** and D of **D** there are isomorphisms

$$\mathbf{GF}(C) \cong C \qquad \mathbf{FG}(D) \cong D$$

in a way compatible with morphisms. Categories are *equivalent* if there is an equivalence between them. Gelfand and Manin overstate an important insight when they call isomorphism 'useless' compared to equivalence:

> Contrary to expectations [isomorphism of categories] appears to be more or less useless, the main reason being that neither of the requirements $\mathbf{GF} = 1_\mathbf{C}$ and $\mathbf{FG} = 1_\mathbf{D}$ is realistic. When we apply two natural constructions to an object, the most we can ask for is to get a new object which is canonically isomorphic to the old one; it would be too much to hope for the new object to be identical to the old one. (Gelfand and Manin, 1996, p. 71)

This is actually not true even in Gelfand and Manin's book. Their central construction is the *derived category* $D(\mathbf{A})$ of any Abelian category **A**. Given **A**, they define $D(\mathbf{A})$ up to a unique isomorphism (1996, §III.2). They use the uniqueness up to isomorphism repeatedly. The notion of isomorphic categories remains central. Yet for many purposes equivalence is enough.[29] The next section returns to it.

14.5.5 Philosophical open problems

We have seen structuralist ideas grow dramatically from Noether through Eilenberg and Mac Lane to Grothendieck. We have alluded to Lawvere's elementary category theory unifying and generalizing the key ideas and to many advances in functorial geometry and number theory. Most of that is quite pure mathematics, but there is physics as well. Sir Michael Atiyah has made functorial tools standard in theoretical particle physics and influential in the search for a general relativistic quantum theory (Atiyah, 2006). Among Lawvere's goals throughout his career has been simpler and yet less idealized continuum mechanics (Lawvere and Schanuel, 1986; Lawvere, 2002). An obvious project for philosophers is to apply the agile morphism-based notion of structure in other fields as, for example, Corfield (2006) looks at cognitive linguistics. That article also describes how categorical foundations put foundations into a wider perspective: We now know that very many widely different formalisms are adequate to interpret mathematics while highlighting many different aspects of practice.

Let us close with the philosophical question of how mathematical objects are identified. On one Zermelo–Fraenkel-based story the set of real numbers

---

[29] Krömer (2007, Chapter 5) discusses Gelfand and Manin's view conceptually but without close comparison to the mathematics in their book. And see Marquis (Forthcoming).

ℝ is identified by choosing specific ZF sets for the natural numbers, for ordered pairs, and so on through, say, Cauchy sequences of rational numbers. Perhaps

$$0 = \phi, 1 = \{\phi\}, \ldots$$

$$\langle x, y \rangle = \{x, \{x, y\}\}$$

But mathematicians rarely make such choices and often contradict the choices they seem to make. Textbooks routinely define a complex number $\langle x_0, x_1 \rangle \in \mathbb{C}$ as an ordered pair of real numbers $x_0, x_1 \in \mathbb{R}$ and yet equate each real number x with a complex number, $x = \langle x, 0 \rangle$, which is false on any ZF definition of ordered pairs I have ever seen in print.[30] Few textbooks define ordered pairs at all.

To be very clear: these facts of practice do not deny the formal adequacy of ZF to reconstruct much mathematics. They show that ZF reconstructions differ from practice on the question of how to identify mathematical objects. The difference is already clear in textbooks while it is greatest in the most highly structured mathematics with the greatest need for rigor.

In practice most structures are defined only up to isomorphism by their morphisms to and from similar structures. Section 14.1.4 did this for quotient rings like $\mathbb{Z}/(3)$. Section 14.2.2 sketched how this way of defining homology groups $H_i(M)$ not only works 'in principle' but is precisely adapted to the key use of homology which is defining homology morphisms $H_i(f) : H_i(M) \to H_i(N)$ for each topological map $f : M \to N$. The morphism-based methods of working up to isomorphism are entirely standard today. Less standardized today, and less thoroughly conceptualized, are the methods of working across many levels of structures-of-structures with the corresponding levels of isomorphism-up-to-isomorphism. These occur in practice and rigorous methods are known but no unified choice of methods is yet standard. The conceptual and foundational issues around them are still debated.[31]

One view of identity has the provocative slogan that 'Every interesting equation is a lie', or more moderately 'behind every interesting equation there lies a richer story of isomorphism or equivalence' (Corfield, 2005, p. 74). An equation between numbers is often an isomorphism between sets. As a trivial example suppose there are as many dogs in a certain sheep pen as there are

---

[30] It is at once easy and pointless to gerrymander a ZF definition with $x = \langle x, 0 \rangle$ for all $x \in \mathbb{R}$. It would only generate other anomalies in other textbook practices.

[31] See fibred categories versus indexed categories in Johnstone (2002) and references there. Bénabou (1985) makes pointed and far-reaching claims for fibred categories.

rams. The equation expresses an isomorphism of sets

$$\#(\text{dogs in this sheep pen}) = \#(\text{rams})$$

$$\{\text{dogs in this sheep pen}\} \cong \{\text{rams}\}$$

and the real interest is the specific correspondence: at this moment my dogs are facing one ram each and that is all the rams. Corfield gives practically important examples from combinatorics and topology, as our Sections 14.2.1 and 14.2.2 raised equations of Betti numbers into isomorphisms of homology groups. Turning every equation into an isomorphism is just the same thing as turning every isomorphism into an equivalence—an 'isomorphism up to isomorphism'. It is not clear how far this can be taken. We have seen that mathematicians cannot entirely dispense with isomorphism in favor of equivalence and so cannot entirely dispense with equality in favor of isomorphism. On the other hand, as Corfield says, philosophers have missed the real importance of equivalence as a kind of sameness of structure (2005, p. 76). Mathematical physicist John Baez has taken this viewpoint very far, originally using *n-categories* or *higher dimensional algebra* as a revealing approach to quantum gravity, but also looking at it conceptually all across mathematics (Baez and Dolan, 2001).

Philosophers are right that structural mathematics raises issues such as: how are purely structural definitions possible? And what is the role of identity versus structural isomorphism? But let us take Resnik's point that 'mathematicians have emphasized that mathematics is concerned with structures involving mathematical objects and not with the "internal" nature of the objects themselves' (Resnik, 1981, p. 529). This is already the rigorous practice of mathematics. That practice offers working answers with powerful and beautiful results.

Acknowledgments. I thank Karine Chemla, José Ferreirós, and Jeremy Gray for discussions of the topics here, and William Lawvere for extensive critique of the article.

Bibliography

ALEXANDROFF, Paul (1932), *Einfachste Grundbegriffe der Topologie* (Berlin: Julius Springer). Translated as *Elementary Concepts of Topology* (New York: Dover, 1962).

ALEXANDROFF, Paul and HOPF, Heinz (1935), *Topologie* (Berlin: Julius Springer). Reprinted 1965 (New York: Chelsea Publishing).

ARTIN, Michael, GROTHENDIECK, Alexander, and VERDIER, Jean-Louis (1972), *Théorie des Topos et Cohomologie Etale des Schémas I*, Séminaire de géométrie algébrique du Bois-Marie, 4 (Berlin: Springer-Verlag). Generally cited as SGA4.

ATIYAH, Michael (1976), 'Bakerian Lecture, 1975: Global geometry', *Proceedings of the Royal Society of London Series A*, 347(1650), 291–299.

—— (2006), 'The Interaction Between Geometry and Physics', in Pavel Etingof, Vladimir Retakh, and I. M. Singer (eds.), *The Unity of Mathematics*, n. 244 in Progress in Mathematics, 1–15 (Boston: Birkhäuser).

BAEZ, John and DOLAN, J. (2001), 'From Finite Sets to Feynman Diagrams', in Björn Engquist and Wilfried Schmid (eds.), *Mathematics Unlimited: 2001 and Beyond*, 29–50 (New York: Springer-Verlag).

BASS, Hyman (1978), 'The Fields Medals II: Solving Geometry Problems with Algebra', *Science*, 202, 505–506.

BÉNABOU, Jean (1985), 'Fibered Categories and the Foundations of Naive Category Theory', *Journal of Symbolic Logic*, 50, 10–37.

BOURBAKI, N. (1958), *Théorie des Ensembles*, 3rd edn (Paris: Hermann).

CARTER, Jessica (Forthcoming), 'Categories for the Working Mathematician: Making the Impossible Possible', *Synthese*.

CORFIELD, David (2005), 'Categorification as a Heuristic Device', in D. Gillies and C. Cellucci (eds.), *Mathematical Reasoning and Heuristics*, 71–86 (London: King's College Publications).

—— (2006), 'Some Implications of the Adoption of Category Theory for Philosophy', in Giandomenico Sica (ed.), *What is Category Theory?* (Milan: Polimetrica) pp. 75–93.

CORRY, Leo (1996), *Modern Algebra and the Rise of Mathematical Structures* (Boston: Birkhäuser).

DELIGNE, Pierre (ed.) (1977), *Cohomologie Étale*, Séminaire de géométrie algébrique du Bois-Marie; SGA 4 1/2 (Berlin: Springer-Verlag). Generally cited as SGA 4 1/2, this is not strictly a report on the Seminar.

—— (1998), 'Quelques idées maîtresses de l'œuvre de A. Grothendieck', in *Matériaux pour l'Histoire des Mathématiques au XXe Siècle (Nice, 1996)* (Soc. Math. France) pp. 11–19.

DIEUDONNÉ, Jean (1981), *History of Functional Analysis* (Amsterdam: North-Holland).

EILENBERG, Samuel and STEENROD, Norman (1945), 'Axiomatic Approach to Homology Theory', *Proceedings of the National Academy of Science*, 31, 117–120.

EISENBUD, David (1995), *Commutative Algebra* (New York: Springer-Verlag).

EISENBUD, David and HARRIS, Joe (1992), *Schemes: The Language of Modern Algebraic Geometry* (New York: Wadsworth).

FALTINGS, Gerd (2001), 'Diophantine Equations', in Björn Engquist and Wilfried Schmid (eds.), *Mathematics Unlimited: 2001 and Beyond* (New York: Springer-Verlag) pp. 449–54.

FOWLER, David (1999), *The Mathematics of Plato's Academy* (Oxford: Oxford University Press).

FREYD, Peter (1970), 'Homotopy is Not Concrete', in F. Peterson (ed.), *The Steenrod Algebra and its Applications* (New York: Springer-Verlag) pp. 25–34.

GELFAND, Sergei and MANIN, Yuri (1996), *Methods of Homological Algebra* (Berlin: Springer-Verlag).

GÖDEL, Kurt (1967), 'On Formally Undecidable Propositions (excerpts)', in J. van Heijenoort (ed.), *From Frege to Gödel* (Cambridge: Harvard University Press) pp. 592–617.

GROTHENDIECK, Alexander (1957), 'Sur quelques points d'algèbre homologique', *Tôhoku Mathematical Journal*, 9, 119–221.

—— (1958), 'The Cohomology Theory of Abstract Algebraic Varieties', in *Proceedings of the International Congress of Mathematicians, 1958* (Cambridge: Cambridge University Press) pp. 103–118.

—— (1985–1987), *Récoltes et Semailles* (Montpellier: Université des Sciences et Techniques du Languedoc). Published in several successive volumes.

GROTHENDIECK, Alexander and DIEUDONNÉ, Jean (1971), *Éléments de Géométrie Algébrique I* (Berlin: Springer-Verlag).

HARTSHORNE, Robin (1977), *Algebraic Geometry* (New York: Springer-Verlag).

HASSE, Helmut (1950), *Vorlesungen über Zahlentheorie* (Berlin: Springer-Verlag).

HELLMAN, Geoffrey (1989), *Mathematics Without Numbers* (Oxford: Oxford University Press).

HERREMAN, Alain (2000), *La Topologie et ses Signes: Éléments pour une Histoire Sémiotique des Mathématiques* (Paris: L'Harmattan).

HILBERT, David and COHN-VOSSEN, S. (1932), *Anschauliche Geometrie* (Berlin: Julius Springer). Translated to English as *Geometry and the Imagination* (New York: Chelsea Publishing, 1952).

HOCKING, John and YOUNG, Gail (1961), *Topology* (Reading: Addison-Wesley).

HUYBRECHTS, Daniel (2005), *Complex Geometry* (Berlin: Springer-Verlag).

IRELAND, Kenneth and ROSEN, Michael (1992), *A Classical Introduction to Modern Number Theory* (New York: Springer-Verlag).

JACOBSON, Nathan (ed.) (1983), *E. Noether: Gesammelte Abhandlungen* (Berlin: Springer-Verlag).

JOHNSTONE, Peter (2002), *Sketches of an Elephant: a Topos Theory Compendium* (Oxford: Oxford University Press). To be finished as three volumes.

KLINE, Morris (1972), *Mathematical Thought from Ancient to Modern Times* (Oxford: Oxford University Press).

KNORR, Wilbur (1993), 'Arithmetike Stoicheiosis', *Historia Mathematica*, 20, 180–192.

KREISEL, Georg (1958), 'Review of *Wittgenstein's Remarks on the Foundations of Mathematics* (Basil Blackwell, Oxford, 1956)', *British Journal for the Philosophy of Science*, 9(34), 135–158.

KRÖMER, Ralf (2007), *Tool and Object: A History and Philosophy of Category Theory* (New York: Birkhäuser).

KRONECKER, Leopold (1887a), 'Ein Fundamentalsatz der allgemeinen Arithmetik', *Crelle, Journal für die reine und angewandte Mathematik*, 100, 209–240.

—— (1887b), 'Über den Zahlbegriff', *Crelle, Journal für die reine und angewandte Mathematik*, 101, 337–355.

LAFFORGUE, Laurent (2003), 'Chtoucas de Drinfeld, Formule des Traces d'Arthur-Selberg et Correspondance de Langlands', in Li Tatsien (ed.), *Proceedings of the International Congress of Mathematicians, Beijing 2002*, vol. 1, 383–400 (Singapore: World Scientific Publishers).

LANG, Serge (1993), *Algebra* (Reading: Addison-Wesley).

LAWVERE, F. William (1963), *Functorial Semantics of Algebraic Theories*, Ph.D. thesis, Columbia University. Republished with commentary by the author in: *Reprints in Theory and Applications of Categories*, No. 5 (2004) pp. 1–121. <http://138.73.27.39/tac/reprints/articles/5/tr5abs.html>.

—— (1965), 'An Elementary Theory of the Category of Sets', Lecture notes of the Department of Mathematics, University of Chicago. Republished with commentary by the author and Colin McLarty in: *Reprints in Theory and Applications of Categories*, No. 11 (2005) pp. 1–35. <http://138.73.27.39/tac/reprints/articles/11/tr11abs.html>.

—— (2002), 'Categorical Algebra for Continuum Micro Physics', *Journal of Pure and Applied Algebra*, 175, 267–287.

LAWVERE, F. William and SCHANUEL, S. (eds.) (1986), *Categories in Continuum Physics. Proceedings of a Workshop held at SUNY, Buffalo, N.Y., May 1982*, n. 1174 in Lecture Notes in Mathematics (New York: Springer-Verlag).

LENG, Mary (2002), 'Phenomenology and Mathematical Practice', *Philosophia Mathematica*, 10(1), 3–25.

MAC LANE, Saunders (1986), *Mathematics: Form and Function* (New York: Springer-Verlag).

—— (1998), *Categories for the Working Mathematician*, 2nd edn (New York: Springer-Verlag).

MCLARTY, Colin (1990), 'The Uses and Abuses of the History of Topos Theory', *British Journal for the Philosophy of Science*, 41, 351–375.

—— (2006a), 'Emmy Noether's 'Set Theoretic' Topology: From Dedekind to the Rise of Functors', in Jeremy Gray and José Ferreirós (eds.), *The Architecture of Modern Mathematics: Essays in History and Philosophy* (Oxford University Press) pp. 211–35.

—— (2007a), 'The Rising Sea: Grothendieck on Simplicity and Generality I', in Jeremy Gray and Karen Parshall (eds.), *Episodes in the History of Recent Algebra* (Providence, RI: American Mathematical Society).

—— (2007b), 'The Last Mathematician from Hilbert's Göttingen: Saunders Mac Lane as a Philosopher of Mathematics', *British Journal for the Philosophy of Science*.

MARQUIS, Jean-Pierre (Forthcoming), *From a Geometrical Point of View: The Categorical Perspective on Mathematics and its Foundations* (Kluwer).

MAZUR, Barry (1991), 'Number Theory as Gadfly', *American Mathematical Monthly*, 98, 593–610.

MILNE, James (1980), *Étale Cohomology* (Princeton: Princeton University Press).
MUMFORD, David (1988), *The Red Book of Varieties and Schemes* (New York: Springer-Verlag).
MUMFORD, David and TATE, John (1978), 'Fields Medals IV. An Instinct for the Key Idea', *Science*, 202, 737–739.
NOETHER, Emmy (1926), 'Abstract', *Jahresbericht DMV*, 34, 104.
──── (1927), 'Abstrakter Aufbau der Idealtheorie in algebraischen Zahl- und Funktionenkörpern', *Mathematische Annalen*, 96, 26–91. I cite this from Jacobson (1983) 493–528.
RESNIK, Michael (1981), 'Mathematics as a Science of Patterns: Ontology and Reference', *Noûs*, 15, 529–550.
──── (1997), *Mathematics as a Science of Patterns* (Oxford: Oxford University Press).
RUDIN, Walter (1953), *Principles of Mathematical Analysis* (New York: McGraw-Hill).
SARKARIA, K. (1999), 'The Topological Work of Henri Poincaré', in I. James (ed.), *History of Topology* (Amsterdam: Elsevier) pp. 123–168.
SERRE, Jean-Pierre (1958), 'Espaces Fibrés Algébriques', in *Séminaire Chevalley*, chapter exposé no. 1 (Secrétariat Mathématique, Institut Henri Poincaré).
SHAPIRO, Stewart (1997), *Philosophy of Mathematics: Structure and Ontology* (Oxford: Oxford University Press).
SOULÉ, Christophe (2003), 'The Work of Vladimir Voevodsky', in Li Tatsien (ed.), *Proceedings of the International Congress of Mathematicians, Beijing 2002*, vol. 1 (Singapore: World Scientific Publishers) pp. 99–104.
TAMME, Gunther (1994), *Introduction to Etale Cohomology* (Berlin: Springer-Verlag).
VAN DALEN, Dirk (1999), 'Luitzen Egbertus Jan Brouwer', in I. James (ed.), *History of Topology* (Amsterdam: Elsevier) pp. 947–964.
VAN DER WAERDEN, Bartel L. (1930), *Moderne Algebra* (Berlin: J. Springer).
WASHINGTON, L. (1997), 'Galois Cohomology', in G. Cornell, J. Silverman, and S. Stevens (eds.), *Modular Forms and Fermat's Last Theorem* (Springer-Verlag) pp. 101–120.
WEBER, Heinrich (1893), 'Leopold Kronecker', *Jahresberichte Deutsche Mathematiker Vereinigung*, 2, 5–31.
WEIL, André (1949), 'Number of Solutions of Equations in Finite Fields', *Bulletin of the American Mathematical Society*, 55, 487–495.
──── (1979), *Oeuvres Scientifiques* (Berlin: Springer-Verlag).

15

The Boundary Between Mathematics and Physics

ALASDAIR URQUHART

The traditional or 'received view' of scientific explanation widely held in the 1960s and 1970s was that scientific theories are applied axiomatic systems, with explanations and predictions taking the form of logical derivations from observational statements. However, this model does not seem to describe accurately some aspects of scientific practice, for example, the use of mean field models in condensed matter physics. Such models are more plausibly described as being mathematical constructs in their own right that may be only loosely related to the phenomena they are designed to model.

This latter view of models, however, has its own difficulties, rooted in the fact that the methods of physicists are more often than not lacking in mathematical rigour. The fruitful tension resulting from this is the main topic of this chapter, together with the problems arising from the recently renewed interaction between mathematics and physics.

15.1 Mathematics and physics

In the 1950s, when the logical positivist approach to the philosophy of science was still a dominating force in North American philosophical circles, a common view of scientific theories was that they were simply applied logical theories. An abstract or purely logical theory was held to consist of a set of logical axioms and rules, in which no fixed interpretation was assigned to the primitive relations and concepts of the theory. An applied theory was one in which observational meanings were assigned to certain primitive terms, the *observation vocabulary*, while the remaining terms (other than purely logical or mathematical notions), the *theoretical terms*, were to be explicitly defined using *correspondence rules* from the observational terms. A description and extended critical discussion of this picture, known as the 'Received View on Theories' can be found in Suppe (1977). Although only small parts of classical physics were in fact given an axiomatic form by the philosophers of science of the day, there was considerable optimism that larger and larger parts of science, or at least the more formal parts, could be formulated as sets of logical postulates and rules.

This approach to the philosophy of science, one that is associated with the names of neo-positivist philosophers such as Carnap, Reichenbach, Hempel, Feigl, and others, is now almost totally out of style. There are good reasons for this. The actual formulation of non-trivial scientific theories as logical edifices is a difficult task, demanding very good knowledge of both the scientific literature and the mathematical concepts needed to make the methods of a given area precise, and it is not surprising that philosophers and logicians never accomplished more than a very small fraction of what they had set out to do.

In any case, the relationship between theory and applications is surely more complicated than the schematic ideas propounded in the philosophical literature of the immediate post-war period. The 'theory and interpretation' view is that the application of a theory can be reduced to the 'theory and model' paradigm beloved of model theorists. So, for example, we can formulate the abstract theory of groups as a first-order theory with a single operation representing group composition. An applied version of this theory would be given by an interpretation in terms of a particular group, say, for example, the group of rigid motions in Euclidean 3-space.

This view is attractive and elegant. However, it is hard to square with the actual practice of scientists. The world is so complex that physicists who are attempting to provide mathematical models of physical reality do not in general begin by a direct attempt to formulate theories in which the primitive terms can be given an immediate empirical interpretation. Rather, they very often construct idealized mathematical models in which the behaviour of certain variables bears at least a qualitative resemblance to the real world—if the resemblance can be quantified, so much the better.[1]

As an example of such a model, let us look at a well-known model of ferromagnetism, the so-called Curie–Weiss model, described in most elementary textbooks of statistical physics (Thompson, 1972, pp. 95–105), (Yeomans, 1992, pp. 50–54). The empirical phenomenon to be explained is the spontaneous magnetization of ferromagnetic materials. If we cool a sample of such a material, while subjecting it to a magnetic field, then there is a sharply

---

[1] Recent work on the role of models and idealization can be found in the two collections Morgan and Morrison (1999) and Jones and Cartwright (2005).

defined critical temperature at which a phase transition takes place. If we turn the external field off above the critical temperature, the material no longer exhibits magnetic properties. However, if we turn it off below the critical temperature, the material retains a residual magnetic field, a phenomenon called 'spontaneous magnetization'.

Spontaneous magnetization appears as a collective property of large numbers of ferromagnetic atoms. Although the physical property presumably arises from the interaction of individual atoms, Pierre Weiss (1907) proposed the simplified model of a 'molecular field' representing the magnetic force η produced by the collection of atoms and postulated the self-consistent equation $\eta = \tanh(\eta/T)$, where T is a variable representing the temperature. This simple and primitive picture, the most elementary example of the class of mean field models, in fact does exhibit some of the key features of the physical system it is intended to depict. It shows a phase transition at a critical temperature, and in general its qualitative features show that we are mathematically on the right track, even though the numerical values we obtain from it are in fact not in accord with physical measurements (Yeomans, 1992, Chapter 4). In spite of the well-known shortcomings of mean field theory, it is the usual starting point for workers in statistical mechanics investigating a new system (Fischer and Hertz, 1991, p. 19).

A partial explanation of the success of Weiss's highly simplified model was provided by a Japanese physicist (Husimi, 1954), who pointed out that Weiss's molecular field can be derived as the large N limit of a collection of classical spin variables $\sigma_i = +1, -1$, interacting through a Hamiltonian in which every spin acts on every other spin—in the large N limit the interaction strength is infinitesimal, that is to say, its strength is of order $1/N$. The mathematical details of Husimi's observation were worked out later in (Kac, 1968)—see also (Thompson, 1972, §4.5).

Physicists use the phrase 'mean field model' to refer to two distinct categories of model. The first is a model of the Weiss type, in which the collective action of a large set of microscopic variables is replaced by a macroscopic quantity representing an average or mean; the second is an interaction model of the type described by Husimi and Kac, in which microscopic variables (such as spins) interact with an unbounded number of other microsopic variables. The two types of model, however, are not completely equivalent mathematically, even in the paradigmatic case of the Curie–Weiss model.

It is clear in the case of mean field models that we are dealing with mathematical caricatures of physical systems, that nevertheless exhibit some of the most important broad features of critical phenomena and phase transitions. It is tempting to say that these models are simply mathematical objects in their own right that are related in only a very rough and ready way to the real world. However, this point of view brings with it considerable difficulties, that are the fundamental subject of this chapter.

The basic difficulty is this: if we construe these models simply as mathematical objects, then we have to face the fact that physicists do not employ normal mathematical methods in investigating them. Rather, the methods that they use are frequently so far from normal mathematical practice that it is sometimes not clear that the objects themselves are even mathematically well defined.

15.2 A renewal of vows

Throughout most of their history, mathematics and physics have been closely intertwined, as can be seen by looking at the history of ancient astronomy (Neugebauer, 1975). Furthermore, if we look at work by mathematicians and physicists from the early 19th century, the standards of mathematical rigour used by both groups of researchers appear indistinguishable. For example, Carl Friedrich Gauss has a high reputation among mathematicians for the rigour of his mathematical reasoning, but a glance at his famous treatise of 1827 on differential geometry (Gauss, 2005) shows that it makes free and uninhibited use of the infinitesimal quantities that later mathematicians were to regard with fear and loathing.

The middle years of the 20th century saw an unprecedented divergence between the two communities of researchers. As Faddeev (2006) remarks, in the earlier part of this century, mathematical physics was not distinguished from theoretical physics—both Henri Poincaré and Albert Einstein were called mathematical physicists. However, with the expansion of theoretical physics in the 1920s and 1930s, a separation between the two areas occurred. Whereas 'mathematical physics' came to be understood as a somewhat restricted area, confined to the study of mathematical techniques such as the solution of partial differential equations and the calculus of variations, the theoretical physicists themselves, caught up in trying to understand the physics of atoms and elementary particles, began to move further and further away from the confines of mathematical rigour. Although there are notable examples of interactions between mathematicians and physicists in the middle part of the 20th century, such as the elucidation of quantum mechanics in terms of linear operators on Hilbert space and the application of fibre bundles to non-Abelian gauge theory, in general the concepts and standards of the two fields showed an ever increasing divergence.

The pressure of experimental results, notably the explosion of new elementary particles discovered during the 1950s and 1960s, forced theoretical physicists into a quasi-empirical mode of operation, and also pushed them in the direction of purely formal calculations. At the same time, the mathematicians were attracted by increasing rigour and abstraction. Emblematic of the period is the inception in the mid-1930s of the series of books by the Franco–American mathematical collective who published under the name of Nicolas Bourbaki. In the foundational volume of his encyclopedia, Bourbaki (2004) writes in the preface: 'This series of volumes,... takes up mathematics at the beginning, and gives complete proofs'. Partly as a result of Bourbaki's influence, the language employed by mathematicians and physicists continued to diverge. The physicists, brought up in the older framework of traditional calculus, with its emphasis on formal manipulations of integrals and infinite series, to a considerable degree no longer shared a common language with the mathematicians. Freeman Dyson (1972) remarked sadly:

> As a working physicist, I am acutely aware of the fact that the marriage between mathematics and physics, which was so enormously fruitful in past centuries, has recently ended in divorce. Discussing this divorce, the physicist Res Jost remarked the other day, 'As usual in such affairs, one of the two parties has clearly got the worst of it.' During the last twenty years we have seen mathematics rushing ahead in a golden age of luxuriant growth, while theoretical physics left on its own has become a little shabby and peevish.

However, in an unexpected and exciting development, the apparently diverse streams of mathematics and physics are beginning to reconverge. It is not quite clear what led to these new developments, but there seem to have been forces acting on both sides of the divide. On the side of physics, the success in the 1970s of the Standard Model of elementary particles, formulated in the language of quantum field theory, led to an astonishingly successful framework in which to fit the unruly zoo of elementary particles that had been discovered experimentally in the decades following the Second World War. The scope and success of this triumph of theoretical physics is remarkable. The $SU(3) \times SU(2) \times U(1)$ model has been confirmed repeatedly, and is consistent with virtually all physics down to the scales probed by current particle accelerators, roughly 10^{-16} cm. The very success of this model has led to a change of emphasis. A great deal of current effort is being devoted to string theory. However, the scale of string theory is roughly 20 orders of magnitude smaller, of the order of the Planck length, 1.6×10^{-33} cm. Consequently, there appears to be no hope of direct experimental tests of this theory. Instead, the physicists are guided to an increasing extent by aesthetic criteria more and more resembling those used by pure mathematicians. On the other side of the divide, mathematicians seem to have grown tired of the abstraction of Bourbaki-style mathematics, and have turned back towards more concrete problems, often using the powerful machinery developed during the period of abstraction to solve long standing classical problems, as the recent solution of the Fermat problem by Andrew Wiles illustrates.

The fresh interactions between mathematicians and physicists have given rise to numerous volumes of proceedings and tutorials in which the two communities have attempted to convey the new insights that they developed during the period of separation. A good example is provided by the series of meetings at Les Houches bringing together number theorists and mathematically minded physicists (Luck *et al.*, 1970; Waldschmidt *et al.*, 1992; Cartier *et al.*, 2006). Another sign of the times is the bulky set of two volumes containing tutorials on quantum fields and strings especially aimed at mathematicians (Deligne *et al.*, 1999), whose stated goal is 'to create and convey an understanding, in terms congenial to mathematicians, of some fundamental notions of physics, such as quantum field theory, supersymmetry and string theory'. Interactions between mathematics and physics are taking place across a broad front, in areas such as low-dimensional topology, conformal field theory, random matrix theory, number theory, higher dimensional geometry, and even theoretical computer science. This is an exciting time!

15.3 Exploring the boundary

Of course, when two partners get together again after a long separation, there are often difficulties and friction. The reunion of mathematics and physics is no exception. The extraordinarily fertile language of the physicists, centred around the Feynman integral, presents enormous difficulties to a purely mathematical understanding. More generally, physicists pay little attention to mathematical rigour in their calculations, and often regard the mathematicians' insistence on well-defined quantities and objects as useless pedantry, holding up progress on basic questions.

Kevin Davey (2003) raises the question of whether mathematical rigour is necessary in physics, and (with appropriate reservations) concludes that it is not. Basing his analysis on the examples of the Dirac delta function and Feynman's path integrals, he points out that physicists are not troubled by the lack of rigour in their own reasoning, because they restrict the inferences involving questionable methods. However, this solution cannot satisfy mathematicians who are interested in adapting ideas from physics in solving their own problems. The inferential restrictions employed by the physicists usually take the form of rules of thumb, and the boundary between secure and questionable inferences is in general far from clear.

All of this, though, should be regarded as an opportunity for philosophers who are not afraid of 'getting their hands dirty'. It is usually held that the emergence of Greek philosophy in Ionia was no accident, since on the eastern shore of the Aegean were flourishing mercantile cities where many cultures met and interacted. In the same way, philosophers of mathematics have the opportunity to study at first hand renewed interactions between two previously separate intellectual communities that developed diverse ideas and even notions of proof during the decades of separation. In his Gibbs lecture, entitled *Missed Opportunities*, Freeman Dyson emphasizes the fruitfulness of the areas in physics where concepts actually clash. After discussing four examples of missed opportunities of interaction from the past, he remarks:

> The past opportunities which I discussed have one important feature in common. In every case there was an empirical finding that two disparate or incompatible mathematical concepts were juxtaposed in the description of a single situation. ... In each case the opportunity offered to the pure mathematician was to create a wider conceptual framework within which the pair of disparate elements would find a harmonious coexistence. I take this to be my methodological principle in looking for opportunities that are still open. I look for situations in which the juxtaposition of a pair of incompatible concepts is acknowledged but unexplained. (Dyson, 1972, p. 645)

The whole passage might be taken as an illustration of the Hegelian dialectic!

Dyson's observations point in a fruitful direction for philosophers of mathematics interested in tracing the origins of mathematical concepts in non-mathematical ideas. In the context of such interactions as Dyson describes, they can study the process of extracting and refining mathematical concepts from the raw material of physics. Furthermore, they can observe in detail the process by which mathematics absorbs recalcitrant and sometimes contradictory material from outside fields.

At this point, it might be as well to forestall a misunderstanding. There are a number of recent excellent studies examining the interaction between mathematics and physics in detail; for example, Batterman (2002), Morrison (2000), Steiner (1998), and Wilson (2000). However, these are largely concerned with the applicability of mathematics to physics. The main theme, both in this introductory chapter, and in the research article that follows, is the converse problem—the applicability of physics to mathematics. The article (Fine and Fine, 1997) is much closer to our current theme; it relates a fascinating story that begins with the experimental physics of neutral mesons, and ends with mathematical developments involving the Atiyah–Singer index theorem.

At least two strategies can be observed in the process of absorbing physical ideas into the fabric of mathematics. One is to treat the ideas of the physicists as purely heuristic in nature, and to establish their mathematical conjectures by conventional means. A second and more radical idea is to attempt to rework the ideas of the physicists and thereby construct new conceptual schemes that validate at least some of their calculations. It is this second idea towards which Dyson points in the quotation above. In the research contribution following this introductory chapter, I shall discuss a number of relevant cases from the history of mathematics, as well as mentioning some open areas where this process of mathematical assimilation is still fragmentary or incomplete.

15.4 Further reading

The topic of the interaction between mathematics and physics is vast, and this chapter discusses only a single aspect of this interaction. In §15.1, mean field models were discussed as examples of models in physics that seem to contradict older received views on scientific explanation. The book by Robert W. Batterman (2002) contains detailed discussion of many other models of a similar type. He points out (Batterman, 2002, p. 13) that science often requires methods that 'eliminate both detail, and, in some sense, precision'. He describes such methods as 'asymptotic methods', and the type of reasoning involved in them as 'asymptotic reasoning'. He draws very interesting and rather heterodox conclusions from this fact concerning models of scientific explanation and theory reduction, as well as the notion of emergent properties. (See also the very interesting review Belot (2005) and the reply by Batterman (2005).) Similar themes in the context of condensed matter physics are discussed in the unusually stimulating monograph by Martin H. Krieger (1996).

In the conclusion to a widely cited essay John von Neumann (1947) wrote:

> At a great distance from its empirical source, or after much 'abstract' inbreeding, a mathematical subject is in danger of degeneration. At the inception the style is usually classical; when it shows signs of becoming baroque, then the danger signal is up. ... Whenever this stage is reached, the only remedy seems to me to be the rejuvenating return to the source: the reinjection of more or less empirical ideas. I am convinced that this was a necessary condition to conserve the freshness and the vitality of the subject and that this will remain equally true in the future.

Mathematics is now in the midst of a period of rapid growth through the absorption of a multitude of new ideas from physics. This renewed interaction between the two fields offers, in addition to the problems already discussed, material for philosophical thought on issues such as the growth of mathematical thought, the porous line between method of discovery and method of demonstration in mathematics and in physics, and issues of concept formation in the mathematical and the physical sciences. Further philosophical reflections on the process of growth of mathematical knowledge are to be found in the collection Grosholz and Breger (2000).

Finally, many thought-provoking ideas on the relationship between mathematics and physics are to be found in the writings of Sir Michael Atiyah, particularly in Volumes 1 and 6 of his *Collected Works* (Atiyah, 1998–2004).

Bibliography

ATIYAH, Michael (1988–2004), *Collected Works* (6 vols) (Oxford: Oxford University Press).

BATTERMAN, Robert W. (2002), *The Devil in the Details: Asymptotic Reasoning in Explanation, Reduction and Emergence* (Oxford: Oxford University Press).

—— (2005), 'Response to Belot's "Whose Devil? Which Details?"', *Philosophy of Science*, 72, 154–163.

BELOT, Gordon (2005), 'Whose Devil? Which Details?', *Philosophy of Science*, 72, 128–153.

BOURBAKI, Nicolas (2004), *Elements of Mathematics: Theory of Sets* (Berlin: Springer-Verlag).

CARTIER, Pierre, JULIA, Bernard, MOUSSA, Pierre, and VANHOVE, Pierre (eds.) (2006), *Frontiers in Number Theory, Physics, and Geometry I: On Random Matrices, Zeta Functions, and Dynamical Systems* (New York: Springer-Verlag).

DAVEY, Kevin (2003), 'Is Mathematical Rigour Necessary in Physics?', *British Journal for the Philosophy of Science*, 54, 439–463.

DELIGNE, Pierre, ETINGOF, Pavel, FREED, Daniel S., JEFFREY, Lisa C., KAZHDAN, David, MORGAN, John W., MORRISON, David R., and WITTEN, Edward (eds.) (1999), *Quantum Fields and Strings: A Course for Mathematicians* (American Mathematical Society and The Institute for Advanced Study).

DYSON, Freeman J. (1972), 'Missed Opportunities', *Bulletin of the American Mathematical Society*, 78, 635–652.

FADDEEV, L.D. (2006), 'What Modern Mathematical Physics is Supposed to Be', in *Mathematical Events of the Twentieth Century* (New York: Springer-Verlag) pp. 75–84.

FINE, Dana and FINE, Arthur (1997), 'Gauge Theory, Anomalies and Global Geometry: The Interplay of Physics and Mathematics', *Studies in History and Philosophy of Modern Physics*, 28, 307–323.

FISCHER, K. H. and HERTZ, J. A. (1991), *Spin Glasses* (Cambridge: Cambridge University Press).

GAUSS, Carl Friedrich (2005), *General Investigations of Curved Surfaces* (New York: Dover Publications Inc.). Edited with an introduction and notes by Peter Pesic.

GROSHOLZ, Emily and BREGER, Herbert (eds.) (2000), *The Growth of Mathematical Knowledge* (New York: Kluwer Academic Publishers).

HUSIMI, Kôdi (1954), 'Statistical Mechanics of Condensation', in *Proceedings of the International Conference of Theoretical Physics, Kyoto and Tokyo, September 1953*, 531–533.

JONES, Martin R. and CARTWRIGHT, Nancy (eds.) (2005), *Idealization XII: Correcting the Model; Idealization and Abstraction in the Sciences* (Rodopi). Poznań Studies in the Philosophy of the Sciences and the Humanities, vol. 86.

KAC, Mark (1968), 'Mathematical Mechanisms of Phase Transitions', in M. Chretien, E. P. Gross, and S. Deser (eds.), *Statistical Physics: Phase Transitions and Superfluidity*, vol. 1 (New York: Gordon and Breach) pp. 241–305. Brandeis University Summer Institute in Theoretical Physics 1966.

KRIEGER, Martin H. (1996), *Constitutions of Matter: Mathematically Modeling the Most Everyday of Physical Phenomena* (Chicago: University of Chicago Press).

LUCK, Jean-Marc, MOUSSA, Pierre, and WALDSCHMIDT, Michel (eds.) (1970), *Number Theory and Physics* (Berlin: Springer-Verlag). Springer Proceedings in Physics, vol. 47.

MORGAN, Mary S. and MORRISON, Margaret (eds.) (1999), *Models as Mediators* (Cambridge: Cambridge University Press).

MORRISON, Margaret (2000), *Unifying Scientific Theories: Physical Concepts and Mathematical Structures* (Cambridge: Cambridge University Press).

NEUGEBAUER, Otto (1975), *A History of Ancient Mathematical Astronomy* (New York: Springer-Verlag).

STEINER, Mark (1998), *The Applicability of Mathematics as a Philosophical Problem* (Harvard: Harvard University Press).

SUPPE, Frederick (ed.) (1977), *The Structure of Scientific Theories*, 2nd edn (Chicago: University of Illinois Press).

THOMPSON, Colin J. (1972), *Mathematical Statistical Mechanics* (New York: Macmillan).

VON NEUMANN, John (1947), 'The Mathematician', in R. B. Heywood (ed.), *The Works of the Mind* (Chicago: University of Chicago Press) pp. 180–196.

WALDSCHMIDT, Michel, MOUSSA, Pierre, LUCK, Jean-Marc, and ITZYKSON, Claude (eds.) (1992), *From Number Theory to Physics* (Berlin: Springer-Verlag).

WEISS, Pierre (1907), 'L'hypothèse du champ moléculaire et la propriété ferromagnétique', *Journal de physique, theórique et appliquée*, 6, 661–690.

WILSON, Mark (2000), 'The Unreasonable Uncooperativeness of Mathematics in the Natural Sciences', *The Monist*, 83, 296–314.

YEOMANS, J. M. (1992), *Statistical Mechanics of Phase Transitions* (Oxford: Oxford University Press).

16

Mathematics and Physics: Strategies of Assimilation

ALASDAIR URQUHART

16.1 Introduction

Philosophers of mathematics and logicians have traditionally shown interest in the foundational controversies of the early part of the 20th century, when non-constructive methods were frowned upon by mathematicians like Brouwer and Weyl. As far as the community of working mathematicians is concerned, this seems to be a dead issue. The vast majority of mathematicians use non-constructive methods quite freely, and seldom seem to worry about their validity. Nevertheless, debates over the methods of proof appropriate in mathematics are by no means dead.

The irruption of the methods of physicists into the realm of pure mathematics, such as Edward Witten's application of topological quantum field theory to the theory of knot invariants, recently gave rise to a heated controversy in the pages of the *American Mathematical Bulletin*. The mathematical physicists Arthur Jaffe and Frank Quinn pointed to some dangers in the use of speculative methods in mathematics, which they dub 'theoretical mathematics' by analogy with theoretical physics (Jaffe and Quinn, 1993). Among the dangers that they see are: a failure to distinguish between conjecture and mathematical speculation, with a resulting drop in the reliability of the mathematical literature, the creation of 'dead areas' in research when full credit is claimed by vigorous theorizers, and unfair distribution of rewards when theoreticians are accorded the lion's share of credit for results that are only proved in a rigorous fashion much later.

Jaffe and Quinn's article gave rise to a vigorous set of responses (Thurston, 1994; Atiyah *et al.*, 1994) from a number of distinguished mathematicians, including Michael Atiyah, Armand Borel, James Glimm, Morris Hirsch,

Saunders Mac Lane, Benoit Mandelbrot, Karen Uhlenbeck, René Thom and William Thurston. The reactions of both mathematicians and physicists to the intentionally provocative article are remarkably varied, and repay close study. One of the most interesting rejoinders came from Sir Michael Atiyah, who remarked:

> My fundamental objection is that Jaffe and Quinn present a sanitized view of mathematics which condemns the subject to an arthritic old age ... if mathematics is to rejuvenate itself and break exciting new ground it will have to allow for the exploration of new ideas and techniques which, in their creative phase, are likely to be as dubious as in some of the great eras of the past. Perhaps we now have high standards of proof to aim at but, in the early stages of new developments, we must be prepared to act in more buccaneering style (Atiyah *et al.*, 1994, p. 1)

Saunders Mac Lane made a vigorous reply to these startling comments of Atiyah, remarking that 'a buccaneer is a pirate, and a pirate is often engaged in stealing. There may be such mathematicians now.... We do not need such styles in mathematics' (Mac Lane, 1997, p. 150).

There is a striking article (Cartier, 2000) by the distinguished French mathematician Pierre Cartier that touches on many of the themes of the debate following the publication of the Jaffe–Quinn article. In the introduction to his paper, Cartier writes:

> The implicit philosophical belief of the working mathematician is today the Hilbert–Bourbaki formalism. Ideally, one works within a *closed system*: the basic principles are clearly enunciated once for all, including (that is an addition of twentieth century science) the formal rules of logical reasoning clothed in mathematical form. The basic principles include precise definitions of all mathematical objects... My thesis is: *there is another way of doing mathematics, equally successful, and the two methods should supplement each other and not fight*. This other way bears various names: symbolic method, operational calculus, operator theory ... Euler was the first to use such methods in his extensive study of infinite series, convergent as well as divergent. ... But the modern master was R. Feynman who used his diagrams, his disentangling of operators, his path integrals... The method consists in stretching the formulas to their extreme consequences, resorting to some internal feeling of coherence and harmony. (Cartier, 2000, p. 6)

These remarks of Cartier are all the more startling if we remember that in the 1950s, he was a core member of the Bourbaki group, contributing perhaps 200 pages a year to the collective work.

A superficial interpretation of the words of Atiyah and Cartier would be that they are advocating a loosening of the standards of mathematical rigour. However, a careful examination of their articles reveals a more nuanced picture. Most of Cartier's paper is devoted to showing how some of the most outrageous claims of Leonard Euler and other formally inclined mathematicians can be given a sensible interpretation, even such crazy-looking assertions as '$\infty! = \sqrt{2\pi}$' (Cartier, 2000, p. 64). Similarly, Atiyah and Thurston, though declaring themselves in opposition to the attitudes of Jaffe and Quinn, which they see as overly restrictive and authoritarian, do not really disagree with them on the question of separating rigorously established results from mathematical speculation. Thus Atiyah remarks: 'I find myself agreeing with much of the detail of the Jaffe–Quinn argument, especially the importance of distinguishing between results based on rigorous proofs and those which have a heuristic basis' (Atiyah *et al.*, 1994, p. 1), and Thurston says: 'I am *not* advocating any weakening of our community standard of proof; I am trying to describe how the process really works. Careful proofs that will stand up to scrutiny are very important' (Thurston, 1994, p. 9).

Nevertheless, it is clear that the debate reflects some major changes taking place in the field of mathematics, changes that are reflected in the concluding remarks of Cartier in a 1997 interview with Marjorie Senechal:

> When I began in mathematics the main task of a mathematician was to bring order and make a synthesis of existing material, to create what Thomas Kuhn called *normal science*. Mathematics, in the forties and fifties, was undergoing what Kuhn calls a solidification period. In a given science there are times when you have to take all the existing material and create a unified terminology, unified standards, and train people in a unified style. The purpose of mathematics, in the fifties and sixties, was that, to create a new era of normal science. Now we are again at the beginning of a new revolution. Mathematics is undergoing major changes. We don't know exactly where it will go. It is not yet time to make a synthesis of all these things—maybe in twenty or thirty years it will be time for a new Bourbaki. I consider myself very fortunate to have had two lives, a life of normal science and a life of scientific revolution. (Senechal, 1998)

It is clear that mathematics, after a period of inner consolidation and abstraction, is now open to influences on many sides, and the resulting opening-up induces anxieties on numerous issues. One of the anxieties relates to the classification of researchers as mathematicians. Benoit Mandelbrot clearly considers himself to be a mathematician, but rails against the evil influence of Bourbaki, and the 'typical members of the AMS,' whom he stigmatizes as 'Charles mathematicians' (since the AMS headquarters is on Charles Street, Providence, RI). According to Mandelbrot, Jaffe and Quinn propose to set up a police state within 'Charles mathematics', and a world cop beyond its borders (Atiyah *et al.*, 1994, p. 16). Charles mathematicians practice a sterile form of research, and were the bane of great creative figures such as Henri Poincaré and Paul Lévy, whom they persecuted because of their lack of rigour. In reply, Saunders Mac Lane dismisses Mandelbrot as 'a notorious non-mathematician' (Mac Lane, 1997, p. 149).

I shall not discuss the problem of demarcation of mathematics from other areas. It is clear that there is a great deal of activity in areas of science that do not fall within the restricted boundaries of rigorous mathematics, but are in some sense mathematical; the work of Mandelbrot certainly falls into this category, as does the seminal work of Mitchell Feigenbaum in the area of dynamical systems. Rather, I shall take for granted that there is a more or less clear distinction to be made between fully rigorous work, and work that is lacking rigour, even though it may be broadly mathematical in some sense. Instead, the topic that I shall discuss in the remainder of this essay is the process whereby the more speculative parts of mathematical work are incorporated into the body of rigorous research.

16.2 Proof in physics and mathematics

If we define a proof as something that conveys conviction that a given assertion is true, then it is obvious that proofs do not have to obey the laws of logic or classical mathematics. In this section, we address the question of what is meant by a proof, as it appears to mathematicians and physicists. The combinatorialist X. G. Viennot sets the scene in the following quotation:

> Usually, in the physics literature, the distinction between establishing a formula from computer experiments and giving a mathematical proof is not clearly stated... When the number of known elements of the sequence is much bigger than the number of elements needed to guess the exact formula, the probability that the formula is wrong is infinitesimal, and thus the formula can be considered as an experimentally true statement. But mathematically, it is just a conjecture waiting for a proof, and maybe more: a crystal-clear [bijective] understanding. (Viennot, 1992, p. 412)

As a result of the evolution of mathematics in the 20th century, the notion of mathematical proof is quite standardized. In a rigorous mathematical argument, all objects are to be defined as set-theoretical objects on the basis of Zermelo–Fraenkel set theory, and their fundamental properties are to be established on the basis of the usual set-theoretical axioms. The resulting consensus on the basis of mathematics gives contemporary mathematics considerable unity and cohesion. The fact that from time to time, mathematicians have advocated alternative foundations such as category theory does not really disturb this basic picture, since, although category theory does in fact provide a more convenient framework than set theory in many cases, there are some general techniques available for reconciling category-theoretic methods with those of set theory (Blass, 1984).

One does not have to go far in the literature of physics to discover that no such consensus on the notion of proof exists in that area. Work in physics runs the gamut from fully rigorous proofs of theorems (most of the literature in general relativity falls into this category) to purely formal manipulations backed up by computer simulations, as in the case of a large part of the literature in condensed matter physics. The important area of quantum field theory occupies a kind of intermediate position, where some of the computations can be made fully rigorous, as a result of the hard work of constructive field theorists (Glimm and Jaffe, 1987; Johnson and Lapidus, 2000), whereas the computations that really matter to the physicists, such as those in quantum electrodynamics and quantum chromodynamics, still seem to languish in a mathematical limbo, being defined only in terms of ill-defined perturbation expansions about the theory of the free field.

The number of physicists who write papers that satisfy the usual standards of mathematical rigour is small, perhaps only a few per cent of the total number of working physicists. However, their work is among the great achievements of the physics of the last century. I might mention here the work of Lars Onsager on the two-dimensional Ising model, and that of Dyson, Lieb, and Thirring on the stability of matter. These are examples of beautiful mathematics as well as beautiful physics, and it is purely a historical accident that this work is not usually considered among the major *mathematical* accomplishments of the last century.

It is easy to understand that classically trained mathematicians might regard with horror this unruly world of purely formal computations, in which it is hard to lay hold of mathematical bedrock in the maelstrom of ill-defined infinite series and objects whose nature is far from clear. The mathematician Yuri Manin conveys this bewilderment very tellingly:

> The author, by training a mathematician, once delivered four lectures to students under the title 'How a mathematician should study physics'. In the lectures he said that modern theoretical physics is a luxuriant, totally Rabelaisian, vigorous world of ideas, and a mathematician can find in it everything to satiate himself except the order to which he is accustomed. Therefore a good method for attuning oneself to the active study of physics is to pretend that you are attempting to induce this very order in it. (Manin, 1981, p. x)

The remainder of this article will be devoted to the topic of just how mathematicians attempt to induce this order.

A time-hallowed form of philosophical writing, that continues to be popular to the present day, is the commentary on a classical text. Typically, the writer proposes an 'interpretation' of this text, and defends it, often with great polemical vigour, against competing interpretations. The commentator frequently seems convinced that the interpretation offered is the only correct one, and all the earlier attempts were completely incorrect and illusory. It is an odd and quite surprising fact that the same philosophers who teach their undergraduate students Quine's sceptical arguments about the notion of meaning, and describe them as great contributions to modern analytic philosophy, also write such polemical articles of historical commentary. The idea that there is a 'correct' view as to what a given philosopher might have meant in a puzzling passage seems taken for granted in such debates.

The rules of the game in such debates between would-be exegetes are far from obvious—it is not clear, for example, just what evidence could be counted as settling the dispute. However, the history of mathematics provides us with some interesting examples where puzzling, even contradictory, computations were later given consistent interpretations. In this case, at least, the rules of the game have greater clarity than in the case of philosophical writings. The criteria of success are fairly simple—to provide a consistent conceptual framework in which the puzzling computations make sense.

16.3 Varieties of infinitesimals

Perhaps the most famous example of such a development lies in the area of calculus. The invention of the calculus is linked indissolubly to that of early modern physics; it is notorious that the early work was lacking in rigour, as new results and ideas appeared in a flood. However, in spite of early attempts at making Newtonian calculus rigorous, such as the work of Colin MacLaurin (1742), a fully satisfactory foundation for the differential and integral calculus did not appear until the mid-19th century. Even so, the older tradition of infinitesimals, even though supposedly banished from rigorous discourse by the new limit concepts, exerted a seemingly irresistible attraction on practical mathematicians, and the folk tradition of more or less naive reasoning with infinitesimals is far from dead even today, as can be seen (for example) from a set of introductory lectures on celestial mechanics by Nathaniel Grossman (1996). In the opening remarks of his first chapter, he says: 'While these objects are indispensable to applied mathematicians and appliers of mathematics, their language has been long banished from the usual calculus books because they were considered to be "unrigorous" ' (Grossman, 1996, p. 1).

Philosophers who are interested in the foundations of mathematics know that in the 1960s, Abraham Robinson created a consistent theory of infinitesimals that has found significant applications in many areas of mathematics. What is not so well known is that there are several theories of infinitesimals, so that competing interpretations of the classical passages involving infinitesimals can be given divergent interpretations. Of these theories, the most interesting from the philosophical point of view is the theory of smooth infinitesimal analysis (Moerdijk and Reyes, 1991; Bell, 1998). In this theory, all functions are continuous—it follows from this that the logic is of necessity non-classical, since with the help of the law of excluded middle we can define the 'blip function' that is 1 at the real number 0, and 0 everywhere else.

Abandoning classical logic may seem a high price to pay, but in fact the resulting theory is very rich. In the 'smooth world' not only are all functions continuous, but in addition, the principle of microlinearity, or microaffineness (Bell, 1998, p. 23) holds. This principle says that if we examine the infinitesimal neighbourhood of a point on a curve in the smooth world, then the curve, restricted to that neighbourhood, is a straight line! In other words, in such a universe, the idea of early writers on the calculus, such as Isaac Barrow, that curves are made out of 'linelets' is here literally true. Smooth infinitesimal analysis provides some retrospective justification for the idea of a curve as a polygon with infinitely many sides that played a central part in the early Leibnizian calculus (Bos, 1974, p. 15).

What is more, the nature of infinitesimals in this theory enables the basic techniques of differential calculus to be reduced to simple algebra. In the early writings on the differential and integral calculus, it was common to neglect higher order infinitesimals in calculations. That is to say, if ϵ is an infinitesimal quantity, then at a certain stage in the calculation, early writers such as Newton treat ϵ^2 as if it were equal to zero (even though at other stages in the calculation, ϵ must be treated as nonzero). This apparent inconsistency was the target of Berkeley's stinging criticisms in *The Analyst* (Berkeley, 1734). Strange to say, these inconsistencies disappear in the smooth worlds invented by the category theorists. In this theory, a *nilpotent infinitesimal* ϵ is one that satisfies $\epsilon^2 = 0$. It follows from the microlinearity postulate that in a smooth world, the axiom $\neg \forall x (x^2 = 0 \Rightarrow x = 0)$ holds (Bell, 1998, pp. 23–24, 103). Here, the fact that the logic is non-classical (intuitionistic) saves us from contradiction, because we cannot infer from this fact that $\exists x (x^2 = 0 \wedge x \neq 0)$; this last formula results in an outright contradiction, whereas the theory of smooth analysis is known to be consistent by means of the models constructed in Moerdijk and Reyes (1991).

The theory of smooth analysis has a good claim to be a consistent interpretation of the work of the classical analysts, and evidence for this is provided in both Moerdijk and Reyes (1991) and Bell (1998). Moerdijk and Reyes give explicit examples from classical synthetic differential geometry, such as the work of Elie Cartan (Moerdijk and Reyes, 1991, pp. 3–6), where intuitive reasoning involving infinitesimal quantities can be interpreted directly in smooth analysis. Similarly, Bell provides a remarkably diverse range of applications in calculus, mathematical physics, and synthetic differential geometry (Bell, 1998, Chapters 3–7) where the informal practice of engineers and physicists can be made fully rigorous. As another example, the computations of Grossman, cited above, seem to fit most naturally into the world of smooth infinitesimal analysis.

An even more striking anticipation of smooth infinitesimal analysis is to be found in the work of the Dutch theologian, Bernard Nieuwentijt; a detailed discussion of his research on the foundations of calculus is to be found in Mancosu (1996, pp. 158–164). In a booklet of 1694, Nieuwentijt attempted to bring rigour to the Leibnizian calculus by proceeding along lines remarkably similar to those described above, since one of his basic assumptions is that there are nonzero quantities ϵ so that $\epsilon^2 = 0$.

Thus, smooth infinitesimal analysis has a good claim to be a satisfying interpretation of the older texts, in spite of its denial of classical logic. This fact should interest philosophers in general, since it affords a significant example where convincing reasons have been offered for a change in the logical basis of a theory, in accordance with Quine's view that no part of science is immune from revision. Geoffrey Hellman (2006) offers further philosophical reflections on this intriguing example of mathematical pluralism.

It remains true, however, that smooth infinitesimal analysis cannot account for all of the classical uses of infinitesimals. Euler, for example, frequently assumes that his infinitesimals have inverses; these are unbounded quantities larger than any finite quantity. Such invertible infinitesimals cannot exist in a smooth world, on pain of contradiction. In fact, Moerdijk and Reyes (1991, pp. v–vi) point out that in the writings of geometers like Sophus Lie and Elie Cartan, there are *two* kinds of infinitesimals, namely, nilpotent infinitesimals and invertible infinitesimals. Robinson's Nonstandard Analysis contains the second kind, but not the first. Hence, there results an interesting bifurcation of the theory, showing that interpretations of anomalous concepts are by no means unique, contrary to a frequent assumption of the philosophical community.

It should be emphasized that if we attempt to make rigorous mathematical sense of earlier inconsistent practices, then it is inevitable that we cannot provide a consistent interpretation of *all* of the earlier calculations and proofs. Indeed, this follows from the logic of the situation. Abraham Robinson (1966, Chapter X) argued strongly that his modern infinitesimal analysis could be seen as providing a retrospective justification for the procedures of Leibniz and his followers. Nevertheless, historians of mathematics such as Bos (1974) have looked askance at this attempt at assimilating Leibnizian methods to modern techniques. However, some of Bos's criticisms of Robinson involve absurd and impossible demands—for example, his first criticism (Bos, 1974, p. 83) is that Robinson proves the existence of his infinitesimals, whereas Leibniz does not! The fact remains—no interpretation of dubious and inconsistent practices inherited from the past can be perfect in every respect. The most we can ask is that a rigorous interpretation reproduce at least some of the most important calculations and concepts of the older technique.

16.4 The umbral calculus

The example of infinitesimal calculus is rather intricate. To illustrate some of the basic methods of assimilation, we shall look at a simpler case, the umbral or symbolic calculus. This calculus first made its appearance as a computational device for manipulating sequences of constants (Blissard, 1861); for the history of the method, see Bell (1938). Later, it was rediscovered by the number theorist and inventor of mathematical games Lucas (1876). The self-taught engineer and theorist of electromagnetism, Oliver Heaviside developed a similar operational calculus in the course of solving the differential equations arising in the theory of electromagnetism. The famous analyst Edmund T. Whittaker rated Heaviside's operational calculus as one of the three most important mathematical discoveries of the late 19th century. However, Heaviside's work was regarded with distrust until Bromwich gave a rigorous interpretation of Heaviside's operators as contour integrals.

All of these developments were regarded as somewhat questionable when they first appeared. To see why, let us use the example of the computation of the Bernoulli numbers B_n in the umbral calculus. These numbers, of which the first few are as follows:

$$B_0 = 1,\ B_1 = -\frac{1}{2},\ B_2 = \frac{1}{6},\ B_3 = 0,\ B_4 = -\frac{1}{30},\ B_5 = 0,\ B_6 = \frac{1}{42}, \ldots,$$

play an important role in number theory and combinatorics. I claim they are defined by the umbral equation: $(B+1)^n = B^n$, for $n \geq 2$, together with the initial condition $B^0 = 1$. By this is meant: expand the left-hand side by the binomial theorem, then replace all of the terms B^k by B_k. Then the result is the usual recursion equation defining the Bernoulli numbers. Blissard's original explanation is not fully satisfactory. He writes:

> According to this method, quantities are considered as divided into two sorts, actual and representative. A representative quantity, indicated by the use of a capital letter without a subindex, as $A, B, \ldots P, Q, \ldots U, V, \ldots$ is such that U^n is conventionally held to be equivalent to, and may be replaced by U_n. (Blissard, 1861, p. 280)

Blissard states clearly that he is operating in a two-sorted logic, with two kinds of variables. Furthermore, the two types of variables are linked by the equation $\mathbf{U}^n = U_n$, where $\mathbf{U}$ is the representative variable corresponding to the actual variable U. This transition between terms involving the two types of variable is known in the literature of the umbral calculus as 'raising and lowering of indices'. The remainder of his proposal, however, is somewhat unclear, because the rules of operation for the representative variables are left tacit, though some of them may be gleaned from Blissard's practice. Furthermore, the nature of the representative quantities was unclear, even when it was plain that the method produced the right answers, used with appropriate caution.

All of this was cleared up in a fully satisfactory way, mostly through the efforts of Gian-Carlo Rota. Rota placed the calculus in the framework of modern abstract linear algebra, and gave a rigorous foundation for the calculations (Rota, 1975). The framework is simple and elegant, and can be explained as follows (Roman, 1984). Let P be the algebra of polynomials in one variable U over the complex numbers, and consider the vector space of all linear functionals over P. Any such linear functional L is determined by its values on the basis polynomials $U^0, U^1, U^2, \ldots$, and consequently can be correlated with the formal power series:

$$f(U) = L(U^0) + L(U^1)U + \frac{L(U^2)}{2}U^2 + \frac{L(U^3)}{6}U^3 + \cdots + \frac{L(U^k)}{k!}U^k + \cdots.$$

If we use the representative term $\mathbf{U}^n$ as an abbreviation for $L(U^n)$, then it is clear that the computation involving the Bernoulli numbers can be justified, and the mysterious equation from which we started can be interpreted as saying $L((U+1)^n) = L(U^n)$. Thus, we have found a way of rewriting the original computations so that every step makes sense. The representative variables are revealed as shorthand expressions for the coefficients of formal power series, and every step in the calculation can be given a clear semantical interpretation in the world of set-theoretical and category-theoretic constructions.

It would be a mistake, though, to see Rota's interpretation as simply bestowing legitimacy on a previously dubious mathematical technique. The interpretation bears fruit in other directions. The embedding of combinatorially defined sequences in the algebra of formal power series is heuristically very fertile, and enables a systematic treatment of special functions. Furthermore, the interpretation makes contact with a large tract of more recent mathematics, such as category theory with its fruitful notion of adjoints. The marriage of 19th-century formal calculation with modern abstract theory produces a sturdy offspring.

Let us look back over this example to see if we can extract some interesting general features. In the case of the umbral calculus, as in many of the other cases we discuss, computations are primary. However, the lack of a clear semantics for the objects manipulated leads mathematicians to treat the methods with distrust, no matter how successful the computations.

Rota's reinterpretation of the umbral calculus has two features that we can detect elsewhere in the work of assimilation. First, the logic is reinterpreted. The umbral calculation

$$(B+1)^2 = B^2 + 2B^1 + 1 = B_2 + 2B_1 + 1$$

is rewritten as

$$L((U+1)^2) = L(U^2 + 2U + 1) = L(U^2) + 2L(U) + L(1) = B_2 + 2B_1 + 1.$$

The introduction of the linear operator L allows the dubious move of raising and lowering indices to be explained simply as a transition between two isomorphic domains of calculation. Second, the reinterpretation involves the introduction of objects of higher type, in this case linear functionals acting on the space of polynomials in one variable.

This reinterpretation of logic and the introduction of higher type entities to provide a semantics is also visible in the theory of infinitesimals. Robinson's nonstandard universe is composed of equivalence classes of higher type objects, since the hyperreals are elements of an ultrapower of a standard model (Goldblatt, 1988). Equations that involve treating infinitesimals as zero quantities (the kind of thing to which Berkeley raised objections), such as $2 + \epsilon = 2$, are reinterpreted in Robinson's framework as $2 + \epsilon \approx 2$, where $x \approx y$ means that x and y are infinitely close (and hence have the same standard part). Interestingly, in the framework of smooth analysis, equations such as $2 + \epsilon^2 = 2$ can be literally true, where ϵ is a nilpotent infinitesimal; the price we have to pay for this literal interpretation of equations is the abandonment of classical logic.

Why are higher types so useful in the process of assimilation? One idea that suggests itself is that there are many more objects in higher types than in lower types. Whereas lower types can always be embedded in higher types, the reverse is not true. In fact, the progress of physics can also be seen as a progress towards more and more abstract representations of physical entities. Whereas the classical notion of a point particle is simply that of a point in Euclidean space, a particle in quantum theory is already represented by a higher type object, a probability distribution over a space of lower type. In reply to Gian-Carlo Rota's question 'What are your views on classical physics versus quantum mechanics?', Stan Ulam gave the following striking response:

> Quantum mechanics uses variables of higher types. Instead of idealized points, or groups of points or little spheres or atoms or bodies, the primitive notion is a probability measure. Quite a logical leap from the classical point of view.
>
> Nevertheless you find in quantum mechanics the strange phenomenon that a theory dealing with variables of higher type has to be imaged on variables of lower type. It is the complementarity between electron and wave. (Cooper, 1989, p. 308)

It is a surprising fact that in all of the cases that we have discussed, impossibility theorems are in the end not a great problem. The criticism of infinitesimal calculus by Berkeley (1734) is clear, concise, and deadly, yet it is easily evaded by appropriate changes in the logic. Similarly, in the theory of Feynman integrals, R. H. Cameron in a paper of 1960 showed that in a certain precise sense there is no 'Feynman measure' underlying the Feynman integral (Johnson and Lapidus, 2000, pp. 82–84). The impossibility theorems in question do not rule out making mathematical sense of the problematic techniques in question, but rather simply rule out certain obvious ways of interpreting them. This should encourage us to look with optimism at the prospects of bringing other fruitful but dubious techniques within the fold of certifiably rigorous mathematics.

16.5 Anomalous objects

The preceding examples both share a common feature—in each case, a calculation takes place that violates certain traditional or classical assumptions. In many cases, the calculus in fact is rather efficient, and this leads to the calculational tradition persisting in communities on the fringe of the (pure) mathematical community, such as engineers and theoretical physicists.

Engineers and physicists are understandably enamoured of calculational devices, and if these formal tricks give the right answer as revealed by experiment or computer simulation, then they are much less inclined than the pure mathematicians to examine the pedigree of a successful piece of machinery for calculation. This appears most flagrantly in the case of quantum field theory. In this case, we have a calculational device (Feynman diagrams) that produces answers that agree to nine decimal places with the measured value of certain physical quantities. As a consequence, the physicists have complete trust in the technique, even though it appears that it still seems to elude a full mathematical formulation (Johnson and Lapidus, 2000).

The situation, though, is mathematically unstable, because the physicists can often compute certain quantities, and 'prove' them to their own satisfaction, while employing the powerful but unrigorous methods peculiar to their own fields. In the introductory chapter, I mentioned two strategies that mathematicians can adopt in dealing with anomalous practices of reasoning and computation. The first is to prove all of the established results by conventional means. This is of course a perfectly reasonable idea, though it still leaves unanswered the question: just why do these outrageous techniques work?

The entities that appear in these calculations, such as infinitesimals or representative variables, are logically anomalous objects. Anomalous objects in my sense are a little like Imre Lakatos' monsters, but anomalous in an even stronger sense, namely not just falling outside a standard definition, but violating fundamental mathematical principles.

An interesting example of an anomalous object is the Dirac delta function. When it first made its appearance in the physics literature, it was explained as a function δ defined on the real line so that it was zero everywhere except at the origin, but the integral of the function was 1:

$$\int \delta(x) \mathrm{d}x = 1.$$

All this is very plausible from the physical point of view. If we think of δ as representing the mass density function of a point particle of mass 1 situated at the origin, then the delta function has exactly the right properties. The trouble is, there is no such function. It's very easy to show using standard calculus that there simply is no such function defined on the classical real line.

We are caught in a dilemma. The physicists have postulated an object that, classically speaking, simply doesn't exist, a logical contradiction. There are various attitudes that one can take to this.

The first is simply to dismiss the unruly objects as nonsense, a common attitude among mathematicians. It is not hard to cite mathematicians who took just such an attitude to objects like the Dirac delta function. For example, in his mathematical autobiography, Laurent Schwartz, who was eventually to find the generally accepted interpretation of the delta function as a distribution, remarks:

> I believe I heard of the Dirac function for the first time in my second year at the ENS. I remember taking a course, together with my friend Marrot, which absolutely disgusted us, but it is true that those formulas were so crazy from the mathematical point of view that there was simply no question of accepting them. (Schwartz, 2001, p. 218)

This attitude is logically impeccable, but it fails to account for the success that the physicists attained with their weird ideas. In other words, it is fruitless and unhelpful.

There is a second attitude, also of an extreme type, that consists in saying—'Well, there's a contradiction, so what? As long as it works, that is all that matters'. This is close to the physicists' own attitudes, since physicists are quite happy to perform purely formal calculations as long as something resembling the right answer comes out in the end. I would also consider the idea of paraconsistent logic to be somewhat along the same line. Although the idea of a true contradiction perhaps makes some kind of sense, the idea on the whole does not appear to be very fruitful.

The third attitude, which I might call the philosophical, or critical attitude, consists in a more nuanced approach to these scandalous invasions of the mathematical universe. The idea here is, on the one hand, to maintain the view that the physicists' calculations are largely correct, but that the objects they *think* they are talking about are in fact of a different nature.

Out of this third, or critical, attitude, applied to the case of the Dirac delta function, comes the modern and very fruitful theory of distributions of Laurent Schwartz. It's interesting to see what this move consists in. The idea here is that the Dirac delta function is not a function at all, but a *distribution*, that is to say, a linear operator on a certain class of functions (the similarity of the strategy followed here to that of Rota's in the case of the umbral calculus is quite striking). The reason this works is that the delta function, as used by physicists, only appears in a certain restricted type of computation, and it is possible to redefine this computation using the operators so that all of the physicists' calculations become perfectly rigorous. In this way, the mathematicians can enjoy the best of both worlds.

Incidentally, the introduction of infinitesimals makes possible the explication of the Dirac delta function in a different and perhaps more intuitive way. We simply define a Dirac delta function as a function with a unit integral, all of whose mass is concentrated within an infinitesimal neighbourhood of the origin. This definition has the advantage of being closer to the original definition of the physicists.

Schwartz's theory of distributions also fits the general pattern that we have detected in our two previous examples. The common feature of the examples of the Dirac delta function, infinitesimals, and the umbral calculus is that the explications given for the anomalous objects and reasoning patterns involving them is what may be described as pushing down higher order objects. In other words, we take higher order objects, existing higher up in the type hierarchy, and promote them to new objects on the bottom level. This general pattern describes an enormous number of constructions.

The process we have just described is closely related to the method of adjoining ideal elements familiar from the history of mathematics. This idea is very clearly described by Martin Davis:

> This is a time-honored and significant mathematical idea. One simplifies the theory of certain mathematical objects by assuming the existence of additional 'ideal' objects as well. Examples are the embedding of algebraic integers in ideals, the construction of the complex number system, and the introduction of points at infinity in projective geometry. (Davis, 1977, p. 1)

Ken Manders (1989) has given a general account of this method as a strategy for unifying and simplifying concepts in mathematics. Here we are emphasizing the role of the method as a means of giving a rigorous interpretation to dubious calculations.

16.6 Strategies of interpretation

Each of the three examples that I discussed above demonstrate the second strategy of assimilation that I described in my introductory chapter. The theory of distributions, nonstandard analysis, and the modern umbral calculus all arose from the idea of constructing new conceptual schemes to validate anomalous calculations, and all of them have proved useful and fruitful techniques that have led to new mathematics.

Sir Michael Atiyah (1995) has given a more complex and nuanced analysis of mathematical strategies. He lists four (rather than two), strategies that mathematicians can adopt towards ideas emerging from the physics community. The first is a version of the first strategy described above, namely that mathematicians should 'take the heuristic results "discovered" by physicists and try to give rigorous proofs by other methods. Here the emphasis is on ignoring the physics background and only paying attention to mathematical results that emerge from physics.' Atiyah's second approach is to 'try to understand the physics involved and enter into a dialogue with physicists concerned. This has great potential benefits since we mathematicians can get behind the scenes and see something of the stage machinery.' It is the dialogue that emerges from this second approach that gives rise to the uneasiness of Jaffe and Quinn.

The third approach mentioned by Atiyah is 'to try to develop the physics on a rigorous basis so as to give a formal justification to the conclusions.' This corresponds quite closely to the second approach that I described in Chapter 15, and of which I gave several examples above. The drawback of this third approach, according to Atiyah, is that 'it is sometimes too slow to keep up with the action. Depending on the maturity of the physical theory and the technical difficulties involved, the gap between what is mathematically provable and what is of current interest to physicists can be immense.' The fourth and most visionary idea is to 'try to understand the deeper meanings of the physics–mathematics connection. Rather than view mathematics as a tool to establish physical theories, or physics as a way of pointing to mathematical truths, we can try to dig more deeply into the relation between them.'

Whatever the attitude that one adopts as a mathematician or logician, it is clear that there is an immense amount of work to be done in this borderline area. The difficulties, both conceptual and mathematical, are severe, but the prizes to be gained through new understanding, are potentially immense.

The last section of this chapter is devoted to the description of an area where the process of assimilation is incomplete, and which consequently poses difficult problems of interpretation.

16.7 The replica method

In this section, I describe an anomalous object in the raw, so to speak. In the physics literature it is known as the Sherrington–Kirkpatrick (SK) model of spin glasses; it belongs to the class of models known generically as mean field models. It is the exact analogue in the field of disordered systems of the Curie–Weiss model of ferromagnetism that I described in my introductory chapter.

The SK model, like the Curie–Weiss model, is a caricature of certain real physical systems. The key feature of such systems is that they involve large collections of magnetic spin variables, where the interaction between these variables is not consistently ferromagnetic (favouring consistent orientation) but rather randomly ferromagnetic and antiferromagnetic (favouring opposite orientation), so that conflict and frustration are inherent elements. A ferromagnet, in its low-temperature behaviour, exhibits spontaneous magnetization, that is to say, it exhibits a tendency for all spins to be aligned consistently. By contrast, a spin glass is expected to show a low-temperature phase in which spins are randomly frozen in a disordered state.

Such systems are in fact quite easy to prepare. If a dilute solution of a magnetic transition metal, such as manganese, is formed in a noble metal, such as copper, then the ferromagnetic atoms, represented by spin variables, are so far apart in the matrix that the dominant interaction is no longer ferromagnetic. The principal interaction force is believed to be the RKKY interaction, first studied in the context of nuclear magnetism by Ruderman and Kittel, and later applied to spin glasses by Kasuya and Yosida. This force oscillates between ferromagnetic and antiferromagnetic, depending on the distance.

The first indications that such alloys have unusual properties appeared in experiments of 1959 and 1960. However, the modern investigation of these materials really took off in the 1970s. Experiments during that decade demonstrated that there appears to be a phase transition in such materials, exactly as for ferromagnets. In the case of spin glasses, the material does not exhibit spontaneous magnetization, but a sharp phase transition appears when the magnetic susceptibility is measured. The physicists interpret this as the transition to a frozen disordered phase of the material. Furthermore, these materials exhibit other unusual experimental properties. For example, they display relaxation and ageing effects on very extended time-scales, perhaps centuries or millenia. In this, they are analogues of ordinary glasses, which also show relaxation times extending over thousands of years; visitors to the Egyptian section of a museum are sometimes rewarded by the sight of a piece of glass that is undergoing crystallization. The terminology of 'spin glass' originates in this analogy. Spin glasses are to magnets as glasses are to crystals, that is to say, they are disordered magnetic materials. For more on the experimental history of these unusual materials, the reader can consult chapter 15 of Fischer and Hertz (1991).

These experimental discoveries led to great excitement in the community of condensed matter physics, as the theorists began trying to understand the spin glass phase revealed by the experiments. Part of the excitement was caused by the fact that the properties revealed by the experiments seemed to be largely independent of the actual materials investigated. Spin glass states have also been found in magnetic insulators and amorphous alloys, where the interactions are of a completely different character (Fischer and Hertz, 1991, p. 2). Thus the experiments seemed to point to some kind of universal behaviour of disordered systems.

The theoretical developments of the 1970s and 1980s were truly remarkable, and led to the development of a theory, in which the replica method played a starring role. The key papers, as well as some very useful background material, are collected in the volume (Mézard et al., 1987).

Before describing what the replica method is, let me give a brief sketch of the SK model. The model consists of a very large number N of sites, each of which has a binary (classical) spin variable (taking the values $+1$ and -1) associated with it. Each of the edges linking the sites is assigned a random value; for simplicity let's suppose that we randomly and independently choose the values $+1$ and -1 for the edges (in the SK model, the edge distribution is Gaussian). In addition, let's say that a positive edge represents friendship and a negative edge enmity. Our problem then is to assign the spins into two camps so that we minimize conflicts—an administrator's problem. If the administrator can divide the people (spins) into two groups so that all the people in one camp are friends, and are enemies of all the people in the other camp, then there is no conflict. But in general, this is impossible, and we have what the physicists (and administrators!) call frustration.

How well can the administrator do in general? Suppose that we have 10,000 people. The physicists, by using their mysterious non-rigorous methods, have computed that on average, the administrator cannot do much better than a situation in which the average person has thirty-eight fewer enemies than friends in their camp (Mézard et al., 1987, p. 2). This is not a great improvement on the situation resulting from the administrator throwing up her hands, and splitting the population in two by tossing a coin.

We are dealing with what is basically a problem in finite combinatorics, like the satisfiability problem. The physicists, though, do not proceed via combinatorics. They translate the problem into the form of a statistical mechanical model, including a parameter called 'temperature', and concentrate their efforts on measuring a crucial quantity called the free energy, representing the degree of frustration in the system. Here is where the notorious replica method makes its appearance.

The physicists make use of a certain expression for the logarithm of a physical quantity, which involves letting a variable n tend to zero. They use the well known classical identity

$$\log Z = \lim_{n \to 0} \frac{Z^n - 1}{n},$$

where Z represents the partition function of the system. This is not puzzling in itself, but their interpretation is. They interpret n as the number of replicas of the system, and then let n (the number of replicas) go to zero. This is already hard to understand, but things get worse.

The physicists predict that there is a critical temperature T_c at which the system undergoes a phase transition. Above T_c, the system is in a single pure state, while below T_c the system undergoes a transition into an infinity of pure states, and the system undergoes a continuous splitting of states down to zero temperature. Above T_c, the physicists consider that they are in the 'replica-symmetric' regime, while below there is 'replica symmetry breaking'. The extent of the gap between the mathematics and physics community can be gauged by the fact that the latter considers the replica-symmetric regime trivial, while the former considers it difficult, and the mathematicians have only succeeded in verifying some of the physicists' predictions with a great deal of difficult effort.

It is below the critical temperature that things get *really* strange. Here the generally accepted solution is due to the outstanding Italian physicist Giorgio Parisi. Parisi introduced what he called a 'replica symmetry breaking matrix'. This is an $n \times n$ matrix that apparently is supposed to correspond to the distances between pure states. The problem is, since n goes to zero, it is a zero by zero matrix. Here I can only quote from a famous paper of Parisi from 1980:

> We face the rather difficult problem of parametrising an $n \times n$ matrix in the limit $n = 0$. To work directly in zero-dimensional space is rather difficult... It is evident that the number of parametrisations is unbounded and the space of $O \otimes O$ matrices with these definitions is an infinite dimensional space. (Parisi, 1980; Mézard *et al.*, 1987, p. 166)

This appears to be mathematical nonsense. But is it complete nonsense? It appears not, because in the first place, the physicists' numbers check out against numerical simulations, and in the second place, many of the predictions made by the physicists using the replica method have been borne out by rigorous proofs. The predictions here are not just numerical predictions, such as the one I mentioned above as a solution to the administrator's problem, but also quite specific and detailed formulas for key quantities such as the free energy.

Even more importantly, the methods developed by the physicists have been applied in a surprising number of different areas, some of them quite far removed from the original physical systems that inspired them. Thus the replica method has been applied in the theory of combinatorial optimization problems, such as the matching problem, the travelling salesman problem, and the graph partitioning problem (this is the problem we described as the administrator's problem above). Other applications have been found in the theory of evolution, neural networks, and the theory of memories, where the

Hopfield model was inspired by ideas from spin glass theory. A large selection of such applications is described in the latter part of the book (Mézard et al., 1987). In each case, the physicists were able to draw surprising and apparently correct conclusions that were not attainable by conventional approaches. Thus, the evidence so far indicates that the physicists have uncovered a broad domain of phenomena, all involving disorder of some kind, that can be described using the nonrigorous tools that they have developed and refined since the 1970s.

So, the evidence is that there is some kind of mathematical sense lurking behind the replica method, just as there was in the case of the Dirac delta function. Although the physicists had some initial misgivings about the method, a quick glance at the literature of the theory of disordered systems shows that in many cases, it is the first tool for which they reach when investigating a new system involving disorder and frustration.

Nevertheless, a rigorous foundation for the method has so far eluded the mathematicians. Remarkable and striking work has been accomplished by probabilists working within the usual mathematical framework of large deviations, concentration inequalities, and the like, and Michel Talagrand, in particular, has made very substantial progress towards verifying the predictions of the physicists. A survey of his work and an introduction to this important and intriguing area is to be found in his monograph (Talagrand, 2003b). Recently, Talagrand, building on the work of Francisco Guerra, achieved a breakthrough by proving Parisi's formula for the free energy of the SK model (Talagrand, 2003a, 2006).

In spite of these major successes, and the verification of the physicists' work, the replica method itself languishes in mathematical limbo. Talagrand remarks:

> At the present time, it is difficult to see in the physicist's replica method more than a way to guess the correct formula. Moreover, the search for arguments making this method legitimate 'a priori' does not look to this author as a promising area of investigation. (Talagrand, 2003b, p. 195)

In spite of these sceptical remarks, the example of the Dirac delta function, the umbral calculus, and the theory of infinitesimals should encourage us to hope that the remarkable success of this method is more than a mere mathematical coincidence. Much remains to be done.

16.8 A plea for nonrigorous mathematics

There are numerous arguments for admitting 'nonrigorous' mathematics into the field of philosophy of mathematics. Even if we identify mathematics with theorems deducible from the axioms of set theory, there still remains the non-formal question of justifying the axioms themselves, including new axioms.

Going beyond this, it would seem that if we define mathematics in this narrow way, that we would exclude from the sphere of mathematics most mathematics from before 1900, surely an unwanted consequence.

I am not arguing for relaxing the standards of rigour in mathematics. On the contrary, I am urging logicians and philosophers to look beyond the conventional boundaries of standard mathematics for mathematical work that is interesting but nonrigorous, with a view to making it rigorous. The examples of calculus, distribution theory, complex function theory, and Brownian motion all show that some of the best mathematics can result from this process. Similarly, philosophers can surely find fruitful areas for their studies in areas lying beyond the usual set-theoretical pale.

The Czech physicist Jan Klima has presented the activities of mathematicians in a rather humiliating light:

> In the fight for new insights, the breaking brigades are marching in the front row. The vanguard that does not look to left nor to right, but simply forges ahead—those are the physicists. And behind them there are following the various canteen men, all kinds of stretcher bearers, who clear the dead bodies away, or, simply put, get things in order. Well, those are the mathematicians. (Zeidler, 1995, p. 373)

By way of contrast, here is a quotation from the great Canadian mathematician, Robert Langlands:

> Field theories and especially conformally invariant field theories are becoming familiar to mathematicians, largely because of their influence on the study of Lie algebras and above all on topology. Nonetheless, in spite of the progress in constructive quantum field theory during recent decades, many analytic problems, especially the existence of the scaling limit, are given short shrift. These problems are difficult and fascinating and merit more attention. ... It is often overlooked that the largely *mathematical* development of Newtonian mechanics in the 18th century was an essential prerequisite to the enormous *physical* advances of the 19th and 20th centuries, that attempts to overcome mathematical obstacles may lead to concepts of physical significance, and that mathematicians, recalling the names of d'Alembert, Lagrange, Hamilton and others, may aspire to nobler tasks than those currently allotted to them. (Langlands, 1996)

I hope that these inspiring words of Langlands may lead philosophers to look beyond the rather restricted range of topics in mathematics that currently preoccupy them.

Bibliography

ATIYAH, Michael (1995), 'Reflections on geometry and physics', in *Surveys in Differential Geometry*, Volume 2 (Cambridge, MA: International Press) pp. 1–6. Reprinted in Atiyah's *Collected Works*, Volume 6 (Oxford: Oxford University Press, 2004), pp. 491–496.

ATIYAH, Michael *et al.* (1994), 'Responses to "Theoretical mathematics: towards a cultural synthesis of mathematics and theoretical physics", by A. Jaffe and F. Quinn', *Bulletin of the American Mathematical Society*, 30, 178–207.

BELL, Eric Temple (1938), 'The History of Blissard's symbolic method, with a sketch of its inventor's life', *American Mathematical Monthly*, 45, 414–421.

BELL, John L. (1998), *A Primer of Infinitesimal Analysis* (Cambridge: Cambridge University Press).

BERKELEY, George (1734), *The Analyst* (London: J. Tonson).

BLASS, Andreas (1984), 'The interaction between category theory and set theory', in *Mathematical Applications of Category Theory* (Providence, RI: American Mathematical Society) pp. 5–29.

BLISSARD, John (1861), 'Theory of generic equations', *Quarterly Journal of Pure and Applied Mathematics*, 4, 279–305.

BOS, H. J. M. (1974), 'Differentials, higher-order differentials, and the derivative in the Leibnizian calculus', *Archive for History of Exact Sciences*, 14, 1–90.

CARTIER, Pierre (2000), 'Mathemagics (A tribute to L. Euler and R. Feynman)', in Michel Planat (ed.), *Noise, Oscillators and Algebraic Randomness: From Noise in Communications Systems to Number Theory* (Berlin: Springer-Verlag). Lectures of a school held in Chapell des Bois, France, April 5–10, 1999.

COOPER, Necia Grant (ed.) (1989), *From Cardinals to Chaos: Reflections on the Life and Legacy of Stanislaw Ulam* (Cambridge: Cambridge University Press). First published in 1987 as a special issue of *Los Alamos Science*.

DAVIS, Martin (1977), *Applied Nonstandard Analysis* (New York: John Wiley & Sons). Reprinted by Dover Publications, 2005.

FISCHER, K. H. and HERTZ, J. A. (1991), *Spin Glasses* (Cambridge: Cambridge University Press).

GLIMM, James and JAFFE, Arthur (1987), *Quantum Physics: A Functional Integral Point of View*, 2nd edn (New York: Springer-Verlag).

GOLDBLATT, Robert (1998), *Lectures on the Hyperreals: An Introduction to Nonstandard Analysis* (Berlin: Springer Verlag). Graduate Texts in Mathematics 188.

GROSSMAN, Nathaniel (1996), *The Sheer Joy of Celestial Mechanics* (New York: Birkhäuser).

HELLMAN, Geoffrey (2006), 'Mathematical pluralism: the case of smooth infinitesimal analysis', *Journal of Philosophical Logic*, 35, 621–651.

JAFFE, Arthur and QUINN, Frank (1993), ' "Theoretical mathematics": Toward a cultural synthesis of mathematics and theoretical physics', *Bulletin of the American Mathematical Society*, 29, 1–12.

JOHNSON, Gerald W. and LAPIDUS, Michel L. (2000), *The Feynman Integral and Feynman's Operational Calculus* (Oxford: Oxford University Press).

LANGLANDS, Robert P. (1996), 'An essay on the dynamics and statistics of critical field theories', in *Canadian Mathematical Society 1945–1995*, Volume 3 (Canadian Mathematical Society).

LUCAS, Éduard (1876), 'Théorie nouvelle des nombres de Bernouilli et d'Euler', *Comptes Rendus de l'Academie des Sciences (Paris)*, 83, 539–541.

MAC LANE, S. (1997), 'Despite physicists, proof is essential in mathematics', *Synthese*, 111, 147–154.

MACLAURIN, Colin (1742), *A Treatise of Fluxions* (Edinburgh: T. W. and T. Ruddimans).

MANCOSU, Paolo (1996), *Philosophy of Mathematics and Mathematical Practice in the Seventeenth Century* (Oxford: Oxford University Press).

MANDERS, Ken (1989), 'Domain extension and the philosophy of mathematics', *Journal of Philosophy*, 86, 553–562.

MANIN, Yu. I. (1981), *Mathematics and Physics* (New York: Birkhäuser). Translated by Ann and Neal Koblitz: Progress in Physics, Volume 3.

MÉZARD, Marc, PARISI, Giorgio, and VIRASORO, Miguel Angel (1987), *Spin Glass Theory and Beyond* (Singapore: World Scientific).

MOERDIJK, I. and REYES, G. E. (1991), *Models for Smooth Infinitesimal Analysis* (New York: Springer-Verlag).

PARISI, Giorgio (1980), 'The order parameter for spin glasses: A function on the interval $0-1$', *Journal of Physics A: Mathematical and General*, 13, 1101–1112.

ROBINSON, Abraham (1966), *Non-standard Analysis* (Amsterdam: North-Holland). Second revised edition 1974. Reprinted with an introduction by Wilhelmus A. J. Luxemburg (Princeton University Press, 1996).

ROMAN, Steven (1984), *The Umbral Calculus* (New York: Academic Press). Reprinted by Dover Publications, 2005.

ROTA, Gian-Carlo (1975), *Finite Operator Calculus* (New York: Academic Press). With the collaboration of P. Doubilet, C. Greene, D. Kahaner, A. Odlyzko, and R. Stanley.

SCHWARTZ, Laurent (2001), *A Mathematician Grappling with his Century* (New York: Birkhäuser). Translated from the French by Leila Schneps.

SENECHAL, Marjorie (1998), 'The continuing silence of Bourbaki—An interview with Pierre Cartier, June 18, 1997', *Mathematical Intelligencer*, 20, 22–28.

TALAGRAND, Michel (2003a), 'The Parisi formula', *Comptes Rendus de l'Academie des Sciences (Paris)*, 337, 111–114.

——— (2003b), *Spin Glasses: A Challenge for Mathematicians* (Berlin: Springer-Verlag).

——— (2006), 'The Parisi formula', *Annals of Mathematics*, 163, 221–263.

THURSTON, William P. (1994), 'On proof and progress in mathematics', *Bulletin of the American Mathematical Society*, 30, 161–177.

VIENNOT, X. G. (1992), 'A survey of polyomino enumeration', in *Actes du 4e Colloque Séries Formelles et Combinatoires Algébrique, 15–19 Juin 1992, Montréal* (Publications du Laboratoire de Combinatoire et d'Informatique).

ZEIDLER, Eberhard (1995), *Applied Functional Analysis: Applications to Mathematical Physics* (Berlin: Springer-Verlag). Applied Mathematical Sciences, Volume 108.

Index of Names

Aigner, M. 337, 351b
Alexandro, P. 381, 402b, 403b
Andradas, C. 163
Apollonius 67, 75–6, 104–105, 132b, 181
Appel, K. 304
Arana, A. 317
Archimedes 67, 181, 193, 195b
Aristotle 83, 97, 179–181, 193–194
Armstrong, D. 282, 283, 299b
Arnauld, A. 141, 185, 194b, 256, 273b
Arnold, V. 128, 132b
Artin, M. 160, 366n, 387, 395, 397n, 403b
Artmann, B. 67, 78b
Aspray, W. 3, 20b
Atiyah, M. 379, 400, 403b, 413–415, 431–432, 438b
Avigad, J. viii, 16, 19–20, 176, 260, 266, 273b, 296, 299b, 314, 315, 331, 332, 338, 340, 350, 351b
Awodey, S. 355n, 367b
Ayer, A. J. 88, 132b
Azzouni, J. 351b

Baez, J. 402, 403b
Bailey, D. 303, 304, 315b
Baker, A. 136–140
Balfe, J. 276
Ballarin C. 343, 351b
Bankoff, L. 191, 194b
Barwise, J. 23, 26, 40b
Bass, H. 382, 403b
Bassler, O. B. 307, 315b
Batterman, R. 135, 138, 148
Bays, T. 9, 20b
Becker, O. 63n, 64b
Beeson, M. 339, 352b
Behboud, A. 77, 78b
Bell, E. T. 425, 438b
Bell, J. viii, 423–424, 438b
Belnap, N. xi
Belot, G. 414, 415b
Bénabou, J. 401, 403b
Benacerraf, P. 1, 276, 283, 286, 299, 354–355, 367b
Bengedem, J. P. van 5, 13
Bergren, J. L. 77, 78b
Berkeley, G. 72, 78b, 84, 423, 427–428, 438b

Bernays, P. 211n, 231, 232, 252, 254, 255
Bertot, Y. 343, 351b
Bessel, F. W. 185n
Beth, E. W. 72
Black, R. 61n
Blackwell, A. 27, 40b
Blanchette, P. 20
Blass, A. 421, 438b
Blissard, H. J. M. 425–426, 438b
Bôcher, M. 72, 78b
Bochnak, J. 162n, 163, 177
Bolzano, B. ix, 141–142, 145, 161, 182–186, 187n, 191, 193, 194b
Boolos, G. 220, 252b
Borel, A. 417
Borwein, J. 303, 304, 315b
Bos, H. 348, 351b, 423, 425, 438b
Bourbaki, N. 361, 367b, 396, 406b, 411
Breger, H. 5, 20b
Bressoud, D. 5, 20b
Bricker, P. 282n, 283n
Brill, A. 277n
Bröcker, L. 177b
Browder, F. 252b
Brouwer, L. viii, 7, 165, 166, 370, 380, 381, 417
Brown, J. R. 23, 28, 40b
Brumfiel, G. W. 158–160, 163n, 165, 166–170, 177b
Brummelen, G. van 77, 78b
Bruner, R. 256
Bundy, A. 315b
Burge, T. 271, 273b

Cameron, R. H. 428
Cantor, G. ix, 48, 146
Cardan, G. 156
Cariani, F. vi, 132
Carnap, R. 6, 408
Carr, D. 328n, 352b
Cartan, E. 424
Carter, J. 355, 357, 364, 367b, 398, 403b
Cartier, P. 412, 415b 418–419, 438b
Cartwright, N. 408, 416b
Casteran, P. 343, 351b
Cauchy, A. 31, 40b, 160
Cayley, A. 342

Cellucci, C. 5, 20b
Chemla, K. 402
Chevalley, C. 262, 273b
Chihara, C. 355n, 367b
Chou, C. C. 348, 352b
Cicero 185, 194b
Clebsch, A. 277n
Cobb, J. 367
Cohn-Vossen, S. 223, 255b, 381, 403b
Collins, J. 276
Colton, S. 313, 315
Colyvan, M. 135
Conrad, B. 256, 297
Conway, J. 357–360, 367b
Coolidge, J. L. 240, 252
Cooper, N. 428, 438b
Corfield, D. 5, 7–10, 18, 19, 20b, 291n, 296n, 299b, 310, 315b, 364, 367b, 400–403, 404b
Cornell, G. 355, 367b
Corry, L. 371, 404b
Coste, M. 177b
Courant, R. 161, 177b
Cournot, A. 141, 144
Cox, D. 260, 265, 266n, 273b, 290n, 299b
Coxeter, H. M. S. 229, 252b
Cunningham, S. 23

Darrigol, O. 293, 299b
Dalen, D. van 380, 406b
Davey, K. 412, 415b
Davis, M. 431, 438b
Davis, P. 3, 20b
de Finetti, B. 309
De Morgan, A. 126
Dean, E. 302
Dedekind, R. 7, 22, 40b, 56, 187, 194b, 216n, 220, 241, 250, 252b, 262n, 263, 267, 271, 273b, 276–278, 293, 295–299, 300b, 354n, 368b, 370, 371
Dehn, M. 207, 221, 252
Deligne, P. 391, 396, 403b
Demopoulos, W. 198
Desargues, G. 76
Descartes, R. 73, 77, 80, 81, 83, 92, 184
Detlefsen, M. v, viii, 15, 270n, 284, 317, 365
Diamond, H. 190, 194b
Dickson, L. E. 342, 352b
Dieudonné, J. 193, 194b, 370n, 387, 390, 403b, 404b
Dirichlet, P. 199, 263, 273b
Docherty, M. 256, 367
Dolan, J. 402, 403b
Donnelly, K. 351b

Drobisch, M. 184n
Dubnov, Y. 95, 102, 124–125, 132b
Dummett, M. 3, 140, 148b
Dyson, F. 411, 413–414, 415b

Eddy, R. 40b
Edmonds, A. 191, 194b
Edwards, H. 260, 266n, 273b, 290, 296n, 300b
Eilenberg, S. ix, 364, 366, 368b, 382, 400, 403b
Einstein, A. 2, 410
Eisenberg, T. 40b
Eisenbud, D. 279n, 300b, 394n, 403b
Elgin, C. 258, 273b
Elliot G. 397
Engel, F. 206, 255b
Enriques, F. 269, 273b
Erdös, P. 15, 189–191, 194b, 337
Etchemendy, J. 23
Euclid 22, 25, 31, 41b, 65–67, 70–71, 75–77, 78b, 80–81, 85–94, 97n, 98, 100–101, 104, 108–110, 113–114, 118, 120, 123, 125–126, 130, 132b, 157, 201, 206–210, 212, 220, 229, 230, 232, 240, 241, 242, 245n, 247, 250, 345, 352b, 354, 357
Eudemus 97
Euler 26, 31, 37–38, 166, 265, 266, 271, 289–292, 309, 311, 347–348
Ewald, W. ix, 183, 186n, 194b, 195b, 198, 212, 214, 230, 252b, 254b, 255b

Faddeev, L. D. 410, 415b
Fallis, D. 307, 315b
Faltings, G. 370, 374, 378, 385, 403b
Feferman, S. 139, 148b
Feigenbaum, M. 420
Feigl, H. 408
Ferneau, C. viii
Fernmüller, C. 176
Ferreiros, J. 5, 18, 20b, 402
Field, H. 281
Fine, A. 413, 415
Fine, D. 413, 415
Fine, K. 270n, 273
Fischer, K. H. 409, 415b, 433, 438b
Flament, D. 359, 368b
Fomenko, A. 23, 41b
Formanek, E. 191, 195b
Forrest, P. 282, 283, 300b
Fowler, D. 97n, 132b, 372, 404b
Fraassen, B. van 143
Freed, D. S. 415b

INDEX OF NAMES 443

Frege, G. ix–x, 7n, 83, 85, 187, 188n, 195b, 210, 211, 212, 214, 215, 218, 220, 221, 249, 252, 253
Freyd, P. 397, 404b
Friberg, B. 191n, 195b
Friedman, H. 340, 351b
Friedman, M. 152, 153, 170, 178b
Frölich, A. 259n, 273b

Gaifman, H. 310–312, 315b
Galileo 76
Gao, X. S. 351b
Gauss, C. F. 184n, 185, 187, 195b, 207, 250, 259–263, 265–266, 271, 273b, 303, 318, 410, 416
Gauthier, Y. 198
Gelfand, S. 400, 404b
Gentner, D. 44, 64b
Gentzen, G. 60
Giaquinto, M. viii, 14, 18, 20, 31, 33, 36, 41b, 53n, 64b, 68
Gillies, D. 5, 20b
Gilmer, R. 191, 195b
Glimm, J. 417, 421, 438b
Glymour, C. 317
Gödel, K. viii–ix, 6, 10–12, 60–62, 64b, 139, 339, 371, 404b
Goldblatt, R. 438b
Goldfarb, W. 323n, 352b
Goldman, J. 260, 273
Gonthier, G. 305
Good, I. J. 309, 311n, 315
Goodman, N. 136, 289, 290
Goodwin, W. 74, 78b
Gouvêa, F. 261n
Gray, D. 351b
Gray, J. 5, 18, 20b, 402
Grialou, P. 27n, 41b
Grosholz, E. 5, 20b, 28
Grossman, N. 422–424, 438b
Grothendieck, A. ix, 17, 366, 384–385, 387, 390, 391, 394, 395, 396n, 397n, 400, 403b, 404b

Hacking, I. 309, 310, 312, 315b
Hadamard, J. 189–190
Hafner, J. v, ix, 15, 142–143, 148, 151, 178b
Hahn, H. 78b, 132b
Haken, W. 304
Hale, B. 3, 272, 274b
Hales, T. 304
Hallett, M. v, ix, 15, 199–201, 204–209, 211n, 212–214, 216–218, 221, 224, 225n, 227–228, 230n, 232n, 233n, 236n, 237n, 240, 241n, 244–245, 246, 247, 249, 250, 252b, 253b
Harari, O. 71, 78b, 141, 148b
Hardy, G. H. 338, 352b
Harris, J. 256, 257, 274b, 279n, 300b, 403b
Hart, W. D. 253b
Hartshorne, R. 67, 78b, 244, 246, 253, 384n, 390, 394n, 404
Heath, T. L. 88n, 89n, 90n, 100, 101, 108, 123, 126, 132n, 230, 240, 241, 245, 253
Heaviside, O. 425
Hecke, E. 262n, 274b
Hegel, G. W. F. 262n
Heis, J. 317
Hellman, G. 139, 147, 355n, 368b, 396, 404b, 424, 438b
Helmholz, H. 293
Hempel, C. G. 152, 153
Herreman, A. 380, 404b
Hersh, R. 3, 20b
Hertz, J. A. 409, 415b, 433, 438b
Hiddleston, E. 256
Hilbert, D. (1) viii–ix, 3, 7, 15, 23, 41b, 60, 67, 78b, 80–82, 85, 87, 120, 132b, 188, 195, 198–251, 253b, 254b, 277, 300b, 345, 370, 381, 404b
Hilbert, D. (2) 281
Hintikka, J. 72, 78b
Hippocrates 97
Hirsch, M. 417
Hjelmslev, J. 93, 132b
Hobson, E. W. 195b
Hocking, J. 363, 368b, 382, 404b
Hodges, W. viii
Hoffman, D. 23, 38–39, 41b
Hoffman, J. 38–39
Hopf, H. 381, 403b
Hunt W. 339, 352b
Husimi, K. 409, 416b
Huybrechts, D. 384n, 404b
Huygens, C. 76

Ingham, A. 190, 193, 195b
Isaacson, D. viii, 249n, 255
Israel, D. 109

Jacobson, N. 404b
Jaffe, A. 417–419, 420, 432, 438b
Jamnik, M. 30, 41b
Jannson, L. 256
Jantzen, B. 302
Jeffrey, L. C. 415b
Jenkins, J. 192, 195b

Johnson, G. W. 421, 428–429, 439b
Johnstone, P. 404
Jones, M. 408, 416b
Joyal, A. 23, 41b

Kac, M. 409, 416b
Kant, I. 8, 22, 41b, 72, 74, 217, 280
Karzel, H. 225, 243
Kazarinoff, D. K. 191, 195b
Kazhdan, D. 415
Keranen, J. 317
Kerkhove, B. van 5, 13, 20b
Kieffer, S. 302
Killing, W. 241, 255
Kim, J. 263, 270, 274
Kirchhoff, G. 213, 255
Kitcher, P. 3, 4–7, 10, 13, 15, 17–18, 20b, 141–142, 144–148, 149b, 151–159, 162–163, 166–168, 169n, 170–176, 178b, 263n, 274b
Kleiman, S. 291n, 300b
Klein, F. 88, 132b, 188, 194, 195, 277n
Kline, M. 3, 20b, 372–374, 404b
Kment, B. 270n
Knorr, W. 88n, 100, 122n, 132b, 372, 404b
Knuth, D. E. 352b
Koetsier, T. 5, 20b
Kosslyn, S. 47n, 54, 64b
Kreisel, G. 63n, 371–372, 404b
Krieger, M. 5, 20b, 363, 368n, 414, 416b
Kripke, S. 323n, 352b
Krömer, R. 356n, 368b, 400, 405b
Kronecker, L. 7, 370, 372–375, 378
Krug, R. B. 352b
Kuhn, S. 5, 271, 274b
Kuratowski, K. 282, 284, 285
Kürschák, J. 246
Kushner, D. 143, 149b, 152
Kyburg, H. 302, 309

la Vallée Poussin, C. de 189–190
Lafforgue, L. 371n, 405b
Lagrange, J. L. 76, 156–158, 182n, 196b
Lakatos, I. 3, 6–10, 14, 17, 18, 20b, 291n, 300b, 429
Laksov, D. 291n, 300b
Lambert, K. 76
Landau, E. 23, 41b
Lang, S. 357, 358–363, 368, 398, 405b
Langlands, R. 364, 437, 439
Lapidus 421, 428–429, 439b
Larvor, B. 4, 5, 20b
Lauda, A. 23, 41b

Lawvere, F. W. 354n, 363n, 368, 396n, 398, 400, 402, 405b
Lefschetz, S. 166n, 372
Legendre, A.M. 76, 258–259, 262–267, 270
Leibniz, G. W. 76, 80, 85, 108, 181, 182, 193, 196b, 271
Leinster, T. 23, 41b
Lemmermeyer, F. 261–262, 274b
Leng, M. 138, 139n, 149b, 397, 405b
Lenstra, H. 259n, 274b
Levi, I. 310, 311n, 316b
Lewis, D. K. 258, 274b, 279–283, 298, 300
Lie, S. 209n, 277n, 424
Lipton, P. 134–135, 149b
Littlewood, J. E. 352b
Livesey, S. 181, 196b
Lloyd, G. E. D. 68, 78b
Locke, J. 72–73
Loemker, L. 80, 132b
Longo, G. 27, 40b
Lucas, E. 439b
Luck, J. M. 412, 416b
Luh, J. 192, 196b
Lyusternik, L. 31

Mac Lane, S. ix, 17, 358, 360n, 363, 366, 368, 372, 376n, 377, 382n, 384n, 395, 397, 398, 400, 405, 418, 420, 439b
Macbeth, D. 70, 78b
MacBride, F. 357–358, 366–367, 368b
MacCarthy, T., viii
MacLaurin, C. 188, 196b, 422, 439b
Maddy, P. 6, 7, 10–13, 19, 20 21b, 140–141, 149b, 272, 356
Majer, U. ix, 198–201, 199–201, 204–209, 211n, 212–214, 216–218, 221, 224, 225n, 227–228, 230n, 232n, 233n, 236n, 237n, 240, 241n, 244–245, 246, 247, 249, 250, 253
Mancosu, P. ix–x, 6, 13, 15, 21b, 27, 36, 41b, 42b, 63n, 64b, 141–143, 145, 148, 149b, 151, 178b, 182n, 196b, 256, 276, 263, 302, 312, 313, 316b, 317, 424, 439b
Mandelbrojt, S. 144–145, 149b
Mandelbrot, B. 416–418
Manders, K. x, 14, 70–71, 78b, 317, 323n, 345, 431, 439b
Manin, Y. 400, 404b, 421, 439b
Marquis, J. P. 400, 405b
Maxwell, E. 2, 95, 102, 125, 133b
May, K. 261, 274b
Mayberry, J. 367
Mazur, B. 291, 300b, 359, 366, 368, 373n, 405b

INDEX OF NAMES 445

McCune, W. 304, 316b
McLarty, C. ix, 16, 19, 256, 349–350, 359n, 362, 377n, 395n, 396, 398, 405b
Meeks, W. 38
Melia, J. 136, 149b
Menelaos, 76
Menn, S. 256
Mézard, M. 434–436, 439b
Miller, N. 25, 26, 42b, 70–71, 78b, 346–348
Milne, J. 256, 262, 384, 406b
Moerdijk, I. 423–424, 439
Momtchiloff, P. vi
Monastyrsky, M. 364, 366n, 368b
Monge, G. 203, 239
Montuchi, P. 34, 42b
Moore, J. 352b
Mordell, L. 142, 149b
Morgan, M. 408, 415b, 416b
Morrison, M. 408, 415b, 416b
Morrow, G. R. 90
Mott, J. 191, 195b
Moussa, P. 415b–416b
Mueller, I. 87–88, 100–101, 125, 133b
Mullin, R. C. 192n, 196b
Mumford, D. 387, 399, 406
Mumma, J. 67, 70–71, 75, 78b, 317, 346–348, 352b

Needham, T. 23, 38, 42b
Nelsen, R. 23, 42b
Nelson, G. 340, 352b
Nerlich, G. 135
Netz, R. 67, 78b
Neugebauer, O. 410, 416b
Neumann, J. von 283, 285
Newton, I. 76, 101, 133b, 271, 423
Nicole, P. 141, 256, 273b
Nieuwentijt, B. 424
Noether, E. ix, 17, 256, 274b, 277n, 355, 364–365, 367, 369b, 370–372, 376–378, 380–381, 396, 400, 406b
Nolan, D. 271, 274b
Norman, J. 23, 28n, 32, 42b, 74, 78b

Okada, M. 40b
Olbers, H. 185n
Onsager, L. 421
Oppen, D. 340, 352b

Page, W. 34, 42b
Palais, R. 23, 42
Panza, M. 317
Pappus 77, 97n, 133b, 181n, 182n, 196b

Parisi, G. 435–436, 439
Pasch, M. 23, 42b, 84, 87, 94, 120, 203, 210, 212, 255b, 345
Paseau, A. 9, 20b
Patashnik, O. 352b
Peacocke, C. 271–272, 274b
Peirce, C. S. 26
Pincock, C. 19–20, 21b
Plato ix, 32–33, 42b, 80, 86, 88n, 97n, 179, 196b
Poincaré, H. ix, 218n, 255b, 319–320, 322, 352b, 366–367, 369b, 410, 419
Pólya, G. 291n, 300b, 302, 309, 311–312, 316b
Poncelet, J. V. 75–7, 78b, 104–105, 108, 133b
Presbuerger, M. 338
Proclus 71, 78b, 88n, 89–90, 100, 112–116, 119–124, 126–127, 130
Proops, I. 256
Protagoras, 126, 128
Putnam, H. ix, 268, 274b, 276, 291n, 300b
Pylyshyn, Z. 27n, 42b

Quine, W. V. O. 6, 10, 136, 141, 356
Quinn, F. 417–419, 432, 438b

Rabin, M. 305
Raff, P. 351b
Ramsey, F. P. 309
Ramus, P. 185n, 196b
Reichenbach, H. 408
Reid, T. 365, 369b
Rescher, N. xi
Resnik, M. 45, 64b, 143, 149b, 151, 178b, 354, 355n, 356–360, 362–363, 369b
Rey, G. 271, 274b
Riemann, B. 15, 146, 190, 199, 257d, 276, 278, 293, 299, 300b
Robinson, A. 191, 423–425, 427, 439
Robinson, R. 269, 274b
Roman, S. 426, 439b
Rota, G. C. 189, 190, 193, 196b, 426–428, 430, 439
Rouse-Ball, W. W. 42b, 78b
Roy, M. F. 177b
Rudin, W. 398n, 406b
Ruiz, J. M. 177b
Russ, S. 186, 194b
Russell, B. x, 3, 6, 23, 42b, 43, 64b
Russo, E. vi
Ryckman T. 20, 62, 64b
Ryle, G. 328–329, 352b

446   INDEX OF NAMES

Saccheri, G. 76
Sagüillo, J. 20
Samuel, P. 229, 252b, 265, 274b
Sandborg, D. 143, 150
Sandifer, E. 290n, 300b
Sanford, D. 263, 274b
Sarkaria, K. 379, 406b
Sawyer, W. 39n, 42b
Schanuel, S. 400, 405b
Schaper, H. von 200
Schlimm, D. 198
Schmidt, A. 230n
Schwartz, L. 430–431, 439b
Scott, D. viii, 176
Seidenberg, A. 160–161, 163–165, 166n, 168–169, 172
Seidenfeld, T. 302, 311n
Selberg, A. 15, 189, 196b
Seligman, J. 86
Senechal M. 419, 439b
Serre, J. P. 371, 390–391, 406
Shabell, L. 74
Shapiro, S. 62, 135, 317, 355n, 356, 359, 360, 369b, 396, 406b
Sherry, D. 84, 133b
Shin, S. J. 26, 42b
Shimojima, A. 27n, 42b
Shoemaker, S. 266, 274b
Sica, G. 351b
Sider, T. 259, 275b, 281–288, 300
Sieg, W. ix, 198, 212, 214, 230, 252b, 254b, 255b
Smith, D. 357–360, 367b
Sober, E. 140, 149b
Solovay, R. 305
Sommer, J. 243, 245, 247, 255
Sorensen, H. 72, 78b
Soulé, 371n, 406b
Spitznagel, E. 192, 196b
Spivak, M. 257, 259, 275b
Stäckel, P. 206, 255b
Stahl, H. 277, 293, 294, 301b
Stalker, D. 257
Stanton, D. 190–191, 196b
Staudt, K. G. C. von 203
Steele, J. M. 338, 352b
Steenrod, N. 364, 366, 368b, 382n, 403b
Steiner, M. 77, 135–136, 138, 139n, 141–144, 147–148, 149b, 150b, 151, 178b
Stevenhagen, P. 259, 274
Stewart, I. 268, 275
Strassen, V. 305

Suppe, F. 408
Sutton, W. 44n, 64b

Tait, W. 323
Talagrand, M. 436, 439b
Tall, D. 268, 275
Tamme, G. 394n, 406b
Tappenden, J. x, 7n, 15, 18, 20, 21b, 36n, 142, 144, 150b, 151, 176, 263n, 264n, 271, 275b, 294n, 296n, 301b, 367
Tarski, A. 6, 67, 74, 78b, 120, 131, 132b, 160–161, 163–165, 166n, 168–169, 172, 178b, 339, 352b
Tate, J. 261
Taussky-Todd, O. 364, 369b
Taylor, B. 258, 275b
Taylor, M. 259n, 273b
Tennant, N. 23, 26, 42b
Thales 185
Thom, R. 418
Thomas, I. 196b
Thompson, C. J. 408–409, 416b
Thurston, W. P. 307, 316b, 417–419, 439b
Tiwari, A. 339, 352b
Tollmien, C. 369b
Tropfke, J. 97, 133b
Tymoczko, T. 3, 21b, 307, 316b

Uhlenbeck, K. 418
Ulam, S. 428
Urquhart, A. x, 17, 302

Vanhove, P. 415
Verhoeven, L. 144, 150b, 151, 178b
Viéte, F. 185, 197b
Viennot, X. G. 420, 440b
Vineberg, S. 256
Virasoro, M. A. 439b
Vitrac, P. 67, 78b

Wagner, S. 276
Waismann, F. 139n
Waldschmidt, M. 412, 416b
Wallis, J. 184, 185n, 188, 193, 197b
Washington, L. 366, 369b, 384, 406b
Wasserman, L. 311n
Watson, G. M. 338, 352b
Weatherson, B. 301b
Weber, H. 144, 150b, 151, 296, 297, 300b, 385, 406
Weierstrass 161, 183n, 186–7, 277
Weil, A. 17, 166n, 261, 290, 291, 292, 297, 301, 364, 371, 384, 385, 406b

Weiner, M. 71
Weiss, P. 409, 416b
Weyl, H. 62, 417
Whittaker, E. T. 338, 352b
Wiedijk, F. 304, 316b, 332, 353b
Wiener, N. 189, 282, 284, 285
Wilson, M. 176, 413, 416b
Wittgenstein, L. 323–328, 343, 353b
Witten, E. 415b, 417
Woo, J. 191, 197b
Woodger, J. H. 6
Wright, C. 3, 272, 274b
Wu, W. T. 358, 353b

Wussing, H. 342, 353b
Wyman, B. 265n

Yeomans, J. M. 408, 409, 416b
Young, G. 363, 368n, 382, 404b

Zeilberger, D. 190–191, 196b
Zeidler, E. 440b
Zermelo, E. 283, 285
Zhang, J. Z. 348, 351b
Ziegler, G. 337, 351b
Zimmermann, W. 23, 42b